# Functional Equations in Probability Theory

RAMACHANDRAN BALASUBRAHMANYAN

*Indian Statistical Institute*
*New Delhi, India*

KA-SING LAU

*Department of Mathematics and Statistics*
*University of Pittsburgh*
*Pittsburgh, Pennsylvania*

ACADEMIC PRESS, INC.

**Harcourt Brace Jovanovich, Publishers**

Boston San Diego New York
London Sydney Tokyo Toronto

This book is printed on acid-free paper. ♾



ACADEMIC PRESS, INC.
1250 Sixth Avenue, San Diego, CA 92101

*United Kingdom Edition published by*
ACADEMIC PRESS LIMITED
24–28 Oval Road, London NW1 7DX

Library of Congress Cataloging-in-Publication Data

Ramachandran, B. (Balasubrahmanyan), date.
Functional equations in probability theory / Ramachandran Balasubrahmanyan, Ka-Sing Lau.
p. cm. — (Probability and mathematical statistics)
Includes bibliographical references and index.
ISBN 0-12-437730-0 (alk. paper)
1. Functional equations. 2. Probabilities. I. Lau. Ka-Sing. II. Title. III. Series.
QA431.R348 1991
515 — dc20 90-28769
CIP

Printed in the United States of America
91 92 93 94 9 8 7 6 5 4 3 2 1

Affectionately dedicated
to
Asha and Ashwin
and
Eveline

Hardcore mathematics is, in some sense, the same as it has always been. It is concerned with problems that have arisen from the actual physical world and other problems inside mathematics having to do with numbers and basic calculations, solving equations. This has always been the main part of mathematics. Any development that sheds light on these is an important part of mathematics.

—M. F. Atiyah

# Contents

# Preface

A more appropriate (and less pretentious) title for this monograph would have been "Some Functional Equations in Probability Theory". We are far from claiming that the material covered includes all or even most such studies; the choice of topics has been dictated by the "active" research interests of the authors. Topics of related interest (mostly nonoverlapping in content) can be found in the two monographs, both originally published in Russian: "Characterization Problems in Mathematical Statistics," by A. M. Kagan, Yu. V. Linnik, and C. R. Rao (Wiley, 1973), and "Characterization Problems Associated with the Exponential Distribution," by T. A. Azlarov and N. A. Volodin (Springer, 1986). Another recent publication of interest to analytical probabilists is "Analytic Methods of Probability Theory," reporting mostly the work of the Leipzig School, by H.-J. Rossberg, B. Jesiak, and G. Siegel (Akademie-Verlag, Berlin, 1985).

The present monograph was projected in 1986, when four areas of interest to either or both of its authors appeared to have reached a definitive stage, warranting their compilation in book form. These areas, to which a fifth has been added, in place of an appendix as originally proposed, and the contents of the various chapters are described in the Introduction.

Several parts of this monograph were presented by one of the authors in a course of lectures (under the same title) delivered in the Winter Semester of 1986–1987 at the Department of Mathematics and Statistics, McGill University, Montreal, while visiting there at the instance of Professor V. Seshadri. A write-up of these lectures was also issued as one of the technical reports of the department. The support extended by the department (and the governments of Canada and Quebec through the NSERC and FCAR grants) is acknowledged here, as is the support of the Indian Statistical Institute during the preparation of this monograph. Part of the organization of the monograph was done while one of the authors was visiting the Delhi

Centre of the Institute under a Fulbright fellowship, Indo–U.S. cultural exchange program; this support, as well as the arrangements from Professors P. Masani and C. R. Rao, and that of the Department of Mathematics and Statistics, University of Pittsburgh, are gratefully acknowledged here.

Professor J. Deny cheerfully came out of retirement to respond to a few queries and Professor C. Berg provided some valuable clarifications. We thank them both for their ready help. We would like to thank Professors C. Lennard, W. B. Zeng and Dr. C. H. Chu for many helpful remarks and corrections. Also, a referee is gratefully acknowledged for his many helpful suggestions concerning improvement of the manuscript.

We are grateful to Professor Z. W. Birnbaum, Founding Editor of this series, and to Charles B. Glaser, Editor, Pascha Gerlinger, Production Editor, and Joseph Clifford, Editorial Assistant, of Academic Press, for their unfailing courtesy, active interest, and help in processing this monograph. The contents of this monograph had been of abiding interest to the late Professor E. Lukacs (who was an editor of this series), and we take this opportunity of paying our tribute to his memory as well as to that of the late, great academician Yu. V. Linnik, whose ideas and methods will continue to inspire succeeding generations of analysts and probabilists.

We thank Mrs. LaVerne Lally of the Department of Mathematics and Statistics, University of Pittsburgh, for typing assistance, undertaken cheerfully and executed promptly and efficiently.

We shall be thankful to be informed of any misprints, obscurities, omissions, and blunders.

B. Ramachandran, New Delhi
K. S. Lau, Pittsburgh

# Introduction

We attempt here to motivate and briefly describe the several problems discussed in this monograph and outline the contents of the various chapters.

The main areas of discussion may be summarized as follows:

**1.** The Integrated Cauchy Functional Equation (ICFE)

$$f(x) = \int_{\mathbb{R}_+} f(x+y)\, d\sigma(y), \qquad x \geq 0, \tag{0.1}$$

where $f \geq 0$, and $\sigma$ is a positive measure; its applications and ramifications.

**2.** The problem of identical distribution of two linear forms in independent and identically distributed random variables, with particular reference to the context of the common distribution of these random variables being normal.

**3.** Characteristic functions $f$ satisfying a functional equation of the form

$$f(t) = \prod_{j=1}^{\infty} (f(b_j t))^{a_j}, \qquad t \in \mathbb{R};\ a_j > 0,\ b_j \text{ real},\ |b_j| < 1. \tag{0.2}$$

**4.** Characterizations of the underlying processes as stable or semistable, through the hypothesis of identical distribution to within a shift (location parameter) of two stochastic integrals defined in the sense of convergence in probability.

**5.** The skew-convolution equation $\mu = \mu \bullet \sigma$ ($\mu, \sigma$ being measures, $\sigma$ given, $\mu$ to be solved for) and the related equation $f = f \bullet \sigma$ ($\sigma$ a given measure, $f$ a real-valued function to be solved for) on subsemigroups of $\mathbb{R}^d$.

We now outline briefly their discussion in this monograph, providing some historical perspective as well.

**1.** The Cauchy functional equation in its basic form is a familiar one: $f: \mathbb{R} \to \mathbb{R}$ is continuous and satisfies the equation

$$f(x + y) = f(x)f(y) \qquad \text{for all } x, y \in \mathbb{R}. \tag{0.3}$$

The solutions, as is well-known, are $f(x) = e^{ax}$ for all $x \in \mathbb{R}$, for some $a \in \mathbb{R}$. The solutions continue to hold on $(0, \infty)$ if (0.3) is assumed to hold only for $x > 0$, $y > 0$. An immediate consequence (in fact, a restatement) is that the "memorylessness property" of the exponential law characterizes it: If a random variable $X \geq 0$ has the property that, for every $x > 0$, $y > 0$, $P\{X > x + y \mid X > y\} = P\{X > x\}$, then $X$ has an exponential distribution. An analogous result holds for the discrete analog of the exponential law, namely the geometric law. Both laws even have a "strong lack of memory": Thus, if $X$ is exponentially distributed and $Y \geq 0$ is a random variable independent of $X$, then $P\{X > Y + x \mid X > Y\} = P\{X > x\}$ for $x > 0$. An investigation of the converse proposition—under what conditions, if any, on $Y$, this property will imply that $X$ has an exponential distribution—is among the first instances leading to what has come to be now called an Integrated Cauchy Functional Equation (ICFE): (0.1); a motivation for this nomenclature is attempted in the introductory section of Chapter 2. This as well as several types of probabilistic questions involving the above-mentioned laws can be answered satisfactorily by the solutions of the ICFE; they are given in Section 2.5 as applications of the ICFE.

The ICFE also arises in Feller's discussion of the "renewal theorem." In Volume I of his work, "An Introduction to Probability Theory and Its Applications," in Chapter XIII, Section 11, the following result is established (Lemma 3, p. 337):

"Let $\{w_n\}$, $n \in \mathbb{Z}$, be a ("doubly infinite") sequence of numbers such that $0 \leq w_n \leq 1$, and such that $w_n = \sum_{k=1}^{\infty} f_k w_{n-k}$ for every $n$, where the $f_n \geq 0$, $n \in \mathbb{N}$, are such that $\sum_{k=1}^{\infty} f_k = 1$, and further the greatest common divisor of those $n$ for which $f_n > 0$ is unity. If $w_0 = 1$, then $w_n = 1$ for all $n$."

In Volume II of the same work, the renewal theorem in the non-arithmetic case—Section XI.2, Lemma 1 (half-line case) and Section XI.9, Lemma 1 and Corollary (whole line case)—is established; Feller cites the paper by Choquet and Deny (1960), of which we have more to say in paragraph (5).

The ICFE (in the discrete case) also arises in the context of a property of the Poisson law. Suppose $X \geq 0$ is an integer-valued random variable, and suppose a damage process (such as birds eating insect eggs on a leaf) reduces $X$ to $Y$ according to a binomial damage model: that is, for some $0 < p < 1$,

$$P\{Y = r \mid X = n\} = \binom{n}{r} p^r (1-p)^{n-r}; \qquad r = 0, 1, \ldots, n;\ n = 1, 2, \ldots. \tag{0.4}$$

If $X$ has a Poisson distribution, an easy computation establishes that

$$P\{Y = r \mid X \text{ damaged}\} = P\{Y = r\} = P\{Y = r \mid X \text{ undamaged}\}, \qquad r = 0, 1, \ldots. \tag{0.5}$$

A natural question then, whether (0.5) characterizes the Poisson law (as the distribution of $X$) subject to the damage model (0.4) holding, was answered affirmatively by H. Rubin and C. R. Rao, who used the Bernstein theorem (see Section 1.2) characterizing (normalized) "completely monotone" functions on $(0, \infty)$ as Laplace–Stieltjes transforms of probability measures supported on $[0, \infty)$. It was later pointed out by Shanbhag (1977) that the problem reduces to solving for $\{v_m\}$ given $\{p_n\}$, $m, n = 0, 1, \ldots$, where both are nonnegative real sequences and satisfy the sequence of relations

$$v_m = \sum_{n=0}^{\infty} p_n v_{m+n}, \qquad m = 0, 1, \ldots,$$

falling under the scope of Theorem 2.1.2 below, established by him for this purpose. This problem—the Rao–Rubin characterization of the Poisson law—is dealt with in Section 2.5. The basic idea of Shanbhag's, of considering $\sup_m(v_{m+1}/v_m)$, underlies the other intensive studies of the ICFE on subsemigroups of $\mathbb{R}$ in the early and mid-1980s aimed at streamlining the disparate, *ad hoc* approaches of earlier papers to solve the ICFE in various guises. In particular, Theorem 2.2.4 was established in Lau and Rao (1982) and its proof simplified in Ramachandran (1982a); it is adequate for dealing with several characterizations of the exponential and geometric laws (Section 2.5).

The ICFE on $\mathbb{Z}_+$ and on $\mathbb{R}_+$ is dealt with in Sections 2.1, 2.2, and 2.3, and the ICFE on $\mathbb{R}_+$ involving signed measures in Section 2.4.

ICFE's on $\mathbb{R}_+$ with error terms arise not only in the context of the "stability" of solutions of exact ICFE's—as is only to be expected—but also in the course of solving "exact" functional equations of a type that may be called "exponentiated ICFE's"; these form the material of Chapter 4. The ICFE on $\mathbb{Z}$ and on $\mathbb{R}$ are dealt with in Chapter 8; a proof similar to Theorem 2.2.4, and a proof using the Krein–Milman theorem are given. A variant of the ICFE given by assuming that (0.1) holds for $x \geq 0$, while the integral is extended over $\mathbb{R}$ is also considered. It makes use of the Wiener–Hopf decomposition of a positive measure; this result together with Theorem 2.2.4 can be used to give yet another approach to the ICFE on $\mathbb{R}$.

**2.** In a classic and monumental study of Yu. V. Linnik's (1953a, b), the problem of identical distribution of two linear forms $\sum_{j=1}^{n} a_j X_j$, $\sum_{j=1}^{n} b_j X_j$ ($a_j$, $b_j$ real) in independent and identically distributed random variables $X_j$ was considered; under the assumption that $\max_j |a_j| \neq \max_j |b_j|$, a set of necessary and sufficient conditions was obtained for the $X_j$ to be normal. A. A. Zinger (1975) completed the study by considering the complementary cases. Zinger also considerably simplified Linnik's original proof, considering at the same time a more general form of the basic functional equation; his proof is to be found in Kagan *et al.* (1973). In Riedel (1985), a proof (applicable to infinite linear forms as well, under certain assumptions) using the Mellin (rather than the Laplace) transform as the basic tool, and appealing only to the elementary theory of meromorphic functions was provided. In Chapter 5, we provide this proof for the "sufficiency" part, and Linnik's proof for the "necessity" part; also given are certain other results concerning related functional equations, with particular reference to the normal law.

**3.** R. Shimizu (1968) and B. Ramachandran and C. R. Rao (1970) considered characteristic functions satisfying a functional equation of the form (0.2)—the former, *per se*, and the latter as arising in discussing a particular case of a regression problem. The infinite divisibility of such an $f$ was established (under some now superfluous conditions on the constants $a_j$, $b_j$) and the form of the Lévy representation obtained; for an account of these, see Kagan *et al.* (1973), Section 5.6.1. In later papers by Shimizu and by Shimizu and Davies, still imposing some conditions on the constants and on the measures concerned, it was established that such an $f$ is, to within a location parameter, a semistable characteristic function. The work on the ICFE in the 1980s made it possible to do away with (most) superfluous restrictions and the "final form" of the solution is the content of Sections 3.3 and 3.4.

**4.** Following on Linnik's work on the normal law, discussed in paragraph (2), it was natural to raise similar questions concerning stable and semistable laws. The work of Shimizu and Shimizu–Davies in this context has already been cited in paragraph (3). In the format that two stochastic integrals (defined in the sense of convergence in probability) with respect to a stochastic process $\{X(t): t \geq 0\}$—assumed to be continuous in probability and to have homogeneous and independent increments—are identically distributed to within a location parameter, M. Riedel (1980b) obtained a set of necessary and sufficient conditions in order that the underlying process be a stable process. These considerations were extended to semistable processes in B. Ramachandran (1991a). These studies form the content of Chapter 6, along with some related material such as a strengthened version of a preliminary result by E. Lukacs in the area.

**5.** The convolution equation $\mu * \sigma = \mu$, where $\mu$ and $\sigma$ are Radon measures on a locally compact abelian group, with $\sigma$ given and $\mu$ to be solved for, was the subject of the papers Choquet and Deny (1960) and Deny (1960). The latter paper in particular remained unknown to most analytical probabilists; the former, detailing the solution to the case where $\mu$ and $\sigma$ are probability measures—usually referred to as the Choquet–Deny theorem—is cited by Feller (1971), Volume II, in his discussion of renewal theory. The *ad hoc* pre-1980s solutions to various forms of the ICFE were obtained in ignorance of Deny (1960). Recently, Deny's discussion was extended to (suitable) semigroups of locally compact abelian groups. These results, specialized to subsemigroups of Euclidean spaces, are presented in Chapter 9.

The contents of the chapters and sections not already cited are briefly described now. Chapter 1 collects together several auxiliary results on the Cauchy functional equations, on analysis, and on distribution functions and characteristic functions on $\mathbb{R}$. The reader will do well to start with Chapter 2 and come back to relevant parts of Chapter 1 as the need arises. Chapter 3 contains a discussion mainly of stable laws, with emphasis on the functional equations approach to obtaining the closed-form formulas for such characteristic functions, and briefly of semistable and "generalized stable" laws. Chapter 7, devoted to "miscellaneous results," deals with three topics that are more appropriately termed "equidistribution problems," or just plain "distribution problems"; even here, Section 7.2 deals with a functional equation with particular reference to normal characteristic functions as solutions. Briefly, these topics are: (1) If $X$ and $Y$ are independent and identically distributed random variables with moments of all orders such that $X + Y$ and $XY$ have the same distribution, identify the common d.f. of $X$ and $Y$. (2) If $X$ and $Y$ are independent and identically distributed random variables, when do assumptions on $(aX + bY)^2$ having a chi-square distribution for one or more choices of the real constants $a$ and $b$ imply that $X$ and $Y$ are standard normal variables? (3) To what extent can the normal law be characterized as the common distribution of the independent and identically distributed random variables concerned, if (i) a quadratic form (of a specified kind) in them has a noncentral chi-square distribution, or (ii) two homogeneous quadratic forms in them both have chi-square distributions (with appropriate degrees of freedom)?

CHAPTER

# 1

# Background Material

In this chapter, we collect together certain auxiliary results required in the following chapters. Proofs are given where they either have not appeared in book form or are not otherwise readily available. For other proofs, references to standard treatises are provided.

## 1.1. CAUCHY FUNCTIONAL EQUATIONS

We will use the following notations:

- $\mathbb{R}$ the set of real numbers,
- $\mathbb{R}_+$ the set of nonnegative real numbers,
- $\mathbb{N}$ the set of natural numbers,
- $\mathbb{Z}$ the set of integers, and
- $\mathbb{Z}_+$ the set of nonnegative integers.

The classical Cauchy functional equation concerns a continuous $f: \mathbb{R} \to \mathbb{R}$ satisfying

$$f(x + y) = f(x)f(y), \qquad \forall\, x, y \in \mathbb{R},$$

the only solutions being the exponential functions. In this section, we will consider some variants of this equation, having $\mathbb{R}_+$ or subsemigroups of $\mathbb{R}_+$ as domains, which will be needed for later developments.

We will use the usual symbol $(p, q) \in \mathbb{N}$ to denote the greatest common divisor (g.c.d.) of $p, q \in \mathbb{Z}$. It is well known that if $(p, q) = d$, then there exist $r, s \in \mathbb{Z}$ such that $rp + sq = d$. It follows that any multiple of $d$ can be represented as $ap + bq$ for suitable $a, b \in \mathbb{Z}$. Of particular interest is a situation where $p$, $q$, $a$, $b$ are all in $\mathbb{Z}_+$, as given by the following result.

**Lemma 1.1.1.** *Let $p, q \in \mathbb{N}$, and let $d = (p, q)$. Then any multiple of $d$ that is $\geq pq$ can be represented in the form $ap + bq$ for some $a, b \in \mathbb{Z}_+$.*

***Proof.*** We may assume without loss of generality that $(p, q) = 1$. For any fixed $n \in \mathbb{N}$, there exist $r, s \in \mathbb{Z}$ such that $n = rp + sq$. Since $n, p, q > 0$, at least one of the numbers $r$ and $s$ is positive, say $s$. Write $s = tp + b$, where $t, b \in \mathbb{Z}_+$ and $0 \leq b < p$, and let $a = (r + tq)$. It follows that $n = ap + bq$. If $n \geq pq$, we have

$$ap = n - bq \geq pq - bq > 0.$$

So $a > 0$, and the lemma is proved.

**Theorem 1.1.2.** *Let $d$ be the g.c.d. of $p_1, \ldots, p_n \in \mathbb{N}$, $n \geq 2$. Then, there exists $r \in \mathbb{N}$ such that, for any $m \geq r$,*

$$md = a_1 p_1 + \cdots + a_n p_n$$

*for some $a_1, \ldots, a_n$ in $\mathbb{Z}_+$.*

***Proof.*** If $n = 2$, the assertion reduces to Lemma 1.1.1. Assume that it holds for $n - 1$. Let $d'$ be the g.c.d. of $p_1, \ldots, p_{n-1}$, and let $d = (d', p_n)$. By Lemma 1.1.1, there exists $m_0 \in \mathbb{Z}_+$ such that, for $m \geq m_0$,

$$md = ad' + bp_n$$

for some $a, b \in \mathbb{Z}_+$. By the induction assumption, there exists $s \in \mathbb{Z}_+$ such that

$$(s + a)d' = \sum_{j=1}^{n-1} a_j p_j,$$

for some $a_1, \ldots, a_{n-1}$ in $\mathbb{Z}_+$. Then, letting $a_n = b$, we have

$$md + sd' = \sum_{j=1}^{n} a_j p_j \qquad \text{for } m \geq m_0.$$

Since $sd'$ on the left-hand side is a multiple of $d$, the assertion follows for $n$ also, and the theorem is proven.

The following is a simple illustration of the use of this theorem.

**Corollary 1.1.3.** *Let $A \subseteq \mathbb{Z}_+$ and let $d$ be the g.c.d. of $A$. Suppose $g: \mathbb{Z}_+ \to \mathbb{R}$ satisfies*

$$g(m + n) = g(m) \qquad \forall\, m \in \mathbb{Z}_+ \text{ and } n \in A.$$

*Then, $g$ has period $d$: $g(m + d) = g(m)$ for all $m \in \mathbb{Z}_+$.*

***Proof.*** Since $d$ is the g.c.d. of $A$, there exist $p_1, \ldots, p_n \in A$ such that $d$ is the g.c.d. of $p_1, \ldots, p_n$. Let $r$ be as in Theorem 1.1.2; then, both $rd$ and $(r+1)d$ can be represented as linear combinations of $p_1, \ldots, p_n$ with coefficients in $\mathbb{Z}_+$. The assumption thus implies that

$$g(m+d) = g((m+d)+rd) = g(m+(r+1)d) = g(m) \qquad \forall\, m \in \mathbb{Z}_+.$$

Let $A \subseteq \mathbb{R}\backslash\{0\}$. We also say that $d > 0$ is the g.c.d. of $A$ if, for each $x \in A$, the quotient $x/d$ is an integer, and $d$ is the largest number with this property. The proof of Theorem 1.1.2 yields the following.

**Theorem 1.1.4.** *Let $d$ be the g.c.d. of $p_1, \ldots, p_n \in \mathbb{R}_+\backslash\{0\}$, $n \geq 2$. Then, there exists $r \in \mathbb{N}$ such that, for any $m \geq r$,*

$$md = a_1 p_1 + \cdots + a_n p_n$$

*for some $a_1, \ldots, a_n \in \mathbb{Z}_+$.*

**Lemma 1.1.5.** *Let $f\colon \mathbb{R}_+ \to \mathbb{R}$ be continuous, and $f(0) = 1$. Suppose the set*

$$T = \{t \geq 0 : f(x+t) = f(x)f(t) \quad \forall\, x \geq 0\}$$

*contains a point other than 0. Then one of the following holds:*

(a) $T = \{0\} \cup [\gamma, \infty)$ *for some $\gamma > 0$, and $f(x) = 0$ for $x \geq \gamma$;*
(b) $T = \mathbb{R}_+$, *and $f(x) = e^{\alpha x}$ for some $\alpha \in \mathbb{R}$;*
(c) $T = \rho\mathbb{Z}_+$ *for some $\rho > 0$, and $f(x) = p(x)e^{\alpha x}$ for some $\alpha \in \mathbb{R}$, and $p$ has period $\rho$.*

***Proof.*** We begin by noting that (i) $s, t \in T$ implies that $s + t \in T$, and (ii) if $s, t \in T$ and $s < t$ with $f(s) \neq 0$, then $t - s \in T$ since

$$f(x + (t-s))f(s) = f(x+t) = f(x)f(t) = f(x)f(t-s)f(s),$$

which implies that $f(x + (t-s)) = f(x)f(t-s)$.

First, consider the case where $f$ vanishes at some point of $T$; let $\gamma = \inf\{t \in T : f(t) = 0\}$. Then, $f(\gamma) = 0$, and $\gamma \in T$ since $f$ is continuous. For $s > \gamma$, $f(s) = f(s-\gamma)f(\gamma) = 0$. $\gamma = 0$ is impossible since $f(0) = 1$, and thus (a) holds in this case.

Suppose $f$ is nonvanishing on $T$; let $\rho = \inf\{t > 0 : t \in T\}$. If $\rho = 0$, then $T$ is a dense semigroup in $\mathbb{R}_+$ (by (i)), and $T$ being closed, $T = \mathbb{R}_+$. Also

$$f(x+t) = f(x)f(t), \qquad \forall\, x, t \in \mathbb{R}_+,$$

implies that $f(x) = e^{\alpha x}$ for some $\alpha \in \mathbb{R}$, and (b) holds. If $\rho > 0$, we claim that $T = \rho\mathbb{Z}_+$. Indeed, since $T$ is closed, $\rho \in T$ and hence (by (i)) $\rho\mathbb{Z}_+ \subseteq T$.

On the other hand, if there exists $t \in T$ such that, for some $n \in \mathbb{Z}_+$, $np < t < (n+1)\rho$, then (since $f(np) \neq 0$) $t - np \in T$ (by (ii)), and $0 < t - np < \rho$. This contradicts the definition of $\rho$; hence, $T \subseteq \rho\mathbb{Z}_+$ and the claim is proved. Now, for any $x \in \mathbb{R}_+$, write $x = n\rho + \bar{x}$, $0 \leq \bar{x} < \rho$. Then,

$$f(x) = f(\bar{x})(f(\rho))^n = f(\bar{x})(f(\rho))^{-\bar{x}/\rho}(f(\rho))^{(x/\rho)} = p(x)e^{\alpha x},$$

for an obvious choice of $\alpha$ and $p(\cdot)$ where $p(x) = p(x + \rho)$. (c) holds in this case.

**Theorem 1.1.6.** *Let $\phi \neq A \subseteq \mathbb{R}_+\backslash\{0\}$, and let $f: \mathbb{R}_+ \to \mathbb{R}$ be continuous and satisfy*

$$f(x + y) = f(x)f(y) \qquad \forall\, x \in \mathbb{R}_+,\, y \in A.$$

*Then one of the following holds:*

(a) *there exists $\gamma \geq 0$ such that $f(x) = 0$ for $x \geq \gamma$,*
(b) *$f(x) = e^{\alpha x}$ for some $\alpha > 0$;*
(c) *$f(x) = p(x)e^{\alpha x}$, where $p(0) = 1$, and $p$ has period $d$, the g.c.d. of $A$.*

***Proof.*** Let $T$ be defined as in Lemma 1.1.5; then, $A \subseteq T$. If $T = \{0\} \cup [\gamma, \infty)$ for some $\gamma > 0$, then (a) holds. Otherwise, either $A$ does not have a g.c.d., thus $A \nsubseteq \rho\mathbb{Z}$ for any $\rho > 0$, hence $T = \mathbb{R}_+$, and (b) holds; or $A$ has a g.c.d. (denoted by) $d$, and then $d$ is the g.c.d. for some $p_1, \ldots, p_n \in A$. Theorem 1.1.4 implies that, for $m$ large,

$$md = a_1 p_1 + \cdots + a_n p_n,$$

where $a_1, \ldots, a_n \in \mathbb{Z}_+$; hence, $md \in T$. Therefore, $T$ can only have the form (c) given by Lemma 1.1.5, $d\mathbb{Z}_+ \subseteq T$, and assertion (c) of the theorem holds.

**Proposition 1.1.7.** *Let $\phi \neq A \subseteq \mathbb{R}_+\backslash\{0\}$, and $f: \mathbb{R}_+\backslash\{0\} \to \mathbb{R}$ be continuous and satisfy*

$$f(x + y) = f(x), \qquad \forall\, x \in \mathbb{R}_+,\, y \in A.$$

*Then, either $f$ is a constant or $f$ has period $d$, where $d$ is the g.c.d. of $A$.*

***Proof.*** We can apply the same argument as in Lemma 1.1.5 and Theorem 1.1.6 with

$$T = \{t \geq 0 : f(x + t) = f(x)\ \forall\, x \in \mathbb{R}_+\}.$$

To conclude this section, we give two more results concerning the exponential functions.

**Proposition 1.1.8.** *Let $S \subseteq \mathbb{R}_+$ be a subsemigroup. Let $g: S \to \mathbb{R}_+\backslash\{0\}$ be nonincreasing and satisfy the relations*

$$g(x + y) \leq g(x)g(y), \qquad g(2y) = g(y)^2, \qquad \forall\, x, y \in S.$$

*Then, $g(x) = e^{-\alpha x}$ for some $\alpha \geq 0$.*

***Proof.*** We first observe that $g(my) = (g(y))^m$ for all $y \in S$. Indeed, by iterating the equality that has been assumed, we have

$$g(2^n y) = g(y)^{2^n}, \qquad n \in \mathbb{Z}_+, y \in S.$$

For any $m \in \mathbb{Z}_+$, let $n \in \mathbb{Z}_+$ be such that $2^n > m$; then,

$$g(2^n y) \leq g(my)g((2^n - m)y) \leq g(y)^m g(y)^{2^n - m} = g(2^n y).$$

Hence, all the inequalities above are actually equalities, and in particular we have $g(my) = g(y)^m$.

If now $x, y \in S$, then, for any $n \in \mathbb{Z}_+$, there exists $m \in \mathbb{Z}_+$ such that

$$mx \leq ny < (m + 1)x.$$

Since $g$ is decreasing on $S$, we have

$$g(mx) \geq g(ny) \geq g((m + 1)x),$$

so that

$$g(x)^m \geq g(y)^n \geq g(x)^{m+1},$$

or, equivalently,

$$g(x) \geq g(y)^{n/m} \geq g(x)^{(m+1)/m}.$$

By letting $n \to \infty$, we see that $n/m \to x/y$, so that

$$g(x)^{1/x} = g(y)^{1/y} \qquad \text{for every } x, y \in S.$$

This implies that there exists $\alpha \geq 0$ such that $g(x) = e^{-\alpha x}$ for all $x \in S$.

**Proposition 1.1.9.** *Let $g: \mathbb{N} \to \mathbb{R}_+$ be nonincreasing and satisfy the functional equation*

$$g(mn) = g(m)g(n) \qquad \forall\, m, n \in \mathbb{N}. \tag{1.1.1}$$

*Then either $g \equiv 0$ or $1$, or there exists $r < 0$ such that $g(m) = m^r$ for all $m \in \mathbb{N}$.*

***Proof.*** Since $g(1) = g(1)^2$, $g(1) = 0$ or $1$. If $g(1) = 0$, then $g(m) = g(m)g(1) = 0$ for all $m$. If $g(1) = 1$, and if $g(2)$ is also 1, then $g(2^k) = 1$ for all $k \in \mathbb{N}$, and since $g$ is nonincreasing, it follows that $g \equiv 1$. Thus, the only nontrivial case is $g(1) = 1$, $g(2) < 1$.

Let $p, q, m \in \mathbb{N}$ be such that $p/q < \log m/\log 2$. Then, $2^p < m^q$, which implies, by (1.1.1) that

$$g(2)^p = g(2^p) \geq g(m^q) = g(m)^q,$$

i.e., $p/q \leq \log g(m)/\log g(2)$. A dual argument shows that any positive rational $p/q > \log m/\log 2$ is also $\geq \log g(m)/\log g(2)$. Hence, we have $\log g(m) = r \log m$ for all $m \in \mathbb{N}$, where $r = \log g(2)/\log 2 < 0$. The assertion of the proposition follows.

***Remark.*** If the monotonicity condition is not imposed, then $g$ may be defined *arbitrarily* at the primes, and (1.1.1) only *defines* $g$ at the other integers, in view of the fact that any $m \in \mathbb{N}$ can be uniquely expressed as a product of powers of primes.

## 1.2. AUXILIARY RESULTS FROM ANALYSIS

Let $\mathcal{B}$ denote the family of Borel subsets of $\mathbb{R}$, and let $\mu$ be a measure on $(\mathbb{R}, \mathcal{B})$. $\mu$ will be called a *Borel measure* on $\mathbb{R}$. The *support* of $\mu$, denoted by supp $\mu$, is the set $\mathbb{R} \setminus \bigcup \{V: \mu(V) = 0, V \text{ open}\}$. $\mu$ is called *$\sigma$-finite* if there exists a sequence $\{E_n\}$ of Borel subsets such that $\bigcup_n E_n = \mathbb{R}$, and $\mu(E_n) < \infty$ for every $n$; we call a positive Borel measure *regular* if, for every $E \in \mathcal{B}$,

$$\mu(E) = \inf\{\mu(V): E \subseteq V, V \text{ open}\}$$

and, for every $E \in \mathcal{B}$ with $\mu(E) < \infty$,

$$\mu(E) = \sup\{\mu(K): K \subseteq E, K \text{ compact}\}.$$

It is easy to show that if $\mu$ is regular, then $\mu$ is $\sigma$-finite; the converse is, however, not true in general: E.g., let $\mu = \delta_0 + \sum_{n=1}^{\infty} \delta_{1/n}$, where $\delta_x$ is the point mass measure at $x$; then, $\mu$ is $\sigma$-finite, but the first condition of regularity is not satisfied for $E = \{0\}$.

A well-known condition for a positive Borel measure on $\mathbb{R}$ to be regular is that $\mu(K) < \infty$ for every compact set $K$ in $\mathbb{R}$ (Rudin, 1974). As a simple corollary of this, we have the following.

**Proposition 1.2.1.** *Let $\mu$ be a positive Borel measure on $\mathbb{R}$. Suppose there exists a positive continuous function $f$ on $\mathbb{R}$ such that $\int_{\mathbb{R}} f\, d\mu < \infty$. Then, $\mu$ is regular.*

***Proof.*** Let $K$ be a compact subset of $\mathbb{R}$. Then, $f$ has a positive minimum $c$ on $K$. Hence,

$$0 < c\mu(K) \leq \int_{\mathbb{R}} f\, d\mu < \infty,$$

and the above criterion applies.

**Proposition 1.2.2.** *Let $\mu$ be a positive $\sigma$-finite Borel measure on $\mathbb{R}$. Suppose $\mu(\mathbb{R}) > a$ for some $a \in \mathbb{R}_+$. Then, there exists a Borel subset $E$ with compact closure, such that $\infty > \mu(E) > a$.*

***Proof.*** Let $\delta > 0$ be such that $\mu(\mathbb{R}) > a + 2\delta$. Let $\{E_n\}$ be a sequence of pairwise disjoint Borel subsets such that $\bigcup E_n = \mathbb{R}$ and $a_n = \mu(E_n) < \infty$. For $n, k \in \mathbb{Z}_+$, let $a_{nk} = \mu([0, k] \cap E_n)$. Then, $\lim_{k\to\infty} a_{nk} = a_n$; hence, there exist $N$ and $K$ large enough such that $\sum_{n=1}^{N} a_{nK} > a + \delta$ (for otherwise, $\sum_{n=1}^{\infty} a_{nk} \le a + \delta$ for all $k$, which implies that $\mu(\mathbb{R}) = \sum_{n=1}^{\infty} \mu(E_n) \le a + \delta$, a contradiction). Then, $E = [0, K] \cap (\bigcup_{n=1}^{N} E_n)$ is a set with the required properties.

Let $g$ be a Borel measurable function on $\mathbb{R}$, and let $\mu$ be a positive regular Borel measure on $\mathbb{R}$. Suppose there exists an $A \ge 0$ such that $\mu\{|g| > A\} = 0$, and $\mu\{|g| > A - \varepsilon\} > 0$ for every $\varepsilon > 0$. Then, $A$ is called the *$\mu$-essential supremum* of $g$; we write $A = \|g\|_{\mu,\infty}$. If no such $A$ exists, then we define $\|g\|_{\mu,\infty} = \infty$.

**Theorem 1.2.3.** *Let $g, \mu$ be defined as above with $\mu(\mathbb{R}) < \infty$, and let*

$$A_r = \left(\int_{\mathbb{R}} |g|^r \, d\mu\right)^{1/r}.$$

*Then* $\lim_{r\to\infty} A_r = \|g\|_{\mu,\infty}$.

***Proof.*** First consider the case where $\|g\|_{\mu,\infty} = A < \infty$. For every $\varepsilon > 0$, we have the obvious estimate:

$$A_r \le (A + \varepsilon)(\mu(\mathbb{R}))^{1/r} \to (A + \varepsilon) \qquad \text{as } r \to \infty.$$

Since $\varepsilon > 0$ is arbitrary, it follows that $\limsup_{r\to\infty} A_r \le A$. Also, for every $\varepsilon > 0$,

$$A_r \ge (A - \varepsilon)(\mu\{|g| > A - \varepsilon\})^{1/r} \to (A - \varepsilon) \qquad \text{as } r \to \infty,$$

and it follows that $\liminf_{r\to\infty} A_r \ge A$. Hence, the assertion holds in this case.

If $A = \infty$, the latter half of the argument, suitably modified, yields the assertion of the theorem.

The following special cases are immediate consequences, on defining $g, \mu$ suitably.

**Corollary 1.2.4.** (a) *Let $\{a_j\}$ and $\{r_j\}$ be two sequences of positive real numbers. Suppose $\sum r_j a_j^s < \infty$ for some $s > 0$. Then,*

$$\lim_{r\to\infty} \left(\sum r_j a_j^r\right)^{1/r} = \sup_j a_j.$$

(b) *Let $I$ be a compact interval, and let $g, h: I \to \mathbb{R}_+$, $h \neq 0$, be continuous. Then,*

$$\lim_{r\to\infty} \left( \int_I (g(t))^r h(t)\, dt \right)^{1/r} = \max_{x\in I} g(x).$$

**Corollary 1.2.5.** *Let $\{a_j\}$, $\{b_j\}$ be nonnegative real sequences such that $\sum a_j^n$, $\sum b_j^n < \infty$ for some $n \in \mathbb{N}$. Suppose $\{a_j\}$ is not a permutation of $\{b_j\}$. Then, there exist at most finitely many $n$ such that $\sum a_j^n = \sum b_j^n$.*

***Proof.*** Suppose the statement is false, and that equality holds for a subsequence $\{k_n\}$, which we shall denote by $\{n\}$ itself for convenience. Since $\{a_j\}$, $\{b_j\}$ converge to 0, there exist $k, l$ such that $a_k = \max_{1\le j<\infty} a_j$, $b_l = \max_{1\le j<\infty} b_j$. Corollary 1.2.4(a) implies that

$$a_k = \lim_{n\to\infty} \left( \sum a_j^n \right)^{1/n} = \lim_{n\to\infty} \left( \sum b_j^n \right)^{1/n} = b_l,$$

and hence

$$\sum_{j\neq k} a_j^n = \sum_{j\neq l} b_j^n.$$

By repeating the argument, we conclude that $\{a_j\}$ is actually a permutation of $\{b_j\}$, which contradicts our assumption.

Let $\mu$ be a $\sigma$-finite signed Borel measure on $\mathbb{R}_+$, and let $|\mu|$ be its total variation measure. Suppose

$$\int_0^\infty e^{-tx}\, d|\mu|(x) < \infty$$

for some $t \in \mathbb{R}$ and let $\sigma_\mu$ denote the infimum of such $t$. Then, $\hat{\mu}$ defined by

$$\hat{\mu}(t) = \int_0^\infty e^{-tx}\, d\mu(x), \qquad t > \sigma_\mu,$$

is called the *Laplace transform* of $\mu$.

Similarly, if $g: \mathbb{R}_+ \to \mathbb{R}$ (or $\mathbb{C}$, the complex plane) is Borel measurable, and if $\int_0^\infty e^{-tx}|g(x)|\, dx < \infty$ for some $t \in \mathbb{R}$, let $\sigma_g$ denote the infimum of such $t$. Then, the Laplace transform $\hat{g}$ of $g$ is defined by

$$\hat{g}(t) = \int_0^\infty e^{-tx} g(x)\, dx, \qquad t > \sigma_g.$$

If $I \subseteq \mathbb{R}$ is an open interval and $f: I \to \mathbb{R}$ has derivatives of all orders there, $f$ is referred to as *completely monotone* on $I$ if $(-1)^n f^{(n)} \geq 0$ on $I$ for all $n \in \mathbb{Z}_+$. A celebrated result in classical analysis concerning such functions is the following.

**Theorem 1.2.6** (S. N. Bernstein). *If $f\colon \mathbb{R}_+ \to \mathbb{R}_+$ is continuous at the origin, with $f(0) = 1$, and if $f$ is completely monotone on $(0, \infty)$, then $f$ is the Laplace transform of a probability measure on $\mathbb{R}_+$, and conversely.*

The converse part is immediate. For a proof of the "direct" part, one may refer to the classic, Widder (1946), p. 160, or to Chung (1974), p. 191 for a probability-theoretic proof (due to W. Feller). Phelps (1966), p. 11, provides an elegant proof using the Krein–Milman theorem, also to be found in Linnik and Ostrovskii (1977), pp. 315–317.

In the definition of the Laplace transform, we may introduce a complex argument $z$ in place of the real $t$. We then have the following result.

**Theorem 1.2.7** (D. V. Widder). *$\hat{\mu}(z)$ is analytic in $\operatorname{Re} z > \sigma_\mu$. If $\mu$ is a positive measure and if $\sigma_\mu > -\infty$, then $\sigma_\mu$ is a singularity for $\hat{\mu}$.*

*Analogous results also hold for $\hat{g}$, the Laplace transform of a function $g$ on $\mathbb{R}_+$.*

A proof may be found in Widder (1946), pp. 57–58. Note that the second statement in the theorem may not be true if $\mu$ is not a positive measure (respectively, if $g$ is not of constant sign).

Finally, we present two classical results in their simplest formulations, referring the reader for definitions, proofs, and more general formulations to Rudin (1974), Hille (1959), or Ahlfors (1966). We recall that a "region" is an open connected subset of $\mathbb{C}$, the complex plane, in our context.

**Theorem 1.2.8** (Cauchy). *Let $\Omega \subseteq \mathbb{C}$ be a simply connected region. If $f\colon \Omega \to \mathbb{C}$ is analytic and if $\gamma$ is a piecewise differentiable closed curve (with range) in $\Omega$, then $\int_\gamma f(z)\,dz = 0$.*

**Theorem 1.2.9** (The Schwarz reflection principle). *Let $D_1$, $D_2$ be two regions contained respectively in the left and right half-planes of the complex plane. Suppose $\bar{D}_1 \cap \bar{D}_2$ contains an interval $J$ of the imaginary axis. If $f_j$ is analytic in $D_j$ and continuous on $D_j \cup J$, for $j = 1, 2$, and $f_1 = f_2$ on $J$, then there exists a function $f$, analytic in $D_1 \cup J \cup D_2$, that agrees with $f_j$ on $D_j$ for $j = 1, 2$.*

We shall also require some auxiliary results of a more technical character—of the Phragmén–Lindelöf type for Chapter 5 and on convergence of measures and on a representation theorem due to Choquet for Chapter 9. These are presented in appendices to these two chapters respectively.

## 1.3. DISTRIBUTION FUNCTIONS AND CHARACTERISTIC FUNCTIONS

We shall assume familiarity with the concepts of a probability space, random variable (r.v.) and distribution function (d.f.). The reader may refer to standard works such as Feller (1971) and Loève (1977) for details.

Let $F$ be a d.f. on $\mathbb{R}$; then, $f: \mathbb{R} \to \mathbb{C}$ given by

$$f(t) = \int_{-\infty}^{\infty} e^{itx}\, dF(x)$$

(the Fourier–Stieltjes transform of $F$) is called the *characteristic function* (ch.f.) of $F$; $F$ is uniquely determined by $f$. In what follows, we shall use one of the letters $F$, $G$, $H$ to denote a d.f. on $\mathbb{R}$ and the corresponding small letter to denote its ch.f. The convolution $H = F * G$ is defined by $H(x) = \int_{-\infty}^{\infty} F(x - y)\, dG(y)$, and also by $\int_{-\infty}^{\infty} G(x - y)\, dF(y)$. If $X$ and $Y$ are two independent r.v.'s with $F$ and $G$ as respective d.f.'s, then $H = F * G$ is the d.f. and $h = f \cdot g$ the ch.f. of $X + Y$. For proofs of these and other basic results on ch.f.'s such as the continuity theorem, the Herglotz–Bochner theorem, etc., we refer the reader to such standard treatises as the classic Gnedenko and Kolmogorov (1964), Lukacs (1970), Feller (1971), and Loève (1977). Since $f$ is continuous and $f(0) = 1$, there exists a neighborhood $N$ of the origin where $f$ does not vanish; there exists a continuous version of the logarithm of $f$ on $N$, denoted by $\phi$, such that $\phi(0) = 0$. We shall always refer to this distinguished branch when we speak of $\log f$ in the following.

$F$ is called a *degenerate* d.f. if it is of the form $\delta_a$ for some $a \in \mathbb{R}$, where $\delta_a(x) = 0$ for $x < a$, and $= 1$ for $x \geq a$ (by an abuse of notation, we also denote by $\delta_a$ the point mass measure at $a$). We shall refer to the ch.f.'s of such d.f.'s also as "degenerate" (or "trivial"). $F$ is called a *lattice distribution* if $F = \sum_{n \in \mathbb{Z}} p_n \delta_{a+nh}$ for some $a$, $h \in \mathbb{R}$, where $p_n \geq 0$ for all $n$, and $\sum p_n = 1$.

A normal d.f. with parameters $\mu$ and $\sigma$ ($\mu$ real, $\sigma > 0$) is given by

$$F(x; \mu, \sigma) = \frac{1}{\sigma\sqrt{2\pi}} \int_{-\infty}^{x} \exp(-(u - \mu)^2/2\sigma^2)\, du, \qquad x \in \mathbb{R}.$$

If $X$ is a r.v. with this $F$ as d.f., then the mean of $X$ is

$$E(X) = \int_{-\infty}^{\infty} x\, dF(x) = \mu,$$

the variance is $\mathrm{Var}(X) = \int_{-\infty}^{\infty} (x - \mu)^2\, dF(x) = \sigma^2$, and the ch.f. of $X$ is given by $f(t) = \exp(it\mu - \frac{1}{2}\sigma^2 t^2)$ for $t \in \mathbb{R}$. For $\mu = 0$, $\sigma = 1$ we have the "standard" normal d.f., denoted by $\Phi$, which has $\exp(-\frac{1}{2}t^2)$ as its ch.f.

A *Poisson* d.f. with parameter $\lambda > 0$ is given by

$$F(\cdot, \lambda) = \sum_{n=0}^{\infty} (e^{-\lambda}\lambda^n/n!)\, \delta_n,$$

and its ch.f. by $f(t; \lambda) = \exp(\lambda(e^{it} - 1))$.

A d.f. $F$ on $\mathbb{R}$ is said to be *absolutely continuous* (with respect to the Lebesgue measure on Borel sets) if there exists a nonnegative Borel measurable $p$ on $\mathbb{R}$ such that

$$F(x) = \int_{-\infty}^{x} p(u)\, du, \qquad \forall x \in \mathbb{R};$$

$p$ is called (a version of) the *probability density function* (p.d.f.) of $F$. Of particular interest and importance in this context is the following.

**Theorem 1.3.1.** *If the ch.f. $f$ of $F$ is (Lebesgue) integrable on $\mathbb{R}$, then $F$ is absolutely continuous, and*

$$p(x) = F'(x) = \frac{1}{2\pi} \int_{-\infty}^{\infty} e^{-itx} f(t)\, dt, \qquad x \in \mathbb{R},$$

*is (a continuous version of) the p.d.f. of $F$.*

**Theorem 1.3.2.** (a) *If, for some $t_0 > 0$, $|f(t_0)| = 1$, then $F$ is a lattice distribution, and conversely.*

(b) *$F$ is degenerate if any one of the following conditions is satisfied:*

(i) $|f(t_1)| = |f(t_2)| = 1$ *for some nonzero $t_1$, $t_2$ with $t_1/t_2$ irrational;*
(ii) $|f(t_n)| = 1$ *for some nonzero sequence $\{t_n\}$ converging to 0;*
(iii) *For some $|a| < 1$, $|f(at)| \leq |f(t)|$ for all $t$ in some neighborhood of the origin.*

***Proof.*** Since $|f(-t)| = |f(t)|$, we may assume that all the $t_n$ are positive.

(a) Since $|f(t_0)| = 1$, we let $f(t_0) = e^{ic}$, where $c$ is a real number; then,

$$\int_{-\infty}^{\infty} \exp(i(t_0 x - c))\, dF(x) = 1,$$

so that

$$\int_{-\infty}^{\infty} (1 - \cos(t_0 x - c))\, dF(x) = 0.$$

It follows that $F$ is purely discrete and can have jumps only at those $x$ that satisfy $t_0 x - c = 2n\pi$, $n \in \mathbb{Z}$. By taking $a = c/t_0$, $h = 2\pi/t_0$, we have the representation $F = \sum_{n \in \mathbb{Z}} p_n \delta_{a+nh}$, and hence $F$ is a lattice distribution. The converse is immediate: $|f(2\pi/h)| = 1$ if $h \neq 0$, $|f(t)| \equiv 1$ if $h = 0$.

(b) If either of the conditions (i) and (ii) is satisfied, then, by part (a), $F$ is a lattice d.f. Suppose $F$ has two points of increase: $a$ and $b$ ($>a$). In case (i), $(b - a)$ is a positive integral multiple of $2\pi/t_1$ as well as of $2\pi/t_2$, and it follows that $t_1/t_2$ is rational, contrary to assumption. In case (ii), $b - a \geq 2\pi/t_n$ for every $n$ and this is impossible since $\{t_n\}$ converges to 0. Thus, $F$ is degenerate in either of these cases. If (iii) holds, then, for some neighborhood $I$ of 0, and for $t \in I$,

$$1 \geq |f(t)| \geq |f(at)| \geq \cdots \geq |f(a^n t)| \to f(0) = 1$$

as $n \to \infty$. This implies that $|f(t)| \equiv 1$ for all $t \in I$, and condition (i) (as well as (ii)) is satisfied. $F$ is thus degenerate.

### *1.3.1. Moments and Characteristic Functions*

We shall give a brief account of some relationships between ch.f.'s and the (absolute) moments of the corresponding d.f.'s. Let $X$ be a r.v. having $F$ as d.f. For $\lambda \in \mathbb{N}$, we shall call $E(X^\lambda) = \int_{-\infty}^{\infty} x^\lambda \, dF(x)$ the *moment of order* $\lambda$ of $F$, and if $\lambda > 0$, $E|X|^\lambda = \int_{-\infty}^{\infty} |x|^\lambda \, dF(x)$ the *absolute moment of order* $\lambda$ for $F$; $F$ is said to have (absolute) moment of order $\lambda$ if $E(|X|^\lambda) < \infty$. If $X$, $Y$ are two independent r.v.'s, then $E|X + Y|^\lambda < \infty$ if and only if $E|X|^\lambda$, $E|Y|^\lambda < \infty$ as well. In particular, if $Y$ also has the same d.f. as $X$, then $E|X - Y|^\lambda < \infty$ if and only if $E|X|^\lambda$ does. In terms of d.f.'s, this means that $F$ has moment of order $\lambda$ if and only if $F_s = F * \tilde{F}$ has, where $\tilde{F}$, the *conjugate* of $F$, is given by $\tilde{F}(x) = 1 - F((-x)_-)$ (the notation $F(x_-)$ denotes the left-hand limit of $F$ at $x$). $\tilde{F}$ is the d.f. of $-Y$, and $F_s$ of $X - Y$. $F$ is called *symmetric* if $\tilde{F} = F$; in this case, the ch.f. of $f$ is real-valued.

The following is a well-known and fundamental result.

**Theorem 1.3.3.** (a) *If $F$ has moment of order $n \in \mathbb{N}$, then $f$ has derivative of order $n$ throughout $\mathbb{R}$ given by*

$$f^{(n)}(t) = \int_{-\infty}^{\infty} (ix)^n e^{itx} \, dF(x).$$

(b) *If $f$ has derivative of order $2k$, $k \in \mathbb{N}$, at the origin, then $F$ has moment of order $2k$; the corresponding result for odd orders is not necessarily true.*

For a proof we refer the reader to Loève (1977) or Lukacs (1970).

The following concerns finite Taylor expansions of ch.f.'s in terms of the moments. In view of what has been said above, by "symmetrization" of d.f.'s, we may confine our attention to symmetric d.f.'s, as we do in what follows.

**Theorem 1.3.4.** (a) *Let $F$ be a symmetric d.f. with moment of order $2n + \lambda$, where $0 < \lambda \leq 2$. Then, $f$ admits the following expansion:*

$$f(t) = 1 + t^2 f^{(2)}(0)/2! + \cdots + t^{2n} f^{(2n)}(0)/(2n)! + O(|t|^{2n+\lambda})$$

as $t \to 0$.

(b) *Let $F$ be a symmetric d.f., and suppose $f$ admits, for some $\lambda \in (0, 2]$, an expansion of the above form in some neighborhood of the origin. Then, $F$ has moments of all orders $<2n + \lambda$ if $0 < \lambda < 2$, and of order $2n + 2$ if $\lambda = 2$.*

For a proof of (a) one can refer to Loève (1977), and of (b), to Ramachandran (1969).

If $F$ has moment of order $2n$, $n \in \mathbb{N}$, then, equivalently, $f^{(2n)}(0)$ exists. Also, $f^{(2n)}/f^{(2n)}(0)$—ignoring the cases of degenerate $F$, which have moments of all orders—is the ch.f. of the d.f. $F_{2n}$ given by

$$F_{2n}(x) = \int_{(-\infty, x]} u^{2n} \, dF(u)/m_{2n},$$

where $m_{2n}$ is the $2n$th moment of $F$. It is clear that $F$ has moment of order $2n + \lambda$, for some $\lambda \in (0, 2)$, if and only if $F_{2n}$ has moment of order $\lambda$.

**Theorem 1.3.5.** *For $0 < \lambda < 2$, $n \in \mathbb{Z}_+$, a d.f. $F$ has moment of order $2n + \lambda$ if and only if $f$ satisfies the conditions:*

(i) *$f^{(2n)}(0)$ exists;*

(ii) *$\int_0^\infty (f^{(2n)}(0) - \operatorname{Re} f^{(2n)}(t)) t^{-1-\lambda} \, dt$ exists finitely.*

*Equivalent to* (ii) *is the condition*

$$\int_0^\infty (|f^{(2n)}(0)| - |f^{(2n)}(t)|) t^{-1-\lambda} \, dt$$

*exists finitely; also equivalent to* (ii) *is the condition*

$$\int_0^c \log |f^{(2n)}(t)/f^{(2n)}(0)| t^{-1-\lambda} \, dt$$

*exists finitely, $c > 0$ being such that $f^{(2n)}$ does not vanish on $[0, c]$.*

***Proof.*** By using the "symmetrization" argument, and by applying the previous remark (replacing $F$ by $F_{2n}$), we can assume that $f$ is real-valued and that $n = 0$. The equivalence of the finiteness of the integrals in (ii) follows from such an assumption and

$$1 - x \leq -\log x \leq 2(1 - x) \qquad \text{for } \tfrac{1}{2} \leq x \leq 1. \tag{1.3.1}$$

Also, we have, by Fubini's theorem,

$$\begin{aligned}\int_0^\infty [1 - \operatorname{Re} f(t)]t^{-1-\lambda}\,dt &= \int_0^\infty \left(\int_{-\infty}^\infty (1 - \cos tx)\,dF(x)\right)t^{-1-\lambda}\,dt \\ &= \int_{-\infty}^\infty \left(\int_0^\infty (1 - \cos tx)t^{-1-\lambda}\,dt\right) dF(x) \\ &= c_\lambda \int_{-\infty}^\infty |x|^\lambda\,dF(x),\end{aligned}$$

where $(0<)\ c_\lambda = \int_0^\infty (1 - \cos u)u^{-1-\lambda}\,du < \infty$ in view of the fact that $0 < \lambda < 2$. The assertion of the theorem then follows at once.

The integral condition on $\log|f^{(2n)}(t)/f^{(2n)}(0)|$ in condition (ii) may be interpreted as providing an estimate of the order of magnitude of that function in the neighborhood of the origin. The following results have more to say on this.

**Theorem 1.3.6.** *Let $f$ be a d.f. with moment of order $2n$ for some $n \in \mathbb{Z}_+$. A necessary condition for $F$ to have moment of order $2n + \lambda$, $0 < \lambda < 2$, is that*

$$\lim_{t\to 0} |t|^{-\lambda} \log|f^{(2n)}(t)/f^{(2n)}(0)| = 0.$$

***Proof.*** As already noted, we need only consider the case $n = 0$, and can also assume that $f$ is real-valued, in particular positive in a neighborhood of the origin.

If $F$ has moment of order $\lambda$, then

$$x^\lambda(1 - F(x)) \leq \int_x^\infty u^\lambda\,dF(u) \to 0 \qquad \text{as } x \to \infty. \tag{1.3.2}$$

Suppose now that $t^{-\lambda}\log f(t)$ does not tend to zero as $t \to 0_+$; then, there exist $c_1 > 0$ and a sequence $\{t_n\} \to 0_+$ such that $t_n^{-\lambda}\log f(t_n) \geq c_1$. By the inequality (1.3.1), we can find $c_2 > 0$ such that $t_n^{-\lambda}(1 - f(t_n)) \geq c_2$ for all large $n$, i.e.,

$$c_3 t_n^{\lambda} \leq \int_0^{\infty} (1 - \cos t_n x)\, dF(x),$$

where $c_3 = c_2/2$. Hence, we have

$$\begin{aligned} c_3 t_n^{\lambda-1} &\leq \int_0^{\infty} (\sin t_n x)(1 - F(x))\, dx \qquad \text{(integrating by parts)} \\ &= \sum_{k=0}^{\infty} t_n^{-1} \int_{k\pi}^{(k+1)\pi} (\sin x)(1 - F(x/t_n))\, dx \\ &\leq t_n^{-1} \int_0^{\pi} (\sin x)(1 - F(x/t_n))\, dx \end{aligned}$$

(the last inequality follows from $\sin(x + \pi) = -\sin x$, and the fact that $1 - F$ is decreasing). Let

$$h(x, t) = (x/t)^{\lambda}(1 - F(x/t));$$

then,

$$c_3 \leq \int_0^{\pi} h(x, t_n) x^{-\lambda} \sin x\, dx.$$

Now, either (a) $h(x, t_n)$ is uniformly bounded for all $x \in (0, \pi)$, and for all $n$, or (b) there exist sequences $\{x_k\}$ and $\{t_{n_k}\}$ such that $h(x_k, t_{n_k}) \to \infty$. In case (a), since $x^{-\lambda} \sin x$ is integrable over $(0, \pi)$ and $h(x, t_n)$ is uniformly bounded, we have, by Fatou's Lemma, that

$$c_3 \leq \int_0^{\pi} \limsup_{n \to \infty} h(x, t_n) x^{-\lambda} \sin x\, dx.$$

Hence $\limsup_{n \to \infty} h(x, t_n) > 0$ for some $x > 0$. This contradicts (1.3.2). In case (b), letting $s_k = x_k/t_{n_k}$ and noting that $1 - F(s_k)$ is bounded, we see that $s_k \to \infty$ as $k \to \infty$, and also that $s_k^{\lambda}(1 - F(s_k)) \to \infty$, which is again impossible by (1.3.2).

The following two theorems give further relations between the existence of the moment of order $\lambda$ for $F$ and the behavior of $\log |f(t)|/|t|^{\lambda}$, for $0 < \lambda < 2$, and for $\lambda = 2$, respectively.

**Theorem 1.3.7.** *Suppose $F$ has moment of order* $0 < \lambda < 2$. Then

$$\lim_{t \to 0} \log |f(t)|/|t|^{\lambda} = 0.$$

*Conversely, suppose there exists a sequence* $\{t_n\} \searrow 0$ *such that* (i) $\sum t_n^{\varepsilon}$ *converges for every* $\varepsilon > 0$; (ii) $\{t_{n-1}/t_n\}$ *is bounded; and* (iii) $\{\log |f(t_n)|/t_n^{\lambda}\}$ *is bounded. Then, $F$ has moments of all orders* $< \lambda$.

***Remark.*** Condition (i) implies that $\lim_{n\to\infty} t_n = 0$. $\{b^n\}$, $0 < b < 1$, is an example of a sequence that satisfies (i) and (ii).

***Proof.*** The first part is a direct corollary of Theorem 1.3.6. Conversely, assume, without loss of generality, that $0 < t_n < t_{n-1} < 1$ for all $n \geq 2$. Let $c$ be a bound of $\log|f(t_n)|/t_n^\lambda$, $n \geq 1$; then,

$$1 - |f(t_n)|^2 \leq 1 - \exp(-ct_n^\lambda) \leq ct_n^\lambda.$$

Since $1 - |f(t)|^2 = 2\int_{-\infty}^{\infty} \sin^2(tx/2)\, dF_s(x)$, by letting $u_n = t_n/2$, we have

$$\int_{-\infty}^{\infty} \sin^2(u_n x)\, dF_s(x) \leq c_1 u_n^\lambda$$

for some $c_1 > 0$. Also, note that $x \sin 1 \leq \sin x$ for $0 \leq x < 1$; we have, on setting $x_n = 1/u_n$, and for $0 < \delta < \lambda$,

$$\begin{aligned}
\int_{x_{n-1}}^{x_n} x^\delta\, dF_s(x) &\leq x_n^\delta x_{n-1}^{-2} \int_{x_{n-1}}^{x_n} x^2\, dF_s(x) \\
&\leq c_2 x_n^\delta x_{n-1}^{-2} u_n^{-2} \int_{x_{n-1}}^{x_n} \sin^2(u_n x)\, dF_s(x) \\
&\leq c_3 (u_{n-1}/u_n)^2 u_n^{\lambda-\delta} \\
&\leq c_4 u_n^{\lambda-\delta} \qquad \text{(by (ii))},
\end{aligned}$$

where $c_2, c_3, c_4 > 0$ are constants. It follows from (i) that

$$\int_0^\infty x^\delta\, dF_s(x) \leq c_4 \sum_{n=1}^\infty u_n^{\lambda-\delta} < \infty,$$

and hence $F_s$ has moments of any order $\delta < \lambda$, and so does $F$.

**Theorem 1.3.8.** (a) *Suppose $F$ has second moment. Then,* $\log|f(t)|/t^2$ *is bounded on some neighborhood of* 0. *Conversely, if* $\log|f(t_n)|/t_n^2$ *is bounded for some nonzero sequence* $\{t_n\}$ *that converges to* 0, *then $F$ has second moment.*

(b) *If* $\log|f(t_n)|/t_n^2 \to 0$ *for some nonzero sequence* $\{t_n\}$ *that converges to* 0, *then $F$ is degenerate.*

***Proof.*** (a) Suppose $F$ has second moment; then, from the inequality $1 - \cos x \leq x^2$, $x \in \mathbb{R}$, we conclude that, for $t > 0$, $(1 - \cos tx)t^{-2} \leq x^2$, and hence

$$(1 - |f(t)|^2)t^{-2} = \int_{-\infty}^{\infty} (1 - \cos tx)t^{-2}\, dF_s(x) \leq \int_{-\infty}^{\infty} x^2\, dF_s(x) < \infty.$$

By the inequality (1.3.1), $\log|f(t)|/t^2$ is bounded on a neighborhood of 0.

For the converse statement, the assumption implies that $|f(t_n)|^2 \geq \exp(-ct_n^2)$ for some $c > 0$, so that, by Fatou's lemma,

$$\begin{aligned}\int_{-\infty}^{\infty} x^2\, dF_s(x) &= 2\int_{-\infty}^{\infty} \liminf_{n\to\infty}(1 - \cos t_n x)t_n^{-2}\, dF_s(x)\\ &\leq 2\liminf_{n\to\infty}\int_{-\infty}^{\infty}(1 - \cos t_n x)t_n^{-2}\, dF_s(x)\\ &= 2\liminf_{n\to\infty}(1 - |f(t_n)|^2)t_n^{-2}\\ &\leq 2\liminf_{n\to\infty}(1 - \exp(-ct_n^2))t_n^{-2}\\ &= 2c.\end{aligned}$$

Hence $F_s$, and therefore $F$, has second moment.

(b) We can apply the above argument to show that $\int_{-\infty}^{\infty} x^2\, dF_s(x) \leq 2c$ for every $c > 0$, and hence $F_s = \delta_0$, so that $F$ is degenerate.

Let $\phi = \log f$ on a neighborhood $I$ of the origin where $f$ does not vanish. We will make a brief study of the relation between the moments of $F$ and the derivatives of $\phi$.

Suppose $F$ has moment of order $n \in \mathbb{Z}_+$; then, $f^{(j)}$ exists on $\mathbb{R}$ for $1 \leq j \leq n$, and $\phi^{(n)}$ on $I$ can be expressed as

$$\frac{f^{(n)}}{f} + c_n \frac{f^{(n-1)}}{f}\frac{f^{(1)}}{f} + g_n,$$

where $c_n$ is a constant, and $g_n$ is a polynomial in the functions $f^{(j)}/f$, $j = 1, \ldots, n-2$, such that its typical terms are of the form

$$c\prod_{j=1}^{n-2}(f^{(j)}/f)^{\alpha_j},$$

where $\sum_{j=1}^{n-2} j\alpha_j = n$; for example,

$$\begin{aligned}\phi^{(2)} &= f^{(2)}/f - (f^{(1)}/f)^2,\\ \phi^{(3)} &= f^{(3)}/f - 3(f^{(2)}/f)(f^{(1)}/f) + 2(f^{(1)}/f)^3.\end{aligned}$$

Recalling Theorem 1.3.5, and the remarks preceding it, we have the following.

**Theorem 1.3.9.** *If $F$ is a symmetric d.f. with $f$ as ch.f., and has moment of order $2n$ for some $n \in \mathbb{Z}$, and $I$ is an interval around the origin where $f$ does* *not vanish, then, for some* $0 < \lambda < 2$ *and for arbitrary* $0 < c \in I$,

$$\int_0^c |f^{(2n)}(t) - f^{(2n)}(0)| t^{-1-\lambda}\, dt < \infty$$

*if and only if*

$$\int_0^c |\phi^{(2n)}(t) - \phi^{(2n)}(0)| t^{-1-\lambda}\, dt < \infty.$$

***Proof.*** Since $f$ is real-valued and continuous with $f(0) = 1$, we may assume that $c$ satisfies $f(t) \geq \frac{1}{2}$ on $[0, c]$. The estimate (1.3.1) of $\log x$ for $\frac{1}{2} \leq x \leq 1$ implies that

$$1 - f(t) \leq -\log f(t) = -\phi(t) \leq 2(1 - f(t)) \tag{1.3.3}$$

and, since $\phi(0) = 0$, the statement is immediate for $n = 0$. Let us consider $n \geq 1$. Since $F$ is symmetric, $f^{(2j-1)}(0) = 0$ for $j = 1, \ldots, n$, and therefore

$$\begin{aligned}\phi^{(2n)}(t) - \phi^{(2n)}(0) &= \frac{f^{(2n)}(t) - f^{(2n)}(0)}{f(t)} + \frac{f^{(2n)}(0)(1 - f(t))}{f(t)} \\ &\quad + c_n \frac{f^{(2n-1)}(t)}{f(t)} \frac{f^{(1)}(t)}{f(t)} + g_n(t) - g_n(0). \end{aligned} \tag{1.3.4}$$

Recall that moments of all orders $\leq 2n$ exist. The assertion follows from the preceding relationship in view of the following three facts: (i) $f(t) \geq \frac{1}{2}$ on $[0, c]$; (ii) $\int_0^c |1 - f(t)| t^{-1-\lambda}\, dt < \infty$ since the moment of order $\lambda (< 2)$ exists; and (iii)

$$\left. \frac{f^{(2n-1)}(t)}{f(t)} \frac{f^{(1)}(t)}{f(t)} \right/ t^2 \to f^{(2n)}(0) f^{(2)}(0)$$

as $t \to 0$, and, for similar reasons, $(g_n(t) - g_n(0))/t^2$ converges to a finite limit as $t \to 0$.

**Theorem 1.3.10.** *If $F$ is a symmetric d.f. with $f$ as ch.f., and has moment of order $2n$ but not of order $(2n + 2)$, then*

$$\lim_{t \to 0} (-1)^{n+1} (\phi^{(2n)}(t) - \phi^{(2n)}(0))/t^2 = \infty.$$

***Proof.*** We use the identity (1.3.4) again. Since $F$ has second moment, $f^{(2)}(0) = \lim_{t \to 0} (1 - f(t))/t^2$ exists. That the moment of order $2n + 2$ does not exist implies that

$$\lim_{t \to 0} (-1)^{n+1} \frac{f^{(2n)}(t) - f^{(2n)}(0)}{t^2} = \lim_{t \to 0} \int_{-\infty}^{\infty} \frac{1 - \cos tx}{t^2} x^{2n}\, dF(x) = \infty.$$

The proof is completed by taking into account facts (ii) and (iii) in the proof of the previous theorem.

### *1.3.2. Infinitely Divisible Laws and Their Ch.f.'s*

Infinitely divisible laws on $\mathbb{R}$ (and special subfamilies such as the stable, self-decomposable, and semistable laws) arise naturally in the investigation of limit distributions of sums of independent random variables. We shall study these special families in Chapters 3, 4, 5, and 6 in the context of functional equations satisfied by them. In the present section, we shall give a short, purely analytical account of infinitely divisible laws without going into their probabilistic origins.

A ch.f. $f$ is said to be *infinitely divisible* (inf. div.) if, for each $n \in \mathbb{N}$, there exists a ch.f. $f_n$ such that $f = (f_n)^n$. The following is a useful criterion:

**Theorem 1.3.11** (DeFinetti's theorem). *A ch.f. $f$ is inf. div. if and only if it has the form*

$$f(t) = \lim_{n \to \infty} \exp(\alpha_n(g_n(t) - 1)),$$

*where $\{\alpha_n\}$ are positive real numbers, and $\{g_n\}$ are ch.f.'s.*

Such an $f$ is necessarily nonvanishing on $\mathbb{R}$, and $\phi = \log f$ admits a representation of the following form, called the *Lévy representation*; we shall denote it by $L(\mu, \sigma^2, M, N)$:

$$\phi(t) = i\mu t - \sigma^2 t^2/2 + \int_{(-\infty, 0)} h(t, u)\, dM(u) + \int_{(0, \infty)} h(t, u)\, dN(u), \quad (1.3.5)$$

where $\mu, \sigma \in \mathbb{R}$, $\sigma \geq 0$, $h(t, u) = e^{itu} - 1 - itu/(1 + u^2)$, $M, N$ are $\sigma$-finite measures on the Borel subsets of $(-\infty, 0)$ and $(0, \infty)$, respectively, such that $N(a, \infty)$ and $M(-\infty, -a)$ are finite for every $a > 0$, and, further,

$$\int_{(-a, 0)} u^2\, dM(u), \qquad \int_{(0, a)} u^2\, dN(u) < \infty, \qquad \forall\, a > 0. \quad (1.3.6)$$

It is often convenient to look upon the integrals in the Lévy representation as Lebesgue–Stieltjes integrals, where $M$ and $N$ are interpreted as "point functions" (rather than as measures), both functions being nondecreasing and right-continuous on their respective domains of definition, and "normalized" so that $M(-\infty) = N(\infty) = 0$.

The Lévy representation for an inf. div. ch.f. is unique. Further, if a function $\phi\colon \mathbb{R} \to \mathbb{C}$ is given by (1.3.5) with $\mu$, $\sigma$, $M$, and $N$ subject to the above restrictions, then $f = e^{\phi}$ is an inf. div. ch.f. For arbitrary inf. div. ch.f.'s, we also have the *Lévy–Khinchin representation* (also unique), given by

$$\phi(t) = i\mu t + \int_{\mathbb{R}} \left(e^{itu} - 1 - \frac{itu}{1+u^2}\right)\frac{1+u^2}{u^2}\,dG(u), \qquad t \in \mathbb{R},$$

where $\mu$ is a real number, $G$ is a finite measure on the Borel subsets of $\mathbb{R}$ (as a point-function, $G$ is bounded, nondecreasing and right-continuous with $G(-\infty) = 0$, $G(\infty) < \infty$), and the integrand is defined at the origin by continuity as $-t^2/2$. For inf. div. ch.f.'s whose d.f.'s have finite second moment, we also have the *Kolmogorov representation*

$$\phi(t) = i\mu t + \int_{\mathbb{R}} \frac{(e^{itu} - 1 - itu)}{u^2}\,dK(u),$$

where $\mu$ is real, $K$ is a finite measure on the Borel subsets of $\mathbb{R}$ (and behaves like the previous $G$, when interpreted as a point-function), and the integrand is again defined at the origin as $-t^2/2$.

The normal ch.f.'s correspond to $M \equiv N \equiv 0$ in the Lévy representation, or $G = K = \sigma^2\,\delta_0$ in the other two representations. The standard Poisson ch.f. with parameter $\lambda > 0$ corresponds to $\sigma = 0$, $M \equiv 0$, $N = \lambda\,\delta_1$, $\mu = \lambda/2$. The degenerate laws correspond to $\sigma = 0$, $M \equiv N \equiv 0$. No d.f. with compact support has an inf. div. ch.f. unless it is degenerate. Concerning the moments of inf. div. laws, we have the following.

**Theorem 1.3.12.** *Let $f$ be an inf. div. ch.f. with the Lévy representation $L(\mu, \sigma^2, M, N)$. Then, $F$ has moment of a given order $\lambda > 0$ if and only if*

$$\int_{(-\infty,\,-1)} |u|^{\lambda}\,dM(u), \qquad \int_{(1,\,\infty)} u^{\lambda}\,dN(u) < \infty.$$

In view of (1.3.6), this can be described succinctly for $\lambda \geq 2$ as: $F$ has moment of order $\lambda$ if and only if both $M$ and $N$ have moment of the same order. For details and proofs, we refer the reader to Ramachandran (1969).

We shall frequently require the following three results pertaining to the normal law. For proofs we refer the reader to Linnik and Ostrovskii (1977), Lukacs (1970), or Ramachandran (1967).

**Theorem 1.3.13** (Lévy–Cramér). *If $f$ is a normal ch.f., and if $f = f_1 f_2$ throughout $\mathbb{R}$, where $f_1$ and $f_2$ are ch.f.'s, then $f_1$ and $f_2$ are necessarily normal (possibly degenerate) ch.f.'s.*

**Corollary 1.3.14.** *If $|f|^2$ is a normal ch.f., so is $f$.*

We remark that in this context, the relation $f = f_1 f_2$ need only hold at some sequence of values of $t$ tending to the origin for us to draw the same conclusion. See Theorem 1.3.15.

Similar results hold for the Poisson law, for the convolution of a normal and a Poisson law, and for certain classes of inf. div. laws of which these are special cases. For a comprehensive account of these fascinating results with deep proofs, we refer the reader to Linnik and Ostrovskii (1977), which also treats multivariate extensions of such results; for results pertaining to univariate inf. div. laws and factorization problems concerning them, one may also refer to Ramachandran (1967).

**Theorem 1.3.15.** *If $G$ is a d.f. having moments of all orders and uniquely determined by its moments, and if $F$ is a d.f. such that $f = g$ at some nonzero sequence $\{t_n\}$ of points converging to 0, then $F = G$.*

A sufficient condition for a d.f. $G$ with moments of all orders to be uniquely determined by them is that the power series $\sum_{n=0}^{\infty} m_n t^n/n!$, where $m_n$ is the moment of order $n \in \mathbb{Z}_+$ for $G$, have nonzero radius of convergence; this requirement is also the same as that the moment-generating function of $G$ exist. A normal law in particular qualifies; thus, if a ch.f. $f$ agrees with $\exp\{i\mu t - \frac{1}{2}\sigma^2 t^2\}$ at some sequence of values of $t$ tending to the origin, then it agrees with the latter on $\mathbb{R}$.

**Theorem 1.3.16.** *Let $P$ be a polynomial in a real variable with possibly complex coefficients. Then,* $\exp P$ *is a ch.f. if and only if $P$ has degree at most two. In this case, the corresponding d.f. is necessarily normal or degenerate.*

**Corollary 1.3.17.** *If a ch.f. is of the form* $\exp P$, *where $P$ is a polynomial, in some neighborhood of the origin, or at a nonzero sequence of points converging to zero, then it is a normal or degenerate ch.f.*

To conclude this section, we give a simple sufficient condition for a ch.f. to have the form $\exp P$, where $P$ is a polynomial.

**Proposition 1.3.18.** *Let $f$ be a ch.f. and let $\phi = \log f$ on a neighborhood $I$ of* 0. *Suppose $\phi$ has derivatives of all orders at* 0, *and there exists $m \in \mathbb{N}$ such that $\phi^{(n)}(0) = 0$ for all $n \geq m$. Then, $\phi$ is a polynomial of degree not greater than $m$ (in fact, at most two, by Corollary* 1.3.17).

***Proof.*** Write $\phi(t) = P(t) + R(t)$ on an interval $I'(\subseteq I)$, neighborhood of 0, where $P$ is a polynomial of degree $m$ and $R(t)/t^n \to 0$ as $t \to 0$ for every

$n > m$ (as implied by the Taylor expansion for $\phi$ up to the power $t^n$). Let $g = e^P$; since $|f(t)| \leq 1$ for real $t$ and $|e^z - 1| \leq |z|e^{|z|}$ for $z \in \mathbb{C}$, we have, for some $c > 0$,

$$|f(t) - g(t)| = |e^{\phi(t)} - e^{P(t)}| \leq |f(t)|\,|1 - e^{-R(t)}|$$
$$\leq |R(t)|e^{|R(t)|} \leq cR(t) \qquad \text{for } |t| \leq 1.$$

Now, for $n \in \mathbb{N}$,

$$|f^{(n)}(0) - g^{(n)}(0)| = \lim_{h \to 0} \sum_{r=0}^{n} (-1)^{n-r}\binom{n}{r}(f(rh) - g(rh))/h^n.$$

Hence, for $n > m$,

$$|f^{(n)}(0) - g^{(n)}(0)| \leq \overline{\lim_{h \to 0}}\, c \sum_{r=0}^{n} \binom{n}{r}|R(rh)|/|h|^n = 0,$$

so that $f^{(n)}(0) = g^{(n)}(0)$ for $n > m$. Clearly,

$$\overline{\lim_{n \to \infty}}\, (f^{(n)}(0)/n!)^{1/n} = \overline{\lim_{n \to \infty}}\, (g^{(n)}(0)/n!)^{1/n} < \infty,$$

and $f$ admits a series expansion in a neighborhood of the origin (*cf.* Feller (1971), p. 514). Since $g$ is an entire function, it follows that so are $f$ and (hence) $\phi$. That $\phi$ also has a power series expansion implies that $\phi = P$.

## NOTES AND REMARKS

Lemma 1.1.1 and Theorem 1.1.2 belong to the folklore of elementary number theory. Lemma 1.1.5 and Theorem 1.1.6 together constitute the content of Marsaglia and Tubilla (1975). Proposition 1.1.8 is abstracted from Ramachandran (1982a). Bergström (1963) cites and uses Proposition 1.1.9 to obtain the Lévy representation for stable laws—our Theorem 3.1.2. Theorems 1.3.5, 1.3.6, 1.3.9, and 1.3.12 are from Ramachandran (1969), Theorems 1.3.7 and 1.3.8 from Ramachandran and Rao (1968), and Theorem 1.3.10 from Linnik (1953a)—also, see Kagan *et al.* (1973), Lemma 2.4.1, and Riedel (1985). Theorem 1.3.13, a conjecture of P. Lévy's, was proved by H. Cramér. Theorem 1.3.15 is implicit in Linnik's work on "$\alpha$-decomposition of probability laws," and is explicitly formulated in Ramachandran and Rao (1968). Theorem 1.3.16 is a special case of a general result due to J. Marcinkiewicz, which states that, if $f$ is an entire ch.f. of finite order $>2$, then the exponent of convergence of its zeros cannot be less than its order; for a proof of the latter, see Linnik and Ostrovskii (1977) or Lukacs (1970), and for a proof by D. Dugué of the former, see Ramachandran (1967).

CHAPTER

# 2

# Integrated Cauchy Functional Equations on $\mathbb{R}_+$

The integral equation

$$\int_S (f(x+y) - f(x)f(y))\,\mathrm{d}\nu(y) = 0, \qquad \forall\, x \in S,$$

where $f: S \to \mathbb{R}$, $S$ is a semigroup of $\mathbb{R}$, and $\nu$ is a positive or a signed Borel measure on $S$, can be considered as an extension of the classical Cauchy functional equation

$$f(x+y) = f(x)f(y), \qquad \forall\, x, y \in \mathbb{R}.$$

If $\int_S f(y)\,d\nu(y)$ is finite and is nonzero, then the above equation reduces to

$$f(x) = \int_S f(x+y)\,\mathrm{d}\sigma(y), \qquad \forall\, x \in S, \tag{*}$$

where $\sigma$ is related to $\nu$ in an obvious manner. Equations of the latter form, with $f$ nonnegative and $\sigma$ a positive measure, repeatedly turn up in analytical probability theory, in characterization problems of mathematical statistics, and in renewal theory. Over the years these equations have been tackled by various authors through different *ad hoc* methods involving elementary real analysis, complex analysis, Fourier and Laplace transforms, and functional analysis (the Krein–Milman theorem and Choquet theory). We will call such an equation an "integrated Cauchy functional equation" (ICFE), and occasionally write the equation (*) as $f = f \bullet \sigma$ on $S$ for simplicity.

In this chapter, our main objective is to characterize the nonnegative solutions of the ICFE on $\mathbb{R}_+$. We begin with a discussion in Section 2.1 of the equation on $\mathbb{Z}_+$. The proof of this special case serves as motivation for the more complicated case of $\mathbb{R}_+$ (Theorem 2.2.4). The proof in both cases requires only elementary real-variable techniques. In Section 2.3, we give an alternative proof of the solution of the ICFE on $\mathbb{R}_+$ by using the technique of exchangable random variables. Section 2.4 concerns the ICFE where $\sigma$ is a signed measure. Some simple applications of the solution of the equation to characterization problems in mathematical statistics are given in Section 2.5; more elaborate applications will follow in the subsequent chapters.

## 2.1. THE ICFE ON $\mathbb{Z}_+$

Let $\{p_n\}_{n=0}^{\infty}$ be a sequence of nonnegative real numbers, not all zero. Our aim is to characterize the nonnegative real squences $\{v_n\}_{n=0}^{\infty}$ that satisfy

$$v_m = \sum_{n=0}^{\infty} v_{m+n} p_n, \qquad \forall\, m \in \mathbb{Z}_+. \tag{2.1.1}$$

We start with the trivial cases corresponding to $p_0 \geq 1$.

**Proposition 2.1.1.** *Let $\{p_n\}_{n=0}^{\infty}$, $\{v_n\}_{n=0}^{\infty}$ be two nonnegative real sequences, satisfying* (2.1.1);

(i) *if $p_0 > 1$, then $v_n = 0$ for all $n \in \mathbb{Z}_+$;*
(ii) *if $p_0 = 1$, let $k = \min\{n \geq 1 : p_n > 0\}$; then, $v_n = 0$ for all $n \geq k$, while $v_0, v_1, \ldots, v_{k-1}$ can be arbitrary.*

***Proof.*** (i) If $p_0 > 1$, then, for any $m \in \mathbb{Z}_+$, (2.1.1) implies that

$$0 \leq \sum_{n=1}^{\infty} v_{m+n} p_n = (1 - p_0) v_m \leq 0,$$

and hence $v_m = 0$ for all $m \in \mathbb{Z}_+$.

(ii) Let $p_0 = 1$. Let $k$ be as in (ii); then $\sum_{n=k}^{\infty} v_{m+n} p_n = 0$ for all $m \in \mathbb{Z}_+$. Since all the terms are nonnegative and $p_k > 0$, by taking $m = 0, 1, 2, \ldots$ in the previous equation, we have $v_k = v_{k+1} = \cdots = 0$. On the other hand, it is clear that (2.1.1) is satisfied for arbitrary $v_0, \ldots, v_{k-1}$.

In view of the above proposition, we shall henceforth consider only the nontrivial cases with $p_0 < 1$.

**Theorem 2.1.2.** *Let $\{p_n\}_{n=0}^{\infty}$ be nonnegative with $p_0 < 1$ and $p_1 > 0$, and let $\{v_n\}_{n=0}^{\infty}$ be a nonnegative solution of Eq.* (2.1.1)*. Then either $v_n = 0$ for all $n$, or $v_n = Bb^n$ for some $B$, $b > 0$, where $b$ satisfies $\sum_{n=0}^{\infty} b^n p_n = 1$.*

***Proof.*** If $\{v_m\}_{m=0}^{\infty}$ is not identically zero, then under our assumptions, it is actually a *positive* sequence: For if $v_{m_0} = 0$ for some $m_0$, then (2.1.1) and $p_1 > 0$ imply that $v_{m_0+1} = 0$; inductively, $v_m = 0$ for all $m \geq m_0$. Equation (2.1.1) then implies that $v_{m_0-1} = v_{m_0-1}p_0$, hence $v_{m_0-1} = 0$, and again by induction $v_m = 0$ for all $m \leq m_0$.

Let, then, $v_m > 0$ for all $m \in \mathbb{Z}_+$; we have, from (2.1.1), that $v_m(1 - p_0) \geq v_{m+1}p_1$, so that for $c = p_1/(1 - p_0)$, we have $cv_{m+1} \leq v_m$. Let, then,

$$b = \sup_m \frac{v_{m+1}}{v_m} (\leq c^{-1}). \tag{2.1.2}$$

The sequence $u_m = bv_m - v_{m+1}$ $(\geq 0)$ also satisfies (2.1.1), and hence $cu_{m+1} \leq u_m$ or, equivalently,

$$c\left(\frac{v_{m+1}}{v_m}\right)\left(b - \frac{v_{m+2}}{v_{m+1}}\right) \leq \left(b - \frac{v_{m+1}}{v_m}\right). \tag{2.1.3}$$

For any $0 < \varepsilon < b^2$, let $\varepsilon_1 = \min\{bc\varepsilon/2, \varepsilon\}$, and let $m$ be such that

$$b - \frac{v_{m+1}}{v_m} < \frac{\varepsilon_1}{2b}\left(\leq \frac{\varepsilon}{2b} < \frac{b}{2}\right).$$

Then, $v_{m+1}/v_m > b/2$, and (2.1.3) implies that

$$b - \frac{v_{m+2}}{v_{m+1}} < \frac{\varepsilon}{2b},$$

so that $b^2 - \varepsilon < v_{m+2}/v_m$. Since $v_{m+2}/v_m \leq b^2$ (by (2.1.2)), it follows from the arbitrariness of $\varepsilon > 0$ that

$$\sup_m \frac{v_{m+2}}{v_m} = b^2.$$

Through an obvious inductive argument, (2.1.3) also implies that given any $\varepsilon > 0$ and $k \in \mathbb{N}$, there exists an $m = m(k, \varepsilon)$ such that

$$b^n - \frac{v_{m+n}}{v_m} < \varepsilon, \qquad n = 1, \ldots, k.$$

It follows that

$$\sup_m \frac{v_{m+n}}{v_m} = b^n, \qquad \forall\, n = 1, 2, \ldots,$$

and

$$\sum_{n=0}^{k} (b^n - \varepsilon)p_n \leq \sum_{n=0}^{k} (v_{m+n}/v_m)p_n \leq 1;$$

hence, $\sum_{n=0}^{\infty} b^n p_n \le 1$. On the other hand, by (2.1.2),

$$1 = \sum_{n=0}^{\infty} (v_{m+n}/v_m)p_n \le \sum_{n=0}^{\infty} b^n p_n \le 1, \qquad \forall\, m \in \mathbb{Z}_+,$$

so that equality must hold throughout in this relation; in particular, since $p_1 > 0$, we have $v_{m+1} = bv_m$ for all $m \in \mathbb{Z}_+$ and $v_m = v_0 b^m$ for all $m \in \mathbb{Z}_+$.

The theorem also implies that if ($p_0 < 1, p_1 > 0$, and) there exists no $b$ satisfying $\sum_{n=0}^{\infty} b^n p_n = 1$, then the only solution is $v_n = 0$ for all $n \in \mathbb{Z}_+$. The general case, not assuming $p_1 > 0$, is given in the next section as a corollary of the main theorem (Corollary 2.2.6).

## 2.2. THE ICFE ON $\mathbb{R}_+$

Let $\sigma$ be a positive $\sigma$-finite Borel measure on $\mathbb{R}_+$, nondegenerate at 0, and let $f: \mathbb{R}_+ \to \mathbb{R}_+$ be locally integrable (with respect to Lebesgue measure) on $\mathbb{R}_+$. The notation ICFE($\sigma$) is used to denote the integral equation

$$f(x) = \int_{\mathbb{R}_+} f(x + y)\, d\sigma(y) \qquad \text{a.e. } x \ge 0 \tag{2.2.1}$$

(a.e. will mean almost everywhere with respect to Lebesgue measure), and $f \in$ ICFE($\sigma$) means that $f$ is a solution of this integral equation. Equation (2.2.1) is also written as $f = f \bullet \sigma$ for brevity. In this section, we are primarily interested in the nonnegative solutions of the ICFE($\sigma$); we first prove a few preliminary results that will lead to the main theorem (Theorem 2.2.4). The notation $\int_a^b$ will mean the integral over the *closed* interval with endpoints $a$, $b$.

**Proposition 2.2.1.** *Let $f \ge 0$ be continuous, and let $\sigma$ be defined as above. Then the following hold*:

(i) *Suppose $f \in$ ICFE($\sigma$), and $f > 0$; then $\sigma^n$, the n-fold convolution of $\sigma$, is again $\sigma$-finite, and $f \in$ ICFE($\sigma^n$).*

(ii) *For any $\beta \in \mathbb{R}$, let $\tilde{f}(x) = e^{-\beta x} f(x)$ and $d\tilde{\sigma}(x) = e^{\beta x}\, d\sigma(x)$; then $\tilde{\sigma}$ is also $\sigma$-finite, and $f \in$ ICFE($\sigma$) if and only if $\tilde{f} \in$ ICFE($\tilde{\sigma}$).*

***Proof.*** We prove (i) for the case $n = 2$; the general case follows by an obvious induction. For $x \geq 0$,

$$\int_{\mathbb{R}_+} f(x + y)\, d\sigma^2(y) = \int_{\mathbb{R}_+} \int_{\mathbb{R}_+} f(x + y + z)\, d\sigma(y)\, d\sigma(z)$$
$$= \int_{\mathbb{R}_+} f(x + z)\, d\sigma(z)$$
$$= f(x).$$

Now, $f > 0$ implies that $\sigma^2$ is $\sigma$-finite, and hence $f \in \text{ICFE}(\sigma^2)$.

Assertion (ii) follows directly from the definition of an ICFE.

We shall refer to $f = 0$ a.e. as the trivial solution in what follows.

**Lemma 2.2.2.** *Let $\sigma$ be a positive $\sigma$-finite Borel measure, nondegenerate at* 0. *Let $f \geq 0$ be a nontrivial locally integrable solution of the* ICFE($\sigma$), *and let $\alpha$ be such that*

$$1 < \int_{\mathbb{R}_+} e^{\alpha y}\, d\sigma(y).$$

*Then, for any $\beta > \alpha$,*

$$\int_x^\infty e^{-\beta y} f(y)\, dy < \infty, \qquad \forall\, x \geq 0. \tag{2.2.2}$$

*If we let $\tilde{f}(x)$ denote the above integral, and let $d\tilde{\sigma}(x) = e^{\beta x}\, d\sigma(x)$, then $\tilde{f}$ is a nonnegative, continuous decreasing solution of the* ICFE($\tilde{\sigma}$).

***Proof.*** By replacing $\sigma$ by $\sigma'$, where $d\sigma'(x) = e^{\alpha x}\, d\sigma(x)$, we may assume without loss of generality that $\alpha = 0$, and hence $\sigma(\mathbb{R}_+) > 1$. By Proposition 1.2.2, there exists a Borel subset $A$ with compact closure such that $\max\{1, \sigma\{0\}\} < \sigma(A) < \infty$.

For any $x \geq 0$, we have, by the Fubini theorem and since $f \in \text{ICFE}(\sigma)$, that

$$\int_0^x f(u)\, du \geq \int_A \left( \int_0^x f(u + y)\, du \right) d\sigma(y) = \int_A \left( \int_y^{y+x} f(u)\, du \right) d\sigma(y),$$

so that

$$\int_A \left( \int_0^{y+x} f(u)\, du \right) d\sigma(y) - \int_0^x f(u)\, du \leq \int_A \left( \int_0^y f(u)\, du \right) d\sigma(y) < \infty,$$

since $f$ is locally integrable and $\sigma(A) < \infty$. Denote the last expression by $a_0$. The fact that $\sigma(A) > 1$ then implies that

$$\int_A \left( \int_x^{x+y} f(u)\, du \right) d\sigma(y) \leq a_0, \qquad \forall\, x \geq 0.$$

Since $\sigma\{0\} < \sigma(A)$, there exists $\eta > 0$ such that $\sigma(A \cap [\eta, \infty)) > 0$. It follows from the preceding inequality that there exists $a_1$ such that

$$\int_x^{x+\eta} f(u)\, du \leq a_1, \qquad \forall\, x \geq 0.$$

Now, for $\beta > 0$,

$$\int_0^\infty e^{-\beta u} f(u)\, du = \sum_{n=0}^\infty \int_{[n\eta,\,(n+1)\eta)} e^{-\beta u} f(u)\, du \leq a_1 \sum_{n=0}^\infty e^{-\beta\eta\cdot n} < \infty.$$

This proves the first part of the theorem. The second part is a direct consequence of the Fubini theorem.

Analogously to Proposition 2.1.1, we have the following.

**Lemma 2.2.3.** *Let $f \in$ ICFE($\sigma$) and be nonnegative. Then the following holds:*

(a) If $\sigma\{0\} > 1$, *then $f \equiv 0$ a.e.*
(b) *If $\sigma\{0\} = 1$, then $f(x) = 0$ a.e. for all $x \geq a$, where*

$$a = \inf\{\operatorname{supp} \sigma \backslash \{0\}\}.$$

(c) *If $\sigma\{0\} < 1$, and $f \not\equiv 0$ a.e., then a strictly positive solution of the ICFE($\sigma$) exists. Conversely if a strictly positive solution $f$ exists, then $\sigma\{0\} < 1$, and $\sigma$ is a regular measure.*

***Proof.*** (a) Let $\sigma\{0\} > 1$. Then,

$$f(x) = \sigma\{0\} f(x) + \int_{(0,\infty)} f(x+y)\, d\sigma(y) \qquad \text{a.e.}, \tag{2.2.3}$$

which implies that $(\sigma\{0\} - 1) f(x) = 0$ a.e., and hence $f = 0$ a.e.

(b) Let $\tilde{f}$ and $\tilde{\sigma}$ be defined as in (2.2.2). Then, $\sigma\{0\} = 1$ implies that $\tilde{\sigma}\{0\} = 1$, and from (2.2.3) we have

$$\int_{(0,\infty)} \tilde{f}(x+y)\, d\tilde{\sigma}(y) = 0. \tag{2.2.4}$$

If $f \neq 0$ a.e. for $x \geq a$, then $\tilde{f}(x) > 0$ for some $x > a$, and because of the decreasing property, $\tilde{f} > 0$ on a neighborhood of $x$. This implies that (2.2.4) cannot hold for $x > a$, which is a contradiction.

(c) We show that $\tilde{f}$ in (2.2.2) (which is a nonnegative, continuous decreasing function) cannot vanish. Suppose $c > 0$ is the smallest value such that $\tilde{f}(x) = 0$ for all $x \geq c$. Choose $0 < \delta < c$ such that $\tilde{\sigma}(0, \delta] < 1 - \tilde{\sigma}\{0\}$ (this can be done since $\lim_{\delta\to 0} \tilde{\sigma}[0, \delta] = \tilde{\sigma}\{0\}$ by a similar argument as in Proposition 1.2.1). Then,

$$\begin{aligned}\tilde{f}(c-\delta) &= \tilde{f}(c-\delta)\tilde{\sigma}\{0\} + \int_{(0,\delta]} \tilde{f}(c-\delta+y)\, d\tilde{\sigma}(y)\\ &< \tilde{f}(c-\delta)\tilde{\sigma}\{0\} + \tilde{f}(c-\delta)(1-\tilde{\sigma}\{0\}) = \tilde{f}(c-\delta),\end{aligned}$$

which is a contradiction. The first statement in (c) is established by noting that $\tilde{f}(0) \neq 0$, and $h(x) = e^{\beta x}\tilde{f}(x)$ is a positive solution of the ICFE($\sigma$).

The second part of statement (c) follows immediately from (a), (b), and Proposition 1.2.1, after readjusting $f$ to be continuous and positive as in Lemma 2.2.2.

The main theorem in this chapter is the following.

**Theorem 2.2.4.** *Suppose $\sigma\{0\} < 1$, and $f$ is a nontrivial, nonnegative locally integrable solution of the* ICFE($\sigma$). *Then,*

$$f(x) = p(x)e^{\alpha x} \qquad a.e.,$$

*where $\alpha \in \mathbb{R}$ and is uniquely determined by*

$$\int_0^\infty e^{\alpha y}\, d\sigma(y) = 1, \tag{2.2.5}$$

*and $p$ satisfies $p(x + y) = p(x)$ for all $y \in$* supp $\sigma$.

Note that if there exists no $\alpha$ satisfying (2.2.5), then $f \equiv 0$ a.e. is the only solution. The following remarks will be used in proving the theorem.

***Remark 1.*** Since $f \neq 0$ a.e., let

$$h(x) = \int_x^\infty \left(\int_y^\infty e^{-2\beta t} f(t)\, dt\right) dy,$$

where $\beta$ is defined as in Lemma 2.2.2 (the existence of the $\alpha$ there is trivial and $\int_0^\infty e^{\beta y}\, d\sigma(y) > 1$). Then, $h$ is positive, convex, and decreasing, and $h \in$ ICFE($\tilde{\sigma}$), where $d\tilde{\sigma}(x) = e^{2\beta x}\, d\sigma(x)$. If $h$ is proven to be of the form $h(x) = p(x)e^{\alpha x}$, then $f$ will also have the same form, as can be seen by taking derivatives.

***Remark 2.*** In view of Remark 1 and Proposition 2.2.1(i), the measure $\nu = \sum_{n=1}^{\infty} 2^{-n}\sigma^n$ is $\sigma$-finite, and $f \in \text{ICFE}(\nu)$. An additional property of $\nu$ is that $\text{supp}\,\nu$ forms a semigroup ($\text{supp}\,\nu = \overline{\bigcup \text{supp}\,\sigma^n}$).

***Proof.*** In view of the above remarks, we assume without loss of generality that $f$ is positive, convex, and decreasing, the support of $\sigma$ is a semigroup, and, furthermore, $\sigma\{0\} = 0$ (otherwise, adjust $\sigma$ as in (2.2.3)). For $y \geq 0$, define

$$g(y) = \sup_{x \geq 0} \frac{f(x+y)}{f(x)}.$$

Then, $0 < g(y) \leq 1$; $g$ is also convex on $(0, \infty)$, and in particular continuous, and satisfies

$$g(y_1 + y_2) \leq g(y_1)g(y_2), \qquad \forall\, y_1, y_2 \geq 0.$$

In particular, $g(2y) \leq g(y)^2$. We claim that $g(2y) = g(y)^2$ for $y \in \text{supp}\,\sigma$. Fix $0 < y \in \text{supp}\,\sigma$ and set $g(y) = c$. For any $\delta > 0$, we have $\sigma(y - \delta, y + \delta) > 0$ and

$$\begin{aligned} c - \frac{f(x+y)}{f(x)} &= \int_{\mathbb{R}_+} \frac{cf(x+t) - f(x+y+t)}{f(x)}\, d\sigma(t) \\ &\geq \int_{(y-\delta,y+\delta)} \frac{cf(x+t) - f(x+y+t)}{f(x)}\, d\sigma(t) \\ &\geq \frac{cf(x+y+\delta) - f(x+2y-\delta)}{f(x)}\, \sigma(y-\delta, y+\delta) \\ &\qquad\qquad \text{(since } f \text{ is decreasing)} \\ &= \frac{f(x+y+\delta)}{f(x)} \left( c - \frac{f(x+2y-\delta)}{f(x+y+\delta)} \right) \sigma(y-\delta, y+\delta). \end{aligned}$$

Now, let $\varepsilon > 0$ be given; $\varepsilon_i$, $i = 1, \ldots, 5$, are positive numbers to be chosen in an obvious manner as some constant multiples of $\varepsilon$; since $g$ is continuous, we can let $\delta > 0$ be such that

$$g(y - \delta) - g(y) < \varepsilon_1, \qquad g(2y - \delta) - g(2y) < \varepsilon.$$

By the definition of $c = g(y)$, there exists $x \geq 0$ such that

$$\frac{f(x+y)}{f(x)} > c - \varepsilon_2 \sigma(y - \delta, y + \delta) = c - \varepsilon_3. \tag{2.2.6}$$

Then, by the convexity of $f$,

$$\frac{f(x+y+\delta)}{f(x)} \geq 2\frac{f(x+y)}{f(x)} - \frac{f(x+y-\delta)}{f(x)} \geq 2(c-\varepsilon_3) - g(y-\delta)$$

$$> c - 2\varepsilon_3 - \varepsilon_1 = c - \varepsilon_4,$$

so that

$$\frac{f(x+2y-\delta)}{f(x+y+\delta)} > c - \varepsilon_2/(c-\varepsilon_4) = c - \varepsilon_5,$$

and

$$\frac{f(x+2y-\delta)}{f(x)} > (c-\varepsilon_4)(c-\varepsilon_5) > c^2 - \varepsilon.$$

This implies that $g(2y-\delta) > c^2 - \varepsilon$, so that $g(2y) > c^2 - 2\varepsilon$; since $\varepsilon > 0$ is arbitrary, we have $g(2y) \geq c^2 = g(y)^2$, and hence $g(2y) = g(y)^2$, $y \in \operatorname{supp} \sigma$.

Now, it follows from Proposition 1.1.8 that, for some $\alpha \geq 0$, $g(y) = e^{-\alpha y}$ for all $y \in \operatorname{supp} \sigma$. Our next claim is that

$$\int_{\mathbb{R}_+} e^{-\alpha y}\, d\sigma(y) \leq 1, \tag{2.2.7}$$

from which equality will follow because

$$1 = \int_{\mathbb{R}_+} \frac{f(x+y)}{f(x)}\, d\sigma(y) \leq \int_{\mathbb{R}_+} e^{-\alpha y}\, d\sigma(y) \leq 1. \tag{2.2.8}$$

Suppose $\int_{\mathbb{R}_+} e^{-\alpha y}\, d\sigma(y) > 1$; by the regularity of $\sigma$ (Lemma 2.2.3(c)) and the fact that $\sigma\{0\} = 0$, there exist $0 < a < b < \infty$, and $\gamma > 1$ such that

$$\int_a^b e^{-\alpha y}\, d\sigma(y) = \gamma.$$

Define $d\bar{\sigma}(y) = e^{-\alpha y}\, d\sigma(y)$, and $p(y) = f(y)e^{\alpha y}$; it follows that

$$p(x) \geq \int_a^b p(x+y)\, d\bar{\sigma}(y) = \gamma p(x+\eta),$$

for some $\eta \in [a, b]$ (the mean value theorem). Let $x_0 \geq 0$, and then let $\{x_n\}$ be a sequence such that

$$x_n \in [x_{n-1} + a, x_{n-1} + b] \qquad \text{and} \qquad p(x_n) \leq p(x_0)\gamma^{-n}. \tag{2.2.9}$$

Note that $x_n \to \infty$ as $n \to \infty$. Set $\delta = \log \gamma (>0)$. From the definition of $p$ and the fact that $f$ decreases, we have, for $x_n \leq z < x_{n+1}$,

$$p(z) \leq p(x_n)e^{b\alpha} \leq p(x_0)e^{b\alpha - n\delta} \leq p(x_0)e^{b\alpha + \delta - \delta(z-x_0)/b}$$

(the last inequality follows from $(n + 1)b \geq x_{n+1} - x_0 > z - x_0$). We thus see that for $x \geq 0$, $y \geq 0$ (taking $x_0 = x$, $z = x + y$),

$$\frac{f(x + y)}{f(x)} = \frac{p(x + y)e^{-\alpha(x+y)}}{p(x)e^{-\alpha x}} \leq ce^{-(\alpha+\delta/b)y},$$

where $c = e^{b\alpha+\delta}$. Hence, $g(y)$ also admits the same estimate. This contradicts (for large $y$) the fact that $g(y) = e^{-\alpha y}$ for $y \in \operatorname{supp} \sigma$, and the second claim follows.

By (2.2.8), $f(x + y) = f(x)e^{-\alpha y}$ for all $x \geq 0$, and for $\sigma$-almost all $y \geq 0$; the continuity of $f$ implies that the equality holds for all $y \in \operatorname{supp} \sigma$ as well. By setting $p(x) = f(x)e^{\alpha x}$, we have $p(x + y) = p(x)$ for all $x \geq 0$, $y \in \operatorname{supp} \sigma$ and the theorem is proved.

**Corollary 2.2.5.** *Let $\sigma$ and $f(x) = p(x)e^{\alpha x}$ be as in Theorem 2.2.4. Then:*

(i) *if $\operatorname{supp} \sigma$ has g.c.d. $\rho > 0$, then $p$ is a periodic function with period $\rho$; and*
(ii) *otherwise, $p$ is a constant.*

***Proof.*** The assertion follows from Proposition 1.1.7 and Theorem 2.2.4.

**Corollary 2.2.6.** *Let $\{p_n\}_{n=0}^{\infty}$, $\{v_n\}_{n=0}^{\infty}$ be nonnegative sequences satisfying*

$$v_m = \sum_{n=0}^{\infty} v_n p_{n+m}, \qquad m \in \mathbb{Z}_+.$$

(i) *If $p_0 \geq 1$, then the statement of Proposition 2.1.1 holds.*
(ii) *If $p_0 < 1$, then either $v_n = 0$ for all $n$, or $v_n = B(n)b^n$, where $\sum_{n=0}^{\infty} b^n p_n = 1$, and $B$ is a periodic function on $\mathbb{Z}_+$ with period equal to the g.c.d. of $\{n: p_n > 0\}$.*

***Proof.*** Let $\sigma\{n\} = p_n$, and $\sigma(E) = 0$ if $E$ is a Borel subset of $\mathbb{R}_+$ disjoint from $\mathbb{Z}_+$. Let $f(x) = p_n$ for $n \leq x < n + 1$. Then, $f$ and $\sigma$ satisfy the ICFE($\sigma$) in Theorem 2.2.4. Now, apply the theorem and Corollary 2.2.5 to obtain $\{v_n\}_{n=0}^{\infty}$ by restricting the solution $f$ to $\mathbb{Z}_+$.

**Corollary 2.2.7.** *Let $\sigma$ be a probability measure, and let $f$ be a bounded, locally integrable solution of the* ICFE($\sigma$). *Then, for each $y \in \operatorname{supp} \sigma$, $f(x + y) = f(x)$ a.e.*

***Proof.*** Let $c$ be such that $|f(x)| < c$ for $x \geq 0$, and let $g = f + c$. Then, $g > 0$ on $\mathbb{R}_+$, and satisfies the ICFE($\sigma$) as well. Theorem 2.2.4 can hence be applied to $g$, and the result follows.

## 2.3. AN ALTERNATIVE PROOF USING EXCHANGABLE R.V.'S

In this section we give an alternative proof of Theorem 2.2.4 by using a simple technique—exchangable random variables.

**Definition 2.3.1.** *Let $\{X_n\}_{n=1}^{\infty}$ be a sequence of real-valued r.v.'s on a probability space $(\Omega, \mathcal{B}, P)$. $\{X_n\}_{n=1}^{\infty}$ is said to be* exchangable *if, for every $n \in \mathbb{N}$, the joint distribution of $(X_{\pi(1)}, \ldots, X_{\pi(n)})$ is the same as that of $(X_1, \ldots, X_n)$ for every permutation $\pi$ on $\{1, \ldots, n\}$.*

**Theorem 2.3.2.** *Let $\{X_n\}_{n=1}^{\infty}$ be a sequence of exchangable r.v.'s. Then, for any Borel subset $A \subseteq \mathbb{R}$,*

$$P\{X_1 \in A\} = \lim_{n\to\infty} E\left(\frac{1}{n}\sum_{i=1}^{n} \chi_{\{X_i \in A\}}\right),$$

*and*

$$P\{X_1 \in A, X_2 \in A\} = \lim_{n\to\infty} E\left(\frac{1}{n}\sum_{i=1}^{n} \chi_{\{X_i \in A\}}\right)^2$$

*($\chi_E$ denotes the indicator function of $E$).*

***Proof.*** The first identity is a trivial consequence of the definition. To prove the second equality, we need only observe that

$$\begin{aligned}
&\lim_{n\to\infty} E\left(\frac{1}{n}\sum_{i=1}^{n} \chi_{\{X_i \in A\}}\right)^2 \\
&\quad = \lim_{n\to\infty} \frac{1}{n^2}\left(\sum_{i=1}^{n} E\chi_{\{X_i \in A\}} + \sum_{i\neq j} E(\chi_{\{X_i \in A, X_j \in A\}})\right) \\
&\quad = \lim_{n\to\infty} \frac{1}{n^2}(nP\{X_i \in A\} + n(n-1)P\{X_1 \in A, X_2 \in A\}) \\
&\quad = P\{X_1 \in A, X_2 \in A\}.
\end{aligned}$$

***An alternative proof of Theorem 2.2.4.*** We still assume without loss of generality that supp $\sigma$ is a semigroup of $\mathbb{R}_+$, $\sigma\{0\} = 0$, and the solution $f$ is a positive, continuous, decreasing function on $\mathbb{R}_+$.

Let $x \geq 0$ be given; we define (as we can) a probability space on which an infinite sequence $\{X_n\}_{n=1}^{\infty}$ of exchangable r.v.'s exists with the following property: For each $n \in \mathbb{N}$, the joint probability distribution of $(X_1, \ldots, X_n)$

is given by

$$P\{X_i \in E_i, i = 1, \ldots, n\}$$
$$= \frac{1}{f(x)} \int_{E_1} \cdots \int_{E_n} f(x + y_1 + \cdots + y_n)\, d\sigma(y_1) \cdots d\sigma(y_n),$$

for any Borel subsets $E_1, \ldots, E_n$ of $\mathbb{R}_+$. Let $0 < y \in \operatorname{supp} \sigma$, and let $A_k = (y - (1/k),\ y + (1/k)) \cap \mathbb{R}_+$, $k \in \mathbb{N}$; then, $\sigma(A_k) > 0$. The continuity of $f$ and Theorem 2.3.2 imply that

$$\frac{f(x + y)}{f(y)} = \lim_{k \to \infty} \frac{P\{X_1 \in A_k\}}{\sigma(A_k)} = \lim_{k \to \infty} \lim_{m \to \infty} \frac{E((1/m) \sum_{i=1}^{m} \chi_{\{X_i \in A_k\}})}{\sigma(A_k)},$$

and

$$\frac{f(x + 2y)}{f(y)} = \lim_{k \to \infty} \frac{P\{X_1 \in A_k, X_2 \in A_k\}}{\sigma(A_k)^2}$$
$$= \lim_{k \to \infty} \lim_{m \to \infty} \frac{E((1/m) \sum_{i=1}^{m} \chi_{\{X_i \in A_k\}})^2}{\sigma(A_k)^2}.$$

The Cauchy–Schwarz inequality hence implies that

$$\left(\frac{f(x + y)}{f(y)}\right)^2 \le \frac{f(x + 2y)}{f(y)}, \qquad \forall\, x \ge 0,\ y \in \operatorname{supp} \sigma;$$

equivalently,

$$\frac{f(x + y)}{f(y)} \le \frac{f(x + 2y)}{f(x + y)}, \qquad \forall\, x \ge 0,\ y \in \operatorname{supp} \sigma.$$

Let $0 < y_0 \in \operatorname{supp} \sigma$ be fixed, and let $g(x) = f(x + y_0)/f(x)$; the above inequality implies that

$$g(x) \le g(x + y_0) \le g(x + 2y_0) \le \cdots, \qquad \forall\, x \ge 0. \tag{2.3.1}$$

Let $x^* \in [0, y_0]$ be such that

$$g(x^*) = \min\{g(x) : x \in [0, y_0]\} \tag{2.3.2}$$

($x^*$ depends on $y_0$). For any $x \ge 0$, write $x = \bar{x} + ny_0$ with $\bar{x} \in [0, y_0]$; then,

$$g(x^*) \le g(\bar{x}) \le g(\bar{x} + ny_0) = g(x).$$

By replacing $x$ by $x^* + x$, and after reshuffling the terms, we have

$$\frac{f(x^* + x)}{f(x^*)} \le \frac{f(x^* + x + y_0)}{f(x^* + y_0)}, \qquad \forall\, x \ge 0.$$

The ICFE($\sigma$) implies that

$$1 = \int_{\mathbb{R}_+} \frac{f(x^* + x)}{f(x^*)} \, d\sigma(x) \leq \int_{\mathbb{R}_+} \frac{f(x^* + x + y_0)}{f(x^* + y_0)} \, d\sigma(x) = 1,$$

and hence

$$\frac{f(x^* + y)}{f(x^*)} = \frac{f(x^* + y + y_0)}{f(x^* + y_0)}, \qquad \forall \, y \in \operatorname{supp} \sigma.$$

By letting $y = ny_0$ and iterating the above equality, we have

$$\frac{f(x^* + ny_0)}{f(x^*)} = \left( \frac{f(x^* + y_0)}{f(x^*)} \right)^n.$$

Then, for any $x \geq 0$,

$$\begin{aligned} \left( \frac{f(x + y_0)}{f(x)} \right)^n &\leq \frac{f(x + ny_0)}{f(x)} \qquad \text{(by (2.3.1))} \\ &\leq \frac{f(x^*)}{f(x)} \frac{f(x^* + (n-1)y_0)}{f(x^*)} \\ &\quad \text{(since } x + y_0 \geq x^* \text{ and } f \text{ is decreasing)} \\ &= \frac{f(x^*)}{f(x)} \left( \frac{f(x^* + y_0)}{f(x^*)} \right)^{n-1}. \end{aligned}$$

Taking the $n$th root on both sides of the inequality, and letting $n \to \infty$, we have

$$\frac{f(x + y_0)}{f(x)} \leq \frac{f(x^* + y_0)}{f(x^*)}, \qquad \forall \, x \geq 0.$$

The choice of $x^*$ implies that equality holds for $x \in [0, y_0]$, and then for all $x \geq 0$. Since $y_0 \in \operatorname{supp} \sigma$ is arbitrary, we have

$$f(0)f(x + y) = f(x)f(y), \qquad \forall \, x \geq 0, y \in \operatorname{supp} \sigma.$$

The proof is hence completed by applying Lemma 1.1.5.

## 2.4. THE ICFE WITH A SIGNED MEASURE

In the previous sections, we have assumed that $\sigma \geq 0$, and that the solutions of the ICFE($\sigma$) are nonnegative. The equation without such an assumption is vastly more complicated (see the notes and remarks at the end of Chapter 9). In this section, we shall deal with a special case that will be needed in the proof of Theorem 3.4.1.

**Theorem 2.4.1.** *Let $\mu$, $\nu$ be positive Borel measures on $\mathbb{R}_+$ such that $\mu + \nu$ is a probability measure. Let $f$ be a bounded, Borel measurable function on $\mathbb{R}_+$ satisfying*

$$f(x) = \int_{\mathbb{R}_+} f(x + y)\, d(\mu - \nu)(y), \qquad a.e.\ x \geq 0. \tag{2.4.1}$$

*Then, $f(x + y) = f(x)$ a.e. $x \geq 0$ for each $y \in \operatorname{supp} \mu$, and $f(x + y) = -f(x)$ a.e. $x \geq 0$ for each $y \in \operatorname{supp} \nu$.*

***Proof.*** Write (2.4.1) as $f = f \bullet (\mu - \nu)$; repeated use of the Fubini theorem yields

$$\begin{aligned} f &= f \bullet \mu - f \bullet \nu = f \bullet \mu - (f \bullet \mu - f \bullet \nu) \bullet \nu \\ &= f \bullet (\mu + \nu^2) - f \bullet (\mu * \nu) \qquad (\text{where } \nu^2 = \nu * \nu) \\ &= \cdots \\ &= f \bullet \left(\mu + \nu^2 * \sum_{i=0}^{n-1} \mu^i\right) - f \bullet (\mu^n * \nu). \end{aligned} \tag{2.4.2}$$

Note that

$$\begin{aligned} |f \bullet (\mu^n * \nu)(x)| &= \left| \int_{\mathbb{R}_+} f(x + y)\, d(\mu^n * \nu)(y) \right| \\ &\leq M \cdot \mu(\mathbb{R}_+)^n \cdot \nu(\mathbb{R}_+) \end{aligned}$$

($M$ is a bound of $|f|$). Since $\mu(\mathbb{R}_+) < 1$ by assumption, the preceding term converges to 0 as $n \to \infty$. We hence have $f = f \bullet \sigma$, where

$$\sigma = \mu + \nu^2 * \left(\sum_{i=0}^{\infty} \mu^i\right).$$

$\sigma$ is a probability measure:

$$\begin{aligned} \|\sigma\| &= \|\mu\| + \|\nu\|^2 \sum_{i=0}^{\infty} \|\mu\|^i = \|\mu\| + \|\nu\|^2 \frac{1}{1 - \|\mu\|} \\ &= \|\mu\| + \frac{(1 - \|\mu\|)^2}{1 - \|\mu\|} = 1 \end{aligned}$$

($\|\sigma\|$ denotes the total variation of $\sigma$). By Corollary 2.2.7, $f(x) = f(x + y)$ for $y \in \operatorname{supp} \sigma$, and in particular for all $y \in \operatorname{supp} \mu$. Equation (2.4.1) then implies that

$$cf(x) = -\int_{\mathbb{R}_+} f(x + y)\, d\nu(y), \qquad \text{a.e. } x \geq 0,$$

where $c = 1 - \mu(\mathbb{R}_+)$. Now, if $y$, $u \in \operatorname{supp} \nu$, then $y + u \in \operatorname{supp} \nu^2 \subseteq \operatorname{supp} \sigma$, so $f(x + y + u) = f(x)$ for almost all $x \geq 0$. It follows that

$$cf(x + y) = -\int_{\mathbb{R}_+} f(x + y + u)\, d\nu(u) = -cf(x), \qquad \forall\, y \in \operatorname{supp} \nu. \tag{2.4.3}$$

This proves the theorem.

The solution in Theorem 2.4.1 can be made more precise as follows:

**Theorem 2.4.2.** *Let $\mu$, $\nu$ and $f$ be defined as in Theorem* 2.4.1. *Then, either*:

(a) *there exists a $\rho > 0$, which we take to be the largest such, such that*

$$\operatorname{supp} \mu \subseteq \{0, 2\rho, 4\rho, \ldots\}, \qquad \operatorname{supp} \nu \subseteq \{\rho, 3\rho, 5\rho, \ldots\},$$

*and in this case $f(x + \rho) = -f(x)$ a.e.; or*

(b) *$f(x) = 0$ a.e. otherwise.*

***Proof.*** Let, for any $\rho > 0$,

$$A(\rho) = \{0, 2\rho, 4\rho, 6\rho, \ldots\}, \qquad B(\rho) = \{\rho, 3\rho, 5\rho, \ldots\}.$$

To prove (a) we assume $\rho$ is taken to be the largest such that $\operatorname{supp} \mu \subseteq A(\rho)$ and $\operatorname{supp} \nu \subseteq B(\rho)$. Let $m\rho \in \operatorname{supp} \mu$, $n\rho \in \operatorname{supp} \nu$ be such that the g.c.d. of $m$, $n$ is 1 ($m$ is even and $n$ is odd). Note that, by Theorem 2.4.1,

$$f(x + m\rho) = f(x), \qquad f(x + n\rho) = -f(x).$$

Lemma 1.1.1 implies that there exists a $k$, necessarily even, such that $m$ divides $k$, and

$$k + 1 = a_1 m + a_2 n,$$

where $a_1$, $a_2$ are nonnegative integers ($a_2$ must be odd in this case); hence,

$$f(x + \rho) = f(x + \rho + k\rho) = f(x + a_1(m\rho) + a_2(n\rho)) = -f(x). \tag{2.4.4}$$

To prove (b) we first observe that if $\operatorname{supp} \mu \cup \operatorname{supp} \nu$ does not generate a lattice, the same is true for $\operatorname{supp} \mu \cup 2 \operatorname{supp} \nu$. The fact that $f$ has every $y \in \operatorname{supp} \mu \cup 2 \operatorname{supp} \nu$ as a period implies that $f$ is a constant (Proposition 1.1.7). Since $f(x + y) = -f(x)$ for $y \in \operatorname{supp} \nu$, $f \equiv 0$. The only nontrivial case is when $\operatorname{supp} \mu \cup \operatorname{supp} \nu$ generates a lattice. Let $0 < \rho$ be the g.c.d. of $\operatorname{supp} \mu \cup \operatorname{supp} \nu$ in such a case. There are two possibilities:

(i) $\operatorname{supp} \mu \not\subseteq A(\rho)$. The g.c.d. of elements in $\operatorname{supp} \mu \cup 2 \operatorname{supp} \nu$ is also $\rho$; hence, by Proposition 1.1.7, $f(x + \rho) = f(x)$ for $x \geq 0$, which contradicts the fact that $f(x + k\rho) = -f(x)$ for some $k\rho \in \operatorname{supp} \nu$, unless $f \equiv 0$.

(ii) $\operatorname{supp} \mu \subseteq A(\rho)$, $\operatorname{supp} \nu \nsubseteq B(\rho)$. The fact that $\rho$ is the g.c.d. of $\operatorname{supp} \mu \cup \operatorname{supp} \nu$ implies that $\operatorname{supp} \nu \cap B(\rho) \neq \emptyset$. Using the same arguments as in (2.4.4), we see that $f(x + \rho) = -f(x)$. On the other hand, $\operatorname{supp} \nu \nsubseteq B(\rho)$ implies that there exists an even $k$ such that $k\rho \in \operatorname{supp} \nu$; hence

$$f(x) = -f(x + \rho) = \cdots = (-1)^k f(x + k\rho) = f(x + k\rho) = -f(x),$$

and $f \equiv 0$.

**Corollary 2.4.3.** *The conclusions of Theorem 2.4.1 and Theorem* 2.4.2 *still hold if the condition that $f$ is bounded is replaced by $f(x + y) - f(x)$ is bounded for every fixed $y > 0$.*

***Proof.*** We only prove the conclusion of Theorem 2.4.2, the other being dealt with in a similar fashion. For each fixed $y$, we apply Theorem 2.4.2 to $f(x + y) - f(x)$, and we have, in case (a),

$$f(x + y + \rho) - f(x + \rho) = -(f(x + y) - f(x)), \qquad \forall\, x \geq 0.$$

This implies that

$$f(x + y + \rho) + f(x + y) = f(x + \rho) + f(x), \qquad \forall\, x \geq 0.$$

Since $y$ is arbitrary, we conclude that $f(x + \rho) + f(x) = c$. Equation (2.4.1) then implies that $c = c(\mu - \nu)(\mathbb{R}_+)$, which yields $c = 0$ (since $|(\mu - \nu)(\mathbb{R}_+)| < 1$); hence, $f(x + \rho) = -f(x)$, $\forall\, x \geq 0$.

For case (b), the previous theorem implies that

$$f(x + y) - f(x) = 0, \qquad \forall\, x \geq 0,\ y \geq 0;$$

hence, $f(x) = c$, and (2.4.1) enables us to conclude that $f \equiv 0$.

Theorem 2.4.2 can be used to study the following simultaneous integral equations.

**Corollary 2.4.4.** *Let $\mu$ and $\nu$ be $\sigma$-finite measures on $\mathbb{R}_+$. Let $g, h \not\equiv 0$ a.e. be nonnegative locally integrable functions on $\mathbb{R}_+$ satisfying*

$$g(x) = \int_{\mathbb{R}_+} g(x + y)\, d\mu(y) + \int_{\mathbb{R}_+} h(x + y)\, d\nu(y),$$

$$h(x) = \int_{\mathbb{R}_+} h(x + y)\, d\mu(y) + \int_{\mathbb{R}_+} g(x + y)\, d\nu(y),$$

*for almost all $x \geq 0$. Then, $g(x) = p(x)e^{\alpha x}$, $h(x) = q(x)e^{\alpha x}$ a.e., where $\alpha$*

*satisfies*

$$\int_{\mathbb{R}_+} e^{\alpha x}\, d(\mu + \nu)(x) = 1, \tag{2.4.5}$$

*and* $p, q \geq 0$ *satisfy either*

(a) $p(x) = q(x)$, *and* $p$ *has every* $y \in \operatorname{supp}(\mu + \nu)$ *as a period*; *or*
(b) $p, q$ *have period* $2\rho$, $p(x + \rho) = q(x)$, $q(x + \rho) = p(x)$, *where* $\rho > 0$ *is such that*

$$\operatorname{supp} \mu \subseteq \{0, 2\rho, 4\rho, \ldots\}, \qquad \operatorname{supp} \nu \subseteq \{\rho, 3\rho, 5\rho, \ldots\}.$$

***Proof.*** We may confine our attention to continuous $g$ and $h$. Let $k = g + h$, so that $k = k \bullet (\mu + \nu)$ as a consequence of the given integral equations. Theorem 2.2.4 implies that there exists an $\alpha$ satisfying (2.4.5) and $k(x) = r(x)e^{\alpha x}$, where $r$ has every $y \in \operatorname{supp}(\mu + \nu)$ as a period; the continuity and periodicity imply that $r$ is bounded. Write

$$g_1(x) = g(x)e^{-\alpha x}, \qquad h_1(x) = h(x)e^{-\alpha x},$$

$$d\tilde{\mu}(x) = e^{\alpha x}\, d\mu(x), \qquad d\tilde{\nu}(x) = e^{\alpha x}\, d\nu(x),$$

and let $s = (g_1 - h_1)$; $s$ is bounded (since $0 \leq g_1, h_1 \leq r$) and satisfies $s = s \bullet (\tilde{\mu} - \tilde{\nu})$, hence has the form stated in Theorem 2.4.2. Let $2p(x) = r(x) + s(x)$, $2q(x) = r(x) - s(x)$; then, $g$, $h$, $p$, $q$ are as described.

## 2.5. APPLICATION TO CHARACTERIZATION OF PROBABILITY DISTRIBUTIONS

Renewal theory is among the foremost as well as historically earliest areas of probability theory in which the ICFE arises; we refer the reader to Feller (1968), p. 337 and Feller (1971), pp. 364, 382 for details. Here we shall consider in outline applications of Theorem 2.2.4 and its discrete analog to certain characterization problems relating to the exponential, geometric, Poisson, and Pareto laws. For details, the reader is referred to the papers cited in the Notes and Remarks.

### *2.5.1. The Lack of Memory Properties of the Exponential and Geometric Laws*

If $X$ is a r.v. having an exponential d.f., then it has the well-known *lack of memory* property, namely

$$P\{X > y + x \mid X > y\} = P\{X > x\}, \qquad \forall\, x, y > 0.$$

Conversely, if this property holds, let $T = 1 - F$, where $F$ is the d.f. of $X$. $T$ satisfies the Cauchy functional equation:

$$T(x + y) = T(x)T(y), \qquad \forall\, x, y \geq 0,$$

so that $F$ has an exponential d.f.

Exponential laws also have the "strong" or "random" lack of memory property (also called the strong Markov property): The $y \geq 0$ in the preceding property can be replaced by a r.v. $Y \geq 0$ independent of $X$. Assuming that $P\{X > Y\} > 0$, we have

$$P\{X > Y + x \mid X > Y\} = P\{X > x\} \qquad \text{for } x \geq 0. \tag{2.5.1}$$

If $G$ is the d.f. of $Y$, then, equivalently,

$$cT(x) = \int_{[0,\infty)} T(x + y)\, dG(y), \qquad \forall\, x \geq 0, \tag{2.5.2}$$

where $c = P\{X > Y\}$. The question then naturally arises whether and when the preceding relations imply that $F$ is exponential. Theorem 2.2.4 then yields the following.

**Theorem 2.5.1.** *Let* (2.5.2) *hold, and let further* $1 > c > G(0)$. *Let* $\lambda(>0)$ *be defined by the relation*

$$\int_0^\infty e^{-\lambda y}\, dG(y) = c,$$

*and let, for* $\rho > 0$, $A(\rho) = \{n\rho \colon n \in \mathbb{Z}_+\}$. *Then*:

(i) *F is an exponential law, with parameter* $\lambda$, *if* $\operatorname{supp} G \nsubseteq A(\rho)$ *for any* $\rho > 0$; *and*

(ii) $F(x) = 1 - p(x)e^{-\lambda x}$ *for all* $x \geq 0$, *where* $p$ *(is right-continuous and) has period* $\rho$ *if* $\operatorname{supp} G \subseteq A(\rho)$ *for some* $\rho > 0$, *which we take to be the largest such.*

The geometric laws, the discrete analogs of the exponential laws, have similar lack of memory properties. Theorem 2.1.2 and Corollary 2.2.6 yield corresponding characterization results for such laws.

### *2.5.2. Record Values and the Exponential and Geometric Laws*

Let $\{X_n\}_{n=1}^\infty$ be a sequence of i.i.d.r.v.'s with $F \neq \delta_0$ as their common d.f. Let $\xi = \sup\{x \colon F(x) < 1\}$ be the right extremity of $F$, and $\eta = \inf\{x \colon F(x) > 0\}$ the left extremity of $F$. $\xi \in \mathbb{R}$ or $\xi = \infty$, while $\eta \in \mathbb{R}$ or

$\eta = -\infty$. We define the sequence $\{R(n)\}_{n=0}^{\infty}$ of the "upper record epochs" of $\{X_n\}$ as: $R(0) = 1$; for $n \in \mathbb{N}$, $R(n) = \min\{k \in \mathbb{N} : X_k > X_{R(n-1)}\}$ if this set is nonempty; and $= \infty$ otherwise. $\{R(n)\}$ is well-defined ($R(n) < \infty$ a.s. $\forall\, n \in \mathbb{N}$) as a strictly increasing sequence of positive integer-valued r.v.'s if either $\xi = \infty$ or, in case $\xi \in \mathbb{R}$, $P\{X_1 = \xi\} = 0$. For a well-defined sequence $\{R(n)\}$, we call $\{X_{R(n)}\}$ the sequence of (*upper*) *record values* of $\{X_n\}$.

The exponential and geometric laws admit the following characterizations through the identical distribution or the independence of certain statistics involving the record values.

**Theorem 2.5.2.** *Let $\{X_n\}$ be a sequence of i.i.d.r.v.'s with a well-defined sequence $\{X_{R(n)}\}$ of record values. If, for some $n \in \mathbb{N}$, $X_{R(n)} - X_{R(n-1)}$ has the same distribution as $X_1$, then $X_1$ has an exponential or a geometric type d.f. (and conversely).*

***Proof.*** Let $T = 1 - F$ (where $F$ is the d.f. of $X_1$), and let $H_j$ denote the d.f. of $X_{R(j)}$, $j \in \mathbb{Z}_+$. Then, our hypothesis implies that $P\{X_1 > 0\} = 1$ (since, by definition, $X_{R(n)} - X_{R(n-1)} > 0$ a.s.), so that the $X_j$ are nonnegative r.v.'s as well, and that

$$\int_0^{\xi} P\{X_{R(n)} - X_{R(n-1)} > x \mid X_{R(n-1)} = y\}\, dH_{n-1}(y) = T(x), \qquad x \geq 0.$$

Now, we note that, for $x \geq 0$ and $y \in \operatorname{supp} H_{n-1}$, the integrand on the left side of the previous relation is equal to

$$\sum_{k=1}^{\infty} P\{X_{R(n-1)+j} \leq y, \text{ for } 1 \leq j < k;\ X_{R(n-1)+k} > x + y \mid X_{R(n-1)} = y\}.$$

By the independence of the sequence $\{X_j\}$, the preceding is equal to

$$\sum_{k=1}^{\infty} F(y)^{k-1} T(x + y) = T(x + y)/T(y). \tag{2.5.3}$$

Hence, our assumption is equivalent to the functional equation

$$\int_0^{\xi} \frac{T(x + y)}{T(y)}\, dH_{n-1}(y) = T(x), \qquad x \geq 0.$$

Now, if $\xi$ is finite, then

$$0 = P\{X_{R(n)} > \xi\} \geq P\{X_{R(n)} - X_{R(n-1)} > 0,\ X_{R(n-1)} > \xi\};$$

and since $P\{X_{R(n)} - X_{R(n-1)} > 0\} = 1$, it follows that $P\{X_{R(n-1)} \geq \xi\} = 0$, so that $P\{X_1 > \xi\} = 0$ as well, in view of $X_1 < X_{R(n-1)}$ a.s. In particular (using as usual the same symbol for a d.f. and for the measure induced by it on $\mathbb{R}$), $H_{n-1}\{\xi\} = F\{\xi\} = 0$. Thus, whether $\xi \in \mathbb{R}$ or not, $dH_{n-1}(y)/T(y)$ defines a $\sigma$-finite Borel measure on $[0, \xi)$, and Theorem 2.4.2 applies. Noting that $H_{n-1}(\{0\}) = 0$, we conclude that $T(x) = p(x)e^{-\lambda x}$ (a.e. $x \geq 0$ and by right-continuity) for all $x \geq 0$, where $p$ has every element of $\operatorname{supp} H_{n-1}$ as a period. Also, $p(0) = 1$ since $T(0) = 1$.

If now the closed subgroup of $\mathbb{R}$ generated by $\operatorname{supp} H_{n-1}$ is $\mathbb{R}$ itself, then $F$ is exponential with parameter $\lambda$. If that subgroup is $d\mathbb{Z}$ for some $d > 0$, then (2.5.3) implies that

$$1 - H_n(x) = \int_0^\xi \frac{T(\max(x, y))}{T(y)} dH_{n-1}(y), \tag{2.5.4}$$

and we conclude that $\operatorname{supp} H_n \subset H_{n-1} \subset d\mathbb{N}$ (recalling that $P\{X_1 = 0\} = 0$). Hence, $X_{R(n)} - X_{R(n-1)} \in d\mathbb{N}$ a.s., so that, by our identical distribution assumption, $X_1 \in d\mathbb{N}$ a.s., i.e., $\operatorname{supp} F \subset d\mathbb{N}$. Taking $d = 1$, without loss of generality, we have, for $x \in [0, 1)$, $0 = F(x) = 1 - p(x)e^{-\lambda x}$, so that

$$p(x) = p(x - [x]) \quad \text{and} \quad F(x) = 1 - e^{-\lambda[x]}, \quad \text{for all } x \geq 0.$$

Thus, $F$ is an exponential or a geometric type d.f. according to whether the closed subgroup generated by $\operatorname{supp} H_{n-1}$ is $\mathbb{R}$ itself or is discrete.

The converse part is easily verified. Hence, the theorem is proven.

**Theorem 2.5.3.** *Let $\{X_n\}$ be a sequence of i.i.d.r.v.'s, with common d.f. $F$, and with a well-defined sequence $\{X_{R(n)}\}$ of record values. Suppose the left extremity $\eta$ of $F$ is in $\mathbb{R}$. Then, $X_{R(n)} - X_{R(n-1)}$ is independent of $X_{R(n-1)}$ if and only if, for some $\lambda > 0$, $d > 0$,*

$$T(x + \eta_{n-1}) = T(\eta_{n-1})e^{-\lambda x} \quad \text{or} \quad T(\eta_{n-1})e^{-\lambda[x/d]}, \quad \text{for } x \geq 0.$$

*Here, $\eta_n$ is the left extremity of $H_n$.*

***Proof.*** We need only prove the "only if" part, as the "if" part is immediate. Since $\eta \in \mathbb{R}$ and $X_{R(j)} > X_1$ a.s., it follows that $\eta_j \in \mathbb{R}$ and $\eta_j \geq \eta$, for all $j \in \mathbb{N}$ ($\eta_0 = \eta$). Arguing as in the derivation of (2.5.3), we have for any $u \in \operatorname{supp} H_{n-1}$, and in particular for $u = \eta$, that

$$\begin{aligned}\frac{T(x + u)}{T(u)} &= P\{X_{R(n)} - X_{R(n-1)} > x \mid X_{R(n-1)} = u\} \\ &= P\{X_{R(n)} - X_{R(n-1)} > x\} \quad \text{(by the independence assumption)} \\ &= \int_{\eta_{n-1}}^\xi \frac{T(x + y)}{T(y)} dH_{n-1}(y). \end{aligned} \tag{2.5.5}$$

Thus, for every $u \in \operatorname{supp} H_{n-1}$, we have

$$T(x + u)/T(u) = T(x + \eta_{n-1})/T(\eta_{n-1}).$$

From this relation, or taking $u = \eta$ in (2.5.5) and applying Theorem 2.2.4 and arguing as in the proof of the preceding theorem, we obtain the conclusions stated in the theorem, the two cases corresponding to the closed subgroup generated by $\operatorname{supp} H_{n-1}$ being $\mathbb{R}$ itself, or the discrete subgroup $d\mathbb{Z}$ (for some $d > 0$), respectively.

**Corollary 2.5.4.** *If, further, $F(\eta) = 0$, then $F$ is necessarily exponential.*

***Proof.*** We have from (2.5.4) that

$$H_1(x) = \int_\eta^\xi \left(1 - \frac{T(\max(x, y))}{T(y)}\right) dF(y).$$

If $F(\eta) = 0$, then since $H_1(\eta + \varepsilon)$ is

$$\geq \int_\eta^{\eta+\varepsilon/2} \left(1 - \frac{T(\eta + \varepsilon)}{T(y)}\right) dF(y),$$

it is positive for every $\varepsilon > 0$, the integrand being positive on $(\eta, \eta + \varepsilon/2)$, and $F(\eta, \eta + \varepsilon/2)$ being $>0$ as well. The relation $\eta_1 \geq \eta$ is obvious, and it then follows that $\eta_1 = \eta$. We establish similarly, by induction, using (2.5.4), that $\eta_j = \eta$ for all $j \in \mathbb{N}$, if $F(\eta) = 0$. Then, Theorem 2.5.3 enables us to assert that

$$T(x + \eta) = e^{-\lambda x} \qquad \text{or} \qquad e^{-\lambda[x/d]}, \qquad \text{for all } x \geq 0,$$

for some $\lambda > 0$, $d > 0$. The second possibility is ruled out, since, for a lattice d.f. with left extremity $\eta \in \mathbb{R}$, $\eta$ is necessarily a point of increase, but this is impossible by our assumption that $F(\eta) = 0$. Hence, the corollary is proven.

### 2.5.3. An Order Statistic Property of the Exponential Law and Related Laws

Let $\{X_i\}_{i=1}^n$ be i.i.d.r.v.'s with $F \neq \delta_0$ as common d.f., and let

$$X_{1:n} \leq X_{2:n} \leq \cdots \leq X_{n:n}$$

be a rearrangement of the $X_i$ in nondecreasing order. Then, $X_{i:n}$ is called the $i$th order statistic (based on a random sample of size $n$ from $F$). The following facts are well known, as well as easily verified: For any $x \in \mathbb{R}$, with $T = 1 - F$ as usual,

$$P\{X_{n:n} \leq x\} = F(x)^n; \qquad P\{X_{1:n} \leq x\} = 1 - T(x)^n;$$

$$F_{r,n}(x) := P\{X_{r:n} \leq x\} = \sum_{j=r}^{n} \binom{n}{j} F(x)^j T(x)^{n-j}.$$

Whatever $F$ is (in particular, purely discrete or continuous), a version of the conditional distribution of $X_{k:n}$ given $X_{r:n}$, $r < k < n$, defined by

$$\int_{\mathbb{R}} P\{X_{k:n} \leq x \mid X_{r:n} = y\}\, dF_{r:n}(y) = F_{k:n}(x),$$

is the same as the distribution of the $(k - r)$th order statistic in a random sample of size $(n - r)$ from the d.f. obtained by "truncating $F$ to the left at $y$," namely,

$$F^*(x) = \begin{cases} \dfrac{F(x) - F(y)}{1 - F(y)} & \text{if } x \geq y, \\ 0 & \text{otherwise.} \end{cases}$$

If $X_1$ has an exponential d.f. with parameter $\lambda(>0)$, a straightforward computation shows that $X_{r+1:n} - X_{r:n}$ has an exponential d.f. with parameter $(n - r)\lambda$, which is also the d.f. of $X_{1:n-r}$. (See the proof of the following theorem for a computation of the d.f. of $X_{r+1:n} - X_{r:n}$ for arbitrary $F$.) We now examine the converse of this proposition.

**Theorem 2.5.5.** *Let $\{X_i\}_{i=1}^n$ be i.i.d.r.v.'s with $F \neq \delta_0$ as common d.f. If $X_{r+1:n} - X_{r:n}$ has the same distribution as $X_{1:n-r}$, for some $r < n$, then*

$$\binom{n}{r} F(0)^r \leq 1,$$

*and there are (only) three possibilities:*

(i) *$F$ is an exponential law with parameter $\lambda$ given by*

$$\binom{n}{r} \int_0^\infty e^{-\lambda x}\, dF^r(x) = 1;$$

(ii) *$F$ is a "two-point" d.f. of the form $p\delta_0 + q\delta_a$, for some $a > 0$, where*

$$\binom{n}{k} p^k = 1 \qquad \text{and} \qquad q = 1 - p;$$

(iii) *$F$ is a mixture of $\delta_0$ and a geometric type d.f.*

***Proof.*** Our hypothesis is equivalent to

$$P\{X_{r+1:n} - X_{r:n} > x\} = T(x)^{n-r}, \qquad \forall\, x \in \mathbb{R}.$$

It follows that $F(0_{-}) = 0$, so that the $X_i$ are necessarily nonnegative random variables. Using the fact about the conditional distribution of a "higher" order statistic given a "lower" one formulated in the preceding, we have (recalling the definition of $F^*$)

$$\begin{aligned}P\{X_{r+1:n} - X_{r:n} > x\} &= \int_0^\infty P\{X_{r+1:n} - X_{r:n} > x \mid X_{r:n} = y\}\, dF_{r:n}(y)\\ &= \int_0^\infty (1 - F^*_{1:n-r}(x+y))^{n-r}\, dF_{r:n}(y)\end{aligned}$$

($F^*_{r:s}$ having the same relationship to $F^*$ as $F_{r:s}$ has to $F$)

$$= \int_0^\infty \left(\frac{T(x+y)}{T(y)}\right)^{n-r} dF_{r:n}(y),$$

which reduces to the (equivalent) form

$$\binom{n}{r} \int_0^\infty T(x+y)^{n-r}\, dF^r(y).$$

Thus, our assumption of identical distribution is equivalent to the functional equation

$$\binom{n}{r} \int_0^\infty T(x+y)^{n-r}\, dF^r(y) = T(x)^{n-r}.$$

$F \neq \delta_0$ rules out the possibility $\binom{n}{r} F^r(0) > 1$. If $\binom{n}{r} F^r(0) = 1$, we have case (ii). If $\binom{n}{r} F^r(0) < 1$, then Theorem 2.2.4 applies and we have the other two cases, depending respectively on whether the closed subgroup generated by supp($F^r$) is $\mathbb{R}$ or discrete. In the latter case, the explicit form of $F$ is as follows:

$$F = \theta\delta_0 + (1-\theta) \sum_{k=1}^{\infty} pq^{k-1}\, \delta_{kp}$$

for some $0 \leq \theta \leq 1$, $0 < p = 1 - q < 1$, where $\binom{n}{r}\theta^r < 1$, and the parameters $\theta$ and $p$ are related according to

$$\binom{n}{r} \int_0^\infty q^{(n-r)x}\, dF^r(x) = 1;$$

i.e.,

$$\binom{n}{r}\left(Q_0 + \sum_{k=1}^{\infty} (Q_k - Q_{k-1}) q^{(n-r)k}\right) = 1,$$

with $Q_k := (1 - (1-\theta)q^k)^r = (F(kp))^r$ for $k \in \mathbb{N}$; $Q_0 = \theta^r$.

### 2.5.4. Constancy of Certain Conditional Moments and the Exponential Law

If $X$ has an exponential d.f. with parameter $\lambda$, then, for any $\alpha > 0$, $EX^\alpha < \infty$ and the conditional expectation

$$E[(X - x)^\alpha \mid X \geq x] = \lambda \int_x^\infty (y - x)^\alpha e^{-\lambda(y-x)}\, dy = \lambda \int_0^\infty u^\alpha e^{-\lambda u}\, du,$$

which is a constant, for all $x \geq 0$.

**Theorem 2.5.6.** *Suppose $X \geq 0$ is a nondegenerate r.v. with $EX^\alpha < \infty$ for some $\alpha > 0$. Suppose further that, for some constant $c > 0$,*

$$E[(X - x)^\alpha \mid X \geq x] = c \qquad \text{for all } x \geq 0. \tag{2.5.6}$$

*Then, $X$ has an exponential d.f. with parameter $\lambda$ given by* (2.5.7).

***Proof.*** Let $F$ be the d.f. of $X$, and $T = 1 - F$. Then, the assumption that $EX^\alpha < \infty$ implies that

$$x^\alpha T(x) \leq \int_x^\infty y^\alpha\, dF(y) \to 0 \qquad \text{as } x \to \infty.$$

Now, (2.5.5) is equivalent to

$$\int_x^\infty (y - x)^\alpha\, dF(y) = cT(x), \quad \forall\, x \geq 0.$$

Integrating by parts and using the previously mentioned fact, we have

$$\int_x^\infty T(y)\, d(y - x)^\alpha = cT(x), \qquad \forall\, x \geq 0,$$

or

$$\int_0^\infty T(x + y)\, d(y^\alpha) = cT(x), \qquad \forall\, x \geq 0.$$

Theorem 2.2.4 implies at once that $T(x) = e^{-\lambda x}$ for $x \geq 0$, where $\lambda$ is given by

$$\int_0^\infty e^{-\lambda y}\, d(y^\alpha) = c. \tag{2.5.7}$$

### 2.5.5. The Pareto Laws

The Pareto laws are important income-distribution models, and are defined by

$$F(x) = \begin{cases} 1 - (a/x)^k & \text{for } x \geq a, \\ 0 & \text{otherwise,} \end{cases}$$

for some $a > 0$, $k > 0$. If $X$ is a r.v. following $F$, it is immediate that $\log(X/a)$ follows an exponential law with parameter $k$. Consequently, every characterization result concerning the exponential law gives rise to a corresponding result on the Pareto law. For formulations of such results, we refer the reader to the papers cited in the Notes and Remarks.

### 2.5.6. The Poisson Law and a Damage Model

Let $X$ be a nonnegative integer-valued r.v., with $P\{X = n\}$ denoted by $p_n$, $n \in \mathbb{Z}_+$. Suppose $X$ is reduced by some damage process to an integer-valued r.v. $Y$ ($X$ could be the number of eggs laid and $Y$ the number of those hatched), according to the damage model

$$P\{Y = r \mid X = n\} = d(r \mid n), \qquad r = 0, 1, \ldots, n.$$

Consider the conditional probabilities

$$P\{Y = r \mid X \text{ undamaged}\} = \frac{p_r d(r \mid r)}{\sum_{n=0}^{\infty} p_n d(n \mid n)},$$

$$P\{Y = r \mid X \text{ damaged}\} = \frac{\sum_{n=r+1}^{\infty} p_n d(r \mid n)}{\sum_{r=0}^{\infty} \sum_{n=r+1}^{\infty} p_n d(r \mid n)}.$$

Straightforward computations show that, under the binomial damage model,

$$d(r \mid n) = \binom{n}{r} p^r q^{n-r}, \qquad r = 0, 1, \ldots, n, \tag{2.5.8}$$

for some $0 < p < 1$, $q = 1 - p$, if $p_n = e^{-\lambda}\lambda^n/n!$, $n \in \mathbb{Z}_+$, for some $\lambda > 0$, then, for $r \in \mathbb{Z}_+$,

$$P\{Y = r\} = P\{Y = r \mid X \text{ undamaged}\} = P\{Y = r \mid X \text{ damaged}\}, \tag{2.5.9}$$

the common value being $= e^{-\lambda p}(\lambda p)^r/r!$. In particular, if $X \sim \text{Poisson}(\lambda)$, and is subject to (2.5.8), then $Y \sim \text{Poisson}(\lambda p)$.

It is then natural to ask: If (2.5.9) holds, then, assuming (2.5.8), does it follow that $X$ must have a Poisson distribution? Theorem 2.1.2 enables us to answer this question in the affirmative. The equality (2.5.9) implies that, for a suitable $c > 0$,

$$p_r = c \sum_{m=0}^{\infty} p_{r+m} \binom{r+m}{r} q^m, \qquad r \in \mathbb{Z}_+,$$

cancelling out $d(r|r) = p^r$ on both sides. Then, $P_r = p_r \cdot r!$ satisfies

$$P_r = c \sum_{m=0}^{\infty} P_{r+m} q^m/m!.$$

Theorem 2.1.2 implies that, for the unique $\lambda > 0$ such that

$$c \sum_{m=0}^{\infty} q^m \lambda^m/m! = 1$$

(i.e., $c = e^{-q\lambda}$), we must have $p_r = b\lambda^r$ for some $b > 0$, so that $p_r = e^{-\lambda}\lambda^r/r!$, $r \in \mathbb{Z}_+$.

A method of proving the above conclusion by appealing to Bernstein's theorem on completely monotone functions is of considerable interest, in addition to being the proof originally devised. The probability generating function (p.g.f.) of an integer-valued r.v. $Z \geq 0$ with $P\{Z = n\} = q_n$ for $n \in \mathbb{Z}_+$ is defined as $\sum_{n=0}^{\infty} q_n t^n$, and converges at least for $|t| \leq 1$ ($t$ real or complex). If now $g$ is the p.g.f. of $X$, then the p.g.f. of $Y$ is easily seen to be equal to

$$\sum_{r=0}^{\infty} t^r \left( \sum_{n=r}^{\infty} p_n \binom{n}{r} p^r q^{n-r} \right) = g(q + pt),$$

and the p.g.f. of the conditional d.f. of $\{Y \mid X \text{ undamaged}\}$ is $g(pt)/g(p)$. Thus, if (2.5.9) holds, we have the functional equation, valid for real $t$ with $|t| \leq 1$ in the first instance,

$$g(q + pt) = g(pt)/g(p).$$

This relation implies in turn that the power series for $g$ converges for all real $t$, and that $g$, as its sum function, is defined on $[0, \infty)$ by $g(x + sq) = g(x)/(g(p))^s$ for $s \in \mathbb{Z}_+$, $0 \leq x \leq q$, and on $(-\infty, 0]$ by $g(x - sq) = g(x)(g(p))^s$ for $s \in \mathbb{N}$, $0 \leq x \leq q$. In particular, $g^{(k)}(x) \geq 0$ for all $k \in \mathbb{Z}_+$ and for all $x \leq 0$ (since every p.g.f. is absolutely monotone on $(0, 1)$, at least). Hence, by Theorem 1.2.6, we conclude that

$$g(x) = g(0) \int_{[0,\infty)} e^{tx}\, dH(t), \qquad x \leq 0,$$

for some d.f. $H$ on $[0, \infty)$. Writing $h(x) = e^{-\lambda x} g(x)$, where $\lambda = -\log g(p)/q$

($>0$), we have, from $h(0) = h(-q) = h(-2q)$, that

$$e^{-\lambda q} = \int e^{-qt}\,dH(t), \qquad e^{-2\lambda q} = \int e^{-2qt}\,dH(t).$$

The conditions for equality in the Cauchy–Schwarz inequality then imply that $H = \delta_\lambda$ necessarily, i.e., $g(x) = e^{\lambda(x-1)}$, so that $X$ has a Poisson($\lambda$) d.f.

## NOTES AND REMARKS

The main results of this chapter are Theorem 2.2.4, supplemented by Theorem 2.1.2. Their counterparts for $\mathbb{R}$, namely, Theorem 8.1.6 supplemented by Theorem 8.1.7, are essentially special cases of the general result by Deny (1960). In the 1970s, several analytical probabilists, unaware of Deny's work, established particular cases of these results (often under superfluous assumptions, using *ad hoc* methods, comprising real, complex, and transform analysis), in the process of solving various characterization problems in mathematical statistics; most of these relate to the exponential laws, and an account of these is to be found in Azlarov and Volodin (1986). Thus, Rao and Rubin (1964) obtained the characterization, discussed in Section 2.5.6, of the Poisson law through a damage model, appealing to Bernstein's theorem on completely/absolutely monotone functions (Theorem 1.2.6). Shanbhag (1977) obtained Theorem 2.1.2, incidentally providing a simple proof of this characterization, a precursor of his work being Feller's discussion of renewal theory. The general result, Theorem 2.2.4, containing all previous *ad hoc* results of its *genre*, was obtained in Lau and Rao (1982); their proof was simplified in Ramachandran (1982a), and a proof using exchangability was provided in Alzaid *et al.* (1987), as in Section 2.3.

In the context of characterizations, particular cases of the general results were obtained in Shimizu (1968) and in Ramachandran and Rao (1970)—see also Kagan *et al.* (1973) for an account of these; in both papers, preliminary versions of the results of Section 2.4 were also obtained, using the complex analysis methods developed in Linnik (1953a, b). Ramachandran (1977b, 1979) discussed the strong memorylessness property of the exponential (and geometric) laws, discussed in Section 2.5.1, using the same methods. Shimizu (1978) provided real analysis methods to discuss special cases of Theorem 2.2.4, and also a proof based on them for the above characterization; in the process, Shimizu also independently established what is known as the Choquet–Deny theorem for $\mathbb{R}_+$ (Corollary 2.2.7 for bounded, real-valued, continuous $f$), and in fact the extended form thereof given by Theorem 2.4.1. In this context, see also the Notes and Remarks for Chapter 3.

Rossberg (1972) and Ramachandran (1982b) used the Wiener–Hopf technique to obtain characterizations of the exponential laws through identical distribution properties relating to order statistics; the latter also provides yet another proof of Theorem 2.5.1, and one of the stated characterizations is the content of our Section 2.5.3. Characterizations of the exponential law through properties of record values, obtained earlier by Ahsanullah, Nayak, and others, were formulated in general form in Lau and Rao (1982) and H.-J. Witte (1988), as in Section 2.5.2, and, through the constancy of certain conditional moments by Sohobov and Geshev (1974), were reformulated in general form, as given in Section 2.5.4, by Lau and Rao (1982). Huang (1981), while providing a proof of Theorem 2.5.1 using the Wiener–Hopf technique, gives an example to show that the condition $c > G(0)$ there cannot be relaxed.

Theorems 2.4.1 and 2.4.2 and Corollary 2.4.3 were first established, under additional assumptions and with different proofs, in Shimizu (1978), following up earlier work of his and of others. The proofs here are essentially as in Ramachandran *et al.* (1988). Various preliminary versions of Corollary 2.4.4 were obtained by Shimizu (1968) and Ramachandran and Rao (1970)—see also Kagan *et al.* (1973), Shimizu (1978), and Davies and Shimizu (1976, 1980); its final form, given here, is from Ramachandran *et al.* (1988).

A recent instance of the use of the Linnik (1953a, b) methods in analytical probability theory is to be found in Ramachandran (1991b), where the following result is established: "If $f$ is a nonvanishing ch.f. (on $\mathbb{R}$), a necessary and sufficient condition for the coresponding d.f. to be symmetric to within a location parameter is that both the functions $f(t)f(-pt)f(-p^2t)$ and $f(t)f(-qt)f(-q^3t)$ should be real-valued throughout $\mathbb{R}$, where $0 < p,\ q < 1$ are the unique real numbers such that $p + p^2 = q + q^3 = 1$."

CHAPTER

# 3

# The Stable Laws, the Semistable Laws, and a Generalization

The stable laws arise in probability theory as limit d.f.'s for normed sums of i.i.d.r.v.'s, i.e., $a_n^{-1} \sum_{i=1}^{n} (X_i - b_n)$ for some norming constants $a_n > 0$, $b_n \in \mathbb{R}$. For details, see for instance Gnedenko and Kolmogorov (1964) or Loève (1977). (The term "law" will be used here and in what follows to denote, indifferently, a d.f. or its ch.f.). In this chapter, we consider the stable laws purely analytically—as ch.f.'s satisfying a functional equation. The derivation of the Lévy representation for these laws, leading in turn to the closed-form formulas for their ch.f.'s, provides an important application of Proposition 1.1.9. A direct derivation of these formulas (without invoking the Lévy representation) leads also to another functional equation of interest (Lemma 3.1.4). These are the content of Section 3.1. In Section 3.2, we study a generalization of stable laws, namely, the semistable laws (of P. Lévy). In Sections 3.3 and 3.4, we consider ch.f.'s satisfying a certain type of functional equation; such laws turn out to be different from semistable laws only by location parameters. In studying them, we invoke the properties of ch.f.'s established in Section 1.3 and information on the ICFE on $\mathbb{R}_+$ developed in Sections 2.2 and 2.4.

## 3.1. THE STABLE LAWS

Two nondegenerate d.f.'s $F_1$ and $F_2$ on $\mathbb{R}$ are said to be of the same *type* if there exist $a, b \in \mathbb{R}$ with $b > 0$ such that $F_2(x) = F_1((x - a)/b)$ for all $x \in \mathbb{R}$. Note that "belongs to the same type" is an equivalence relation on the class of all d.f.'s on $\mathbb{R}$ (and so sets up a partition thereof). All normal laws, for example, belong to one type.

A nondegenerate d.f. $F$ (more accurately, the "type" determined by it) is called a *stable law* (respectively, a *stable type*) if its type is closed under convolution; i.e., for every choice of $a_1, a_2, b_1, b_2 \in \mathbb{R}$ with $b_1, b_2 > 0$, there exist $a, b \in \mathbb{R}$ with $b > 0$ such that

$$F((\cdot - a_1)/b_1) * F((\cdot - a_2)/b_2) = F((\cdot - a)/b).$$

In terms of ch.f.'s, this is equivalent to: For every choice of $b_1, b_2 > 0$, there exist $a, b \in \mathbb{R}$ with $b > 0$ (both depending on $b_1, b_2$) such that

$$f(b_1 t)f(b_2 t) = f(bt)e^{iat}, \qquad t \in \mathbb{R}. \tag{3.1.1}$$

It follows that, for every choice of $b_1, \ldots, b_n$, all positive, $n \in \mathbb{N}$, there exist $\beta, \gamma \in \mathbb{R}$ with $\beta > 0$ (depending of course on the $b_j$) such that

$$\prod_{j=1}^{n} f(b_j t) = f(\beta t)e^{i\gamma t}, \qquad t \in \mathbb{R}. \tag{3.1.1'}$$

***Remarks.*** As we shall see, all stable laws are absolutely continuous with respect to Lebesgue measure. However, closed-form formulas for the probability density functions (p.d.f.'s) thereof are known only for three families, namely:

(a) the normal d..f.'s, with p.d.f.

$$p(x) = \frac{1}{\sigma\sqrt{2\pi}} \exp(-(x - \mu)^2/2\sigma^2), \qquad x \in \mathbb{R}; \mu \in \mathbb{R}, \sigma > 0,$$

and ch.f.

$$f(t) = \exp(i\mu t - \tfrac{1}{2}\sigma^2 t^2), \qquad t \in \mathbb{R};$$

(b) the Cauchy type laws with p.d.f.

$$p(x) = \sigma/(\sigma^2 + (x - \mu)^2), \qquad x \in \mathbb{R}; \mu \in \mathbb{R}, \sigma > 0,$$

and ch.f.

$$f(t) = \exp(i\mu t - \sigma|t|), \qquad t \in \mathbb{R};$$

and

(c) the Lévy–Smirnov type determined by the d.f. with p.d.f.

$$p(x) = \frac{1}{\sqrt{2\pi}} e^{-1/2x} x^{-3/2} \qquad \text{for } x > 0, \qquad \text{zero otherwise},$$

and ch.f.

$$f(t) = \exp(-|t|^{1/2}(1 - i \operatorname{sgn} t)), \qquad t \in \mathbb{R}.$$

More or less straightforward computations establish the correspondences in (a) and (b); as for the third, see the appendix to this chapter (what has been derived there is the ch.f. $f(t/2)$ corresponding to the density $2p(2x)$).

**Theorem 3.1.1.** *A stable law is infintely divisible.*

***Proof.*** Taking $b_1 = \cdots = b_n = 1$ in (3.1.1′), we have

$$f(t)^n = f(\beta_n t)\exp(i\gamma_n t), \qquad t \in \mathbb{R},$$

for some $\beta_n$, $\gamma_n \in \mathbb{R}$ with $\beta_n > 0$. Hence,

$$f(t) = f(t/\beta_n)^n \exp(i\delta_n t) = f_n(t)^n, \qquad t \in \mathbb{R},$$

where $\delta_n$, $f_n$ are defined in an obvious manner and $f_n$ is a ch.f. Hence the theorem is proven.

We proceed to obtain the Lévy representation, through which closed-form formulas for stable ch.f.'s can be obtained. In the latter half of this section, we shall obtain these formulas directly.

**Theorem 3.1.2.** *A ch.f. $f$ is stable if and only if it is nonvanishing on $\mathbb{R}$ and $\phi = \log f$ admits a Lévy representation $L(\mu, \sigma^2, M, N)$ of one of the following forms:*

(i) $M \equiv N \equiv 0$; *in this case,* $\sigma > 0$, *and $F$ is a normal d.f.*;
(ii) *at least one of $M$ and $N$ is* $\not\equiv 0$; *in this case,* $\sigma = 0$, *and*

$$M(u) = c_1|u|^{-\alpha}, \qquad \forall\, u < 0, \qquad N(u) = -c_2 u^{-\alpha}, \qquad \forall\, u > 0,$$

*for some $\alpha$ with $0 < \alpha < 2$, and for some $c_1$, $c_2$ both $\geq 0$, with $c_1 + c_2 > 0$ (i.e., not both zero).*

***Proof.*** The sufficiency of the above conditions is clear. We need only prove the necessity. Since a stable ch.f. is inf. div., it admits a Lévy representation $L(\mu, \sigma^2, M, N)$, which is also unique (Theorem 1.3.12). The uniqueness and (3.1.1) imply that, for every choice of $b_1, b_2 > 0$, there exists a $b > 0$ such that the three relations below are satisfied:

$$\begin{aligned} &\sigma^2(b_1^2 + b_2^2 - b^2) = 0; \\ &M(u/b_1) + M(u/b_2) = M(u/b), \qquad \forall\, u < 0; \\ &N(u/b_1) + N(u/b_2) = N(u/b), \qquad \forall\, u > 0. \end{aligned} \tag{3.1.2}$$

We first observe that if for some $c > 1$ and all $u < 0$, $M(u) \leq M(cu)$, then $M \equiv 0$, and dually for $N$. For, then $M(u) \leq M(c^n u)$ for all $n \in \mathbb{N}$, so that $M(u) = M(-\infty) = 0$.

Suppose now that $M \not\equiv 0$; iterating the second relation in (3.1.2) and taking $b_1 = \cdots = b_n = 1$, we see that, for some $\beta_n > 0$,

$$nM(u) = M(u/\beta_n), \qquad \forall\, u < 0,\ n \in \mathbb{Z}_+. \tag{3.1.3}$$

It follows that $(0 \leq)\ M(u/\beta_n) \leq M(u/\beta_{n+1})$ for all $u < 0$ and then, from the previous observation, that $\beta_n \leq \beta_{n+1}$, i.e., $\{\beta_n\}$ is an increasing sequence. Also, we have from (3.1.3) that

$$M(\beta_m \beta_n u) = M(\beta_{m \cdot n} u), \qquad \forall\, u < 0.$$

Again, by the previous observation, it follows from this relation that $\beta_{m \cdot n} = \beta_m \beta_n$ for all $m, n \in \mathbb{N}$. Therefore, by Proposition 1.1.9, there exists an $\alpha > 0$ such that $\beta_n = n^{1/\alpha}$ for all $n \in \mathbb{N}$. Equation (3.1.3) and the right-continuity of $M$ then imply that $M(u) = M(1)/|u|^\alpha$ for $u < 0$ (if $M \not\equiv 0$) and, similarly, $N(u) = N(1)/u^\alpha$ for $u > 0$ (if $N \not\equiv 0$). Now, setting $b_1 = b_2 = 1$ in (3.1.2), we have that, for the same $b > 0$,

$$\sigma^2(2 - b^2) = 0, \qquad 2 - b^\alpha = 0,$$

so that $\sigma = 0$ if $\alpha \neq 2$. Also, the condition (of the Lévy representation)

$$\int_{(-1,0)} u^2\, dM(u) + \int_{(0,1)} u^2\, dN(u) < \infty$$

implies that, if $M$ or $N$ is not identically zero, then $0 < \alpha < 2$. Hence, $M \equiv N \equiv 0$ if $\alpha \geq 2$; but, again, $\alpha > 2$ is impossible since in such a case we must have $\sigma = 0$ also, implying that $f$ is a degenerate ch.f. (Recall that, by definition, a stable law is nondegenerate.) The case $\alpha = 2$ corresponds to the normal laws.

The $\alpha$ (with $0 < \alpha \leq 2$) appearing in the above theorem is called the *exponent* of the stable law. Now the integrals appearing in the Lévy representation for a stable ch.f. in the cases $0 < \alpha < 2$ can be explicitly evaluated, and we have the following closed-form formulas.

**Theorem 3.1.3.** *$f$ is a stable ch.f. if and only if it is nonvanishing on $\mathbb{R}$ and $\phi = \log f$ is of one of the following forms:*

(i) *$\phi$ is the log ch.f. of a normal law (corresponding to $\alpha = 2$);*

(ii) *$\phi(t) = iat - c|t|^\alpha(1 + i(\operatorname{sgn} t)\theta)$, $t \in \mathbb{R}$, where $0 < \alpha < 2$, $\alpha \neq 1$, and $a, c, \theta \in \mathbb{R}$, $c > 0$, $|\theta| \leq |\tan(\pi\alpha/2)|$;*

(iii) *$\phi(t) = iat - c|t|(1 + i\lambda(\operatorname{sgn} t)\log|t|)$, $t \in \mathbb{R}$, where $a, c, \lambda \in \mathbb{R}$, $c > 0$, and $|\lambda| \leq 2/\pi$ (corresponding to $\alpha = 1$).*

For details of the evaluations, we refer the reader to Gnedenko and Kolmogorov (1964), Loève (1977), or Lukacs (1970).

We may however note here three facts which follow almost immediately from Theorem 3.1.3:

(a) For every $c > 0$, and $0 < \alpha \le 2$, $\exp(-c|t|^\alpha)$ is a ch.f. (These ch.f.'s are the "symmetric stable" ch.f.'s.)

(b) A stable d.f. is absolutely continuous (with respect to Lebesgue measure), since $|f(t)| = \exp(-c|t|^\alpha)$ is integrable over $\mathbb{R}$ (see Theorem 1.3.1).

(c) A stable law with exponent $\alpha < 2$ has (absolute) moments of all orders $<\alpha$, but not of order $\alpha$ itself. This follows from Theorem 1.3.12 and Theorem 3.1.2, or from the fact that (Theorem 1.3.5)

$$\int_0^\infty \frac{1 - |f(t)|}{t^{1+\lambda}}\, dt < \infty$$

if and only if $\lambda < \alpha$, since $|f(t)|$ is of the form $\exp(-c|t|^\alpha)$.

The direct derivation of the closed-form formulas we have given for stable ch.f.'s is of some interest from the functional equation viewpoint. We devote the rest of this section to this topic. We shall establish the necessity part of Theorem 3.1.3, except for the restriction $|\lambda| \le 2/\pi$ for the case $\alpha = 1$. The Lévy representation approach appears to be the only one available for establishing that.

We begin with an auxiliary result on functional equations, of some independent interest.

**Lemma 3.1.4.** *Let $A\colon \mathbb{R}_+ \to \mathbb{R}$, a continuous function and $\beta > 0$, a constant, be such that, for every $n \in \mathbb{N}$, there exists a $\gamma_n \in \mathbb{R}$ satisfying*

$$nA(t) = A(n^\beta t) + \gamma_n t, \qquad t \ge 0. \tag{3.1.4}$$

*Then, for some $k$, $\gamma$, $b \in \mathbb{R}$,*

$$A(t) = \begin{cases} \gamma t + k t^{1/\beta} & \text{if } \beta \ne 1, \\ kt + bt \log t & \text{if } \beta = 1. \end{cases}$$

***Proof.*** Suppose $\beta \ne 1$. Let $m, n \in \mathbb{N}$. Equation (3.1.4) implies that

$$\begin{aligned} mnA(t) &= A(m^\beta n^\beta t) + \gamma_{mn} t \\ &= mA(n^\beta t) - \gamma_m n^\beta t + \gamma_{mn} t \\ &= m(nA(t) - \gamma_n t) - \gamma_m n^\beta t + \gamma_{mn} t, \end{aligned}$$

so that $\gamma_{mn} = m\gamma_n + \gamma_m n^\beta$, and, by symmetry, equals $n\gamma_m + \gamma_n m^\beta$ as well, whence

$$(n - n^\beta)\gamma_m = (m - m^\beta)\gamma_n, \qquad \forall\, m, n \in \mathbb{N}.$$

We conclude that there exists $\gamma$ such that

$$\gamma_n = (n - n^\beta)\gamma, \qquad \forall\, n \in \mathbb{N}.$$

Setting $B(t) = A(t) - \gamma t$, we have $nB(t) = B(n^\beta t)$ for $n \in \mathbb{N}$, $t > 0$. Since $B$ is continuous, it follows by a standard argument that, for some $k \in \mathbb{R}$, we must have $B(t) = kt^{1/\beta}$ for $t > 0$, and hence that $A(\cdot)$ has the stated form.

If $\beta = 1$, then we have from (3.1.4) the identity

$$\frac{A(mnt)}{mnt} - \frac{A(mt)}{mt} - \frac{A(nt)}{nt} + \frac{A(t)}{t} = 0, \qquad t > 0.$$

For fixed $m \in \mathbb{N}$, let $B_m(t) = (A(mt)/mt) - (A(t)/t)$. Then, $B_m(nt) = B_m(t)$ for $n \in \mathbb{N}$, $t > 0$. By Proposition 1.1.7 (using a simple change of variable), this implies that there exists a constant $c_m$ such that $B_m(t) = c_m$ for all $t > 0$ (or simply $B_m(p/q) = B_m(1)$ for all $p, q \in \mathbb{N}$). Thus,

$$\frac{A(mt)}{mt} - \frac{A(t)}{t} = \frac{A(m)}{m} - A(1), \qquad m \in \mathbb{N},\ t > 0.$$

Now, let $C(t) = (A(t)/t) - A(1)$. Then, $C(mt) = C(m) + C(t)$ for all $m \in \mathbb{N}$ and $t > 0$. This implies in turn that the same relation holds for all positive rationals $m$ and $t > 0$, and then by continuity that $C(st) = C(s) + C(t)$ for all $s, t > 0$. Hence, $C(t) = b \log t$ for $t > 0$, for some $b \in \mathbb{R}$, so that $A(\cdot)$ is of the stated form.

***An alternative proof of the necessity part of Theorem 3.1.3.*** If $f$ is a stable ch.f., then, as already seen, for every $n \in \mathbb{N}$, there exists $\beta_n > 0$ such that

$$|f(t)| = |f(t/\beta_n)|^n, \qquad \forall\, t \in \mathbb{R}, \tag{3.1.5}$$

and since $|f(t)| \le 1$, it follows that $|f(\beta_{n+1}t)| \le |f(\beta_n t)|$. Since $F$ is nondegenerate, Theorem 1.3.2(iii) implies then that $\beta_{n+1} \ge \beta_n$. Equation (3.1.5) also implies that $f(\beta_{mn}t) = f(\beta_m\beta_n t)$, $t \in \mathbb{R}$. By Theorem 1.3.2(iii) again, neither $\beta_{mn} > \beta_m\beta_n$ nor $\beta_{mn} < \beta_m\beta_n$ can hold, so that $\beta_{mn} = \beta_m\beta_n$. Hence, by Proposition 1.1.9, $\beta_n = n^{1/\alpha}$ for some $\alpha > 0$. Now, $\alpha > 2$ is impossible, for, setting $t_n = n^{-1/2}$, we have from (3.1.5) that $\log|f(t_n)|/t_n^2 \to 0$ as $n \to \infty$, implying by Theorem 1.3.8(b) that $F$ is degenerate. Thus, $0 < \alpha \le 2$.

Now, if $\alpha = 2$, then $\log|f(t_n)|/t_n^2 = \log|f(1)|$ for all $n \in \mathbb{N}$, $t_n$ as before, and we have from Theorem 1.3.15 and Corollary 1.3.14 that $f$ is a normal ch.f.

We pass to the case $0 < \alpha < 2$. Set $\beta = 1/\alpha$. Then,

$$f(t)^n = f(n^\beta t)\exp(it\gamma_n), \qquad t \in \mathbb{R},$$

for some real $\gamma_n$. Let $A(\cdot)$ be defined by

$$\phi(t) = \log|f(t)| + iA(t), \qquad t \in \mathbb{R}.$$

Then, $A$ is an odd function, continuous on $\mathbb{R}$. Also, $|f(t)|^n = |f(n^\beta t)|$, $n \in \mathbb{N}$, $t \in \mathbb{R}$, implies that, for all $m, n \in \mathbb{N}$, $|f(m^\beta/n^\beta)| = |f(m^\beta)|^{1/n} = |f(1)|^{m/n}$ so that, by continuity, $|f(t)| = \exp(-c|t|^\alpha)$, $t \in \mathbb{R}$, for some $c > 0$.

It then follows from Lemma 3.1.4 and the fact that $\phi(-t) = \overline{\phi(t)}$ that, for $t \in \mathbb{R}$,

$$\phi(t) = \begin{cases} iat - c|t|^\alpha(1 + i(\operatorname{sgn} t)\theta) & \text{if } \alpha \neq 1, \\ iat - c|t|(1 + i\lambda(\operatorname{sgn} t)\log|t|) & \text{if } \alpha = 1. \end{cases}$$

An appeal to the series expansions for stable densities in the cases $\alpha \neq 1, 2$ enables us to establish the restriction on $\theta$ as follows: (The proofs for these expansions, following Feller (1971), pp. 581–583, are outlined in the appendix to this chapter.) Let $\theta = \tan(\pi\gamma/2)$, so that $|\gamma| < 1$ and, choosing $a$ and $c$ (for simplicity) such that $\phi(t) = -|t|^\alpha \exp((i\gamma\pi/2)(\operatorname{sgn} t)$, we have the expansions, in the stated cases, for the p.d.f. of $F$: for $0 < \alpha < 1$, $x > 0$,

$$p(x; \alpha, \gamma) = (\pi x)^{-1} \sum_{k=1}^{\infty} \frac{\Gamma(k\alpha + 1)}{k!} (-x^{-\alpha})^k \sin(k\pi(\gamma - \alpha)/2),$$

$$p(-x; \alpha, \gamma) = p(x; \alpha, -\gamma); \tag{3.1.6}$$

for $1 < \alpha < 2$, $x > 0$,

$$p(x; \alpha, \gamma) = (\pi x)^{-1} \sum_{k=1}^{\infty} \frac{\Gamma(1 + k/\alpha)}{k!} (-x)^k \sin(k\pi(\gamma - \alpha)/2\alpha),$$

$$p(-x; \alpha, \gamma) = p(x; \alpha, -\gamma). \tag{3.1.7}$$

In the case $0 < \alpha < 1$, for large $x > 0$, the dominant term in the expansion (3.1.6) has the same sign as $-\sin(\pi(\gamma - \alpha)/2)$, and so is negative if $\alpha < \gamma < \min(1, 3\alpha)$. Now, we note that the set of "admissible" values of $\theta$, i.e., those for which $\exp\{-|t|^\alpha(1 + i(\operatorname{sgn} t)\theta)\}$ is a ch.f., is obviously a symmetric convex set, and so an interval around the origin; the same is therefore true of the corresponding values of $\gamma$. Since the latter does not contain the interval $(\alpha, \min(1, 3\alpha))$, we must have $|\gamma| \leq \alpha$ for "admissible" values of $\gamma$; equivalently, we must have $|\theta| \leq \tan(\pi\alpha/2)$—in the case $0 < \alpha < 1$.

If $1 < \alpha < 2$, then, it follows from (3.1.6) and (3.1.7) that, for $y > 0$,

$$p(y; \alpha, \gamma) = y^{-1-\alpha} p(y^{-\alpha}, 1/\alpha, \gamma^*),$$

with $\alpha(\gamma^* + 1) = \gamma + 1$. It follows in turn that $\gamma$ is admissible for $\alpha \in (1, 2)$ if and only if $\gamma^*$ is admissible for $1/\alpha \in (\frac{1}{2}, 1)$, and so if and only if $-\gamma^*$ is admissible for such $\alpha$. Consider $\gamma = \alpha - 2 - \delta$ for small $\delta > 0$; then, for large $y > 0$, $p(y, 1/\alpha, -\gamma^*)$ has the same sign as $-\sin(\pi\delta/2\alpha)$, so that such values of $\gamma$ are inadmissible, and we must have $|\gamma| \leq 2 - \alpha$ for $\gamma$ to be admissible, so that $|\theta| \leq |\tan(\pi\alpha/2)|$ in the case $1 < \alpha < 2$.

***Remark 1.*** A more elementary proof that $f$ is a normal ch.f. if $\alpha = 2$ can be obtained as follows: The relation $\log|f(t_n)|/t_n^2 = \log|f(1)|$ implies that $F$ has finite second moment; thus, so does $G = F * \tilde{F}$, with ch.f. $g = |f|^2$. $g$ satisfies the relation $g(t) = g(t/\sqrt{2})^2$, $t \in \mathbb{R}$, and by iteration, we have

$$g(t) = g(t/2^n)^{2^{2n}}, \qquad t \in \mathbb{R},\ n \in \mathbb{N}.$$

Now, $g$ admits a representation of the form

$$g(t) = 1 - \sigma^2 t^2/2 + R(t), \qquad t \in \mathbb{R},$$

where $R(t)/t^2 \to 0$ as $t \to 0$. Hence, noting that $(1 + x_n/n)^n \to e^x$ as $n \to \infty$ if $x_n \to x$, it follows that

$$g(t) = \lim_{n\to\infty} (1 - \sigma^2 t^2/2^{2n+1} + R(t/2^n))^{2^{2n}} = e^{-\sigma^2 t^2/2}$$

and, by Corollary 1.3.14, $f$ is a normal ch.f. as well.

***Remark 2.*** The restriction on $\theta$ for $0 < \alpha < 1$ can also be obtained without using the series expansion: Since $f$ is integrable over $\mathbb{R}$, a continuous version of the probability density function is given (Theorem 1.3.1) by

$$p(x) = \frac{1}{2\pi}\int e^{-itx} f(t)\, dt.$$

In particular (taking $c = 1$ for simplicity), we have

$$\pi p(0) = \operatorname{Re}\int_0^\infty e^{-(1+i\theta)t^\alpha}\, dt.$$

Writing $\tan^{-1}\theta = \pi\gamma/2$, so that $|\gamma| < 1$, we have

$$\pi p(0) = \Gamma(1 + 1/\alpha)\cos(\pi\gamma/2\alpha)\{\cos(\pi\gamma/2)\}^{1/\alpha}.$$

Consider $0 < \alpha < 1$; $p(0)$ is negative if $\alpha < \gamma < 3\alpha$, so that, by the same argument as above, $|\theta| \leq \tan(\pi\alpha/2)$.

## 3.2. THE SEMISTABLE LAWS

A nondegenerate ch.f. $f$ is said to be *semistable* if it does not vanish on $\mathbb{R}$ and satisfies a functional equation of the form

$$f(t) = f(bt)^c, \qquad t \in \mathbb{R}, \tag{3.2.1}$$

for some suitable real $b \neq 0$ and $c > 0$. The *assumption* that $f$ is nonvanishing may be omitted if $c \in \mathbb{N}$. We may assume without loss of generality (interchanging the roles of the two sides if needed) that $c \geq 1$. Then, $|f(t)| \leq |f(bt)|$; since $F$ is nondegenerate, Theorem 1.3.2(iii) implies that $|b| > 1$ is impossible. If $b = 1$ and $c = 1$, nothing at all can be said about $F$, while $b = -1$, $c = 1$ merely implies that $F$ is a symmetric d.f. Thus, the only nontrivial case is where $c > 1$, $|b| < 1$. We therefore shall make this assumption part of our definition of semistable laws.

We proceed to establish the analogs of the results already established in the context of stable laws.

**Theorem 3.2.1.** *A semistable law is infinitely divisible.*

***Proof.*** If $1 < c \in \mathbb{N}$, then (3.2.1) implies that $f$ cannot vanish; then, $f$ is the $k$th power of some ch.f. $f_k$ for $k = c, c^2, \ldots$, and the infinite divisibility of $f$ is immediate. For other $c > 1$, $f$ is nonvanishing by definition. We then have, for $\phi = \log f$,

$$\phi(t) = \lim_{n \to \infty} c^n(f(b^n t) - 1), \qquad t \in \mathbb{R},$$

and Theorem 1.3.8 implies that $f$ is an inf. div. ch.f.

**Theorem 3.2.2** *A ch.f. $f$ is semistable if and only if it is nonvanishing on $\mathbb{R}$ and admits a Lévy representation $L(\mu, \sigma^2, M, N)$ (for $\phi = \log f$) of one of the following forms:*

(i) $M \equiv N \equiv 0$; *in this case, $\sigma > 0$ and $F$ is a normal d.f.*
(ii) *at least one of $M$ and $N$ is $\not\equiv 0$; in this case, $\sigma = 0$ and*

$$M(u) = \xi(u)/|u|^\alpha \quad \textit{for } u < 0; \qquad N(u) = -\eta(u)/u^\alpha \quad \textit{for } u > 0,$$

*where $\xi, \eta$ are nonnegative functions, defined on $(-\infty, 0)$ and $(0, \infty)$, respectively, with*

$$\begin{aligned} \xi(u) &= \xi(u/b), & \eta(u) &= \eta(u/b) & &\textit{if } b > 0, \\ \xi(u &= \eta(-u/b), & \eta(u) &= \xi(-u/b) & &\textit{if } b < 0. \end{aligned} \tag{3.2.2}$$

***Proof.*** The converse part is almost trivial, so that we need only check the "direct" part. If $f$ is a semistable ch.f., it is inf. div. by Theorem 3.2.1, and the uniqueness of its Lévy representation and (3.2.1) imply the following three relations:

$$\left.\begin{array}{lll} & \sigma^2(1 - cb^2) = 0, & \\ N(u) = cN(u/b), & M(u) = cM(u/b) & \text{if } b > 0, \\ N(u) = -cM(-u/b), & M(u) = -cN(-u/b) & \text{if } b < 0. \end{array}\right\} \tag{3.2.3}$$

Let now $\alpha$ be the unique (positive) number such that $c|b|^\alpha = 1$. Set

$$M(u) = \xi(u)/|u|^\alpha, \qquad N(u) = -\eta(u)/u^\alpha;$$

then, $\xi$ and $\eta$ satisfy (3.2.2) and, obviously,

$$\xi(u) = \xi(u/b^2), \qquad \eta(u) = \eta(u/b^2). \tag{3.2.4}$$

$\xi$ and $\eta$ are therefore bounded. Also, $\eta(u_0) = 0$ for some $u_0 > 0$ implies that $\eta(u_0 b^{2n}) = 0$ for all $n \in \mathbb{N}$, so that $N(0+) = 0$ and hence $N \equiv 0$. Thus, if $N \not\equiv 0$, then $\eta$ is strictly positive. It easily follows that $\rho = \inf\{\eta(u): b^2 \le u \le 1\} > 0$, and that $\rho = \inf\{\eta(u): u > 0\}$ as well. Dual statements hold with respect to $\xi$ and $M$. It then follows that, if at least one of $M$ and $N$ is $\not\equiv 0$, then $0 < \alpha < 2$: for, if, say $N \not\equiv 0$, then

$$\begin{aligned} \int_{b^2 u_0}^{u_0} t^2\, dN(t) &\ge (b^2 u_0)^2 (N(u_0) - N(b^2 u_0)) \\ &\ge \rho b^4 u_0^{2-\alpha}(|b|^{-2\alpha} - 1), \end{aligned}$$

and so the right-hand side should tend to zero as $u_0 \to 0$, since the left-hand side does in view of the condition $\int_{(0,1)} u^2\, dN(u) < \infty$ of the Lévy representation. It follows that, if $\alpha \ge 2$, then $M \equiv N \equiv 0$; further, $\alpha > 2$ is impossible since $c|b|^\alpha = 1$ and $\sigma^2(1 - cb^2) = 0$ together imply that $\sigma = 0$, so that $f$ is a degenerate ch.f. Thus, $0 < \alpha \le 2$, with $\alpha = 2$ corresponding to the normal laws. The same argument shows that $\sigma = 0$ if $0 < \alpha < 2$.

As an immediate consequence of Theorem 3.1.2 and 3.2.2, we have the following.

**Corollary 3.2.3.** *Every stable law is also semistable.*

We also note two facts that follow almost at once from the definition of semi-stable laws, noting that if $\psi(t) = -\log|f(t)|/|t|^\alpha$ (for $t \ne 0$), then $\psi(t) = \psi(b^2 t)$ and, consequently, if $(0<)\ \delta,\ \gamma$ denote the minimum and maximum values of $\psi$ *on* $[b^2, 1]$, then we have, for all $t \in \mathbb{R}$,

$$\exp(-\gamma|t|^\alpha) \leq |f(t)| \leq \exp(-\delta|t|^\alpha),$$

and hence:

(a) A semistable d.f. is absolutely continuous (with respect to Lebesgue measure), in view of the right inequality and Theorem 1.3.1.

(b) A semistable law with exponent $\alpha < 2$ has (absolute) moments of all orders $<\alpha$, but not of order $\alpha$. This follows from the above two inequalities and Theorem 1.3.7, also from Theorem 1.3.12 and Theorem 3.2.2.

We remark finally that (3.2.1) is a particular case of (3.4.1), which will follow, and therefore the representation given by Theorem 3.4.1 applies to semistable laws.

## 3.3. THE GENERALIZED SEMISTABLE LAWS AND THE NORMAL SOLUTIONS

Let $\{\beta_j\}$, with $0 < |\beta_j| < 1$ for all $j$, be a real sequence and $\{\gamma_j\}$ a positive sequence. Consider the equation

$$f(t) = \prod_{j=1}^{\infty} f(\beta_j t)^{\gamma_j}, \qquad |t| < \delta, \tag{3.3.1}$$

where $\delta > 0$ is such that the ch.f. $f$ does not vanish on $(-\delta, \delta)$. Then, we have the following.

**Thorem 3.3.1.** *Let $f$ be a nondegenerate ch.f. satisfying* (3.3.1). *Then,* $\sum \gamma_j \beta_j^2 \leq 1$, *and $f$ is a normal ch.f. if and only if* $\sum \gamma_j \beta_j^2 = 1$.

***Proof.*** Note that $g = |f|^2$ satisfies the equation

$$g(t) = \prod g(b_j t)^{\gamma_j}, \qquad |t| < \delta,$$

where $b_j = |\beta_j|$. Suppose, if possible, that $\sum \gamma_j b_j^2 > 1$; then, there exists $N \in \mathbb{N}$ such that $\sum_{j=1}^{N} \gamma_j b_j^2 = \gamma > 1$. If $\psi(t) = -\log g(t)/t^2$ for $0 < |t| < \delta$, then, by the intermediate value theorem, for $0 < t < \delta$,

$$\psi(t) \geq \sum_{j=1}^{N} \gamma_j b_j^2 \psi(b_j t) = \gamma\psi(tb(t)),$$

where

$$(0<) \min\{b_j: 1 \leq j \leq N\} \leq b(t) \leq \max\{b_j: 1 \leq j \leq N\}.$$

Thus, for any $u_0 \in (0, \delta)$, there exists a nonzero sequence $\{u_n\} \to 0$, such that $\psi(u_n) \le \gamma^{-n}\psi(u_0) \to 0$ as $n \to \infty$. By Theorem 1.3.8(b), $G$ has to be degenerate, contrary to assumption. Hence, we must have $\sum \gamma_j b_j^2 \le 1$.

It is obvious that if $f$ is a nondegenerate normal ch.f., then $\sum \gamma_j b_j^2 = 1$. To establish the converse, defining $\psi$ as before, we have

$$\psi(t) = \sum p_j \psi(b_j t) \qquad \text{for } 0 < t < \delta,$$

where $p_j = \gamma_j b_j^2$, so that $\sum p_j = 1$. Rewrite the above equality as

$$\sum p_j(\psi(t) - \psi(b_j t)) = 0, \qquad 0 < t < \delta. \tag{3.3.2}$$

Let $t_0 \in (0, \delta)$ be fixed, and let

$$S_0 = \{t\colon 0 < t < t_0,\ \psi(t) \le \psi(t_0)\}.$$

Equation (3.3.2) implies that $\psi(b_j t_0) \le \psi(t_0)$ for at least one $j$; hence, $S_0$ is nonempty. Let $\tau = \inf S_0$. For the same reason, $\tau = 0$. Thus, there exists a sequence $\{t_n\}$ such that $\psi(t_n) \le \psi(t_0)$ for all $n$, and $G$ has finite variance, by Theorem 1.3.8(a). This fact in turn implies that $\lim_{t\to 0} \psi(t)$ exists, by Theorem 1.3.4(a). Denote this limit by $\psi(0)$; then, $\psi(0) \le \psi(t_0)$. By applying a similar argument to

$$T_0 = \{t\colon 0 < t < t_0,\ \psi(t) \ge \psi(t_0)\},$$

we conclude that $\psi(t_0) \le \psi(0)$ as well. Thus, $\psi$ is constant on $(0, \delta)$, and it follows from Corollary 1.3.17 and Corollary 1.3.14 that $f$ is a normal ch.f.

The following is a simple application of Theorem 3.3.1.

**Theorem 3.3.2.** *Let $\{X_j\}$ be a sequence of nondegenerate i.i.d.r.v.'s, and $\{a_j\}$ a sequence of real numbers such that $\sum a_j X_j$ converges a.s. Suppose further that $\sum a_j X_j$ has the same distribution as $X_1$. Then, the $X_j$ are normally distributed if and only if $\sum a_j^2 = 1$.*

***Proof.*** Let $f$ be the common ch.f. of the $X_j$; we have

$$f(t) = \prod_{j=1}^{\infty} f(a_j t), \qquad t \in \mathbb{R}.$$

If $|a_1| > 1$, then $|f(t)| \le |f(a_1 t)|$ for $t \in \mathbb{R}$, and if $|a_1| = 1$, then $\prod_{j=2}^{\infty} |f(a_j t)| = 1$ in any neighborhood of the origin where $f$ does not vanish. In either case, Theorem 1.3.2 implies that $F$ is degenerate, contrary to assumption. Hence, $|a_1| < 1$ and, similarly, $|a_j| < 1$ for all $j$. Theorem 3.3.1 then yields the desired conclusion.

## 3.4. THE GENERALIZED SEMISTABLE LAWS AND THE NONNORMAL SOLUTIONS

In this section, we continue the investigation of Eq. (3.3.1), now assumed to hold throughout $\mathbb{R}$. It will be convenient to group the $\beta$'s according to sign and rewrite (3.3.1) in the form

$$f(t) = \prod_{j=1}^{\infty} f(\beta_{2j}t)^{\gamma_{2j}} \prod_{j=1}^{\infty} f(-\beta_{2j-1}t)^{\gamma_{2j-1}}, \qquad t \in \mathbb{R}, \tag{3.4.1}$$

where $0 < \beta_j < 1$ and $\gamma_j > 0$ for all $j$. Let

$$B_1 = \{|\log \beta_{2j}| : j \in \mathbb{N}\}, \qquad B_2 = \{|\log \beta_{2j-1}| : j \in \mathbb{N}\}.$$

Then, the following result provides a complete classification and representation for the ch.f.'s satisfying (3.4.1).

**Theorem 3.4.1.** *Let $f$ be a nondegenerate ch.f., nonvanishing on $\mathbb{R}$ and satisfying* (3.4.1). *Then, there exists an $\alpha$, with $0 < \alpha \le 2$, such that $\sum \gamma_j \beta_j^{\alpha} = 1$. Further:*

(a) *if $\alpha = 2$, $f$ is a normal ch.f.;*
(b) *if $0 < \alpha < 2$, then*

$$f(t) = \exp\{iat - |t|^{\alpha}(\Gamma(\log|t|) + i(\operatorname{sgn} t)\Delta(\log|t|))\}, \qquad t \in \mathbb{R},$$

*where $a = 0$ if $0 < \alpha \le 1$, and if $\alpha > 1$, then*

$$a(1 - \sum \gamma_{2j}\beta_{2j} + \sum \gamma_{2j-1}\beta_{2j-1}) = 0.$$

*Also, $\Gamma$ has every $y \in B_1 \cup B_2$ as period, while $\Delta$ has the following properties:*

(i) *if $B_2$ is empty, then $\Delta$ has every $y \in B_1$ as period;*
(ii) *if $B_2$ is nonempty and if there exists a $\rho > 0$, which we take to be the largest such, such that*

$$B_1 \subseteq \{2\rho, 4\rho, \ldots\}, \qquad B_2 \subset \{\rho, 3\rho, \ldots\},$$

*then $\Delta(t + \rho) = -\Delta(t)$ for all $t \in \mathbb{R}$;*
(iii) *$\Delta \equiv 0$ in all other cases.*

*In particular, $F$ differs from a semistable law only by a location parameter.*

**Corollary 3.4.2.** *If $f$ satisfies* (3.4.1) *with the previous assumptions on the $\beta$'s and $\gamma$'s, and if $B_1 \cup B_2$ is not contained in a set of the form $\{\rho, 2\rho, 3\rho, \ldots\}$ for any $\rho > 0$, then $f$ is a symmetric stable ch.f. to within* *a location parameter, i.e.,*

$$f(t) = \exp(iat - c|t|^\alpha), \qquad t \in \mathbb{R},$$

*for some $a, c \in \mathbb{R}$ with $c > 0$, and $\alpha$ with $0 < \alpha \leq 2$ satisfying $\sum \gamma_j \beta_j^\alpha = 1$.*

***Proof of Theorem 3.4.1.*** Theorem 3.3.1 implies that $\sum \gamma_j \beta_j^2 \leq 1$, and assertion (a) holds if $\sum \gamma_j \beta_j^2 = 1$. If $\sum \gamma_j \beta_j^2 < 1$, writing

$$\phi(t) = \log f(t) = \log |f(t)| + iA(t) = -\psi(t) + iA(t), \qquad t \in \mathbb{R},$$

where $\psi$ and $A$ are both continuous, and $A$ is an odd function while $\psi$ is positive and even, we have, from (3.4.1),

$$\psi(t) = \sum \gamma_j \psi(\beta_j t), \quad A(t) = \sum \gamma_{2j} A(\beta_{2j} t) - \sum \gamma_{2j-1} A(\beta_{2j-1} t), \quad t > 0. \tag{3.4.2}$$

Using an obvious change of variable and applying Theorem 2.2.4, we see that there exists an $\alpha(<2)$ such that $\sum \gamma_j \beta_j^\alpha = 1$, and such that $|f(t)| = \exp(-|t|^\alpha \Gamma(\log|t|))$ for $t \in \mathbb{R}$, where $\Gamma$ has every element of $B_1 \cup B_2$ as a period.

To obtain the stated expression for $A(\cdot)$, we need two auxiliary results.

**Lemma 3.4.3.** *Let $g$ be the ch.f. of a symmetric d.f. $G$. Suppose that, for some sequence $\{t_n\}$ of positive numbers, such that for all $n \in \mathbb{N}$, $0 < a < t_{n+1}/t_n < b < 1$, where $a$ and $b$ are constants, the numbers $\{1 - g(t_n)\}/t_n^\alpha$ are bounded. Then, $g \equiv 1$ if $\alpha > 2$; and, for $0 < \alpha \leq 2$, $\{1 - G(x)\}x^\alpha$ is bounded for $x > 0$, and $\{1 - g(t)\}/t^\alpha$ is bounded for $t > 0$.*

***Proof.*** Theorem 1.3.8(b) yields the assertion regarding the case $\alpha > 2$. We consider $0 < \alpha \leq 2$ in what follows. Let $u_n = 1/t_n$, so $u_n \to \infty$. $c_1, c_2, \ldots$ will denote positive constants in what follows. Our assumption implies that

$$\begin{aligned} c_1 t_n^\alpha \geq 1 - g(t_n) &= \int (1 - \cos t_n x)\, dG(x) \\ &= 2 \int \sin^2(t_n x/2)\, dG(x) \geq c_2 t_n^2 \int_0^{u_n} x^2\, dG(x) \end{aligned}$$

(using the fact that $\sin \theta/\theta \geq 2\sin(\frac{1}{2})$ for $0 \leq \theta \leq \frac{1}{2}$, so that

$$\int_0^{u_n} x^2\, dG(x) \leq c_3 t_n^{\alpha - 2}.$$

Let $T = 1 - G$. Then, for $n \in \mathbb{N}$ (we may take $u_0 = 0$),

$$T(u_{n-1}) - T(u_n) = \int_{u_{n-1}}^{u_n} dG(x) \leq (t_n^2/b^2) \int_{au_n}^{u_n} x^2\, dG(x) \leq c_4 t_n^\alpha,$$

in view of the above inequality. Hence,

$$T(u_n) \leq c_4 \sum_{j=1}^{\infty} t_{n+j}^\alpha \leq c_4 t_n^\alpha (1 + b^\alpha + b^{2\alpha} + \cdots) = c_5 t_n^\alpha.$$

If now $u_n \leq x < u_{n+1}$ for some $n$, then $T(x) \leq T(u_n) \leq c_5 t_n^\alpha \leq c_6 t_{n+1}^\alpha \leq c_6 x^{-\alpha}$. Hence, the first assertion of the lemma is proven.

The second assertion of the lemma follows from Theorem 1.3.8(a) for $\alpha = 2$. For $0 < \alpha < 2$, we have

$$\begin{aligned} 1 - g(t) &= 2\int_0^\infty (1 - \cos tx)\, dG(x) = 4\int_0^\infty \sin^2(tx/2)\, dG(x) \\ &\leq 4\int_0^{1/t} \sin^2(tx/2)\, dG(x) + 4\int_{1/t}^\infty dG(x) \\ &\leq t^2 \int_0^{1/t} x^2\, dG(x) + 4T(1/t), \end{aligned}$$

and by an integration by parts, this is

$$\leq 2t^2 \int_0^{1/t} xT(x)\, dx + 3T(1/t) \leq c_7 t^\alpha,$$

in view of $T(x) \leq c_6 x^{-\alpha}$.

**Lemma 3.4.4.** *Let $f$ be a ch.f. such that $1 - |f(t)| = 0(|t|^\alpha)$ as $t \to 0$, for some $\alpha \in (0, 2)$. We have:*

(i) *if $0 < \alpha < 1$, then $1 - f(t) = O(|t|^\alpha)$ as $t \to 0$;*
(ii) *if $1 < \alpha < 2$, then there exists an $a \in \mathbb{R}$ such that*

$$1 + iat - f(t) = O(|t|^\alpha)$$

*as $t \to 0$; and*
(iii) *if $\alpha = 1$, then, for every $0 < \varepsilon < 1$ and $T > 0$,*

$$\sup_{0 < t \leq T} \left| \frac{1 - f(t)}{t} - \frac{1 - f(\varepsilon t)}{\varepsilon t} \right| < \infty.$$

***Proof.*** (i) Since $|f|^2$ is the ch.f. of $F_s = F * \tilde{F}$, our assumption implies that $1 - |f(t)|^2 = O(|t|^\alpha)$ as $t \to 0$, so that, by Lemma 3.4.3,

$$1 - F_s(x) + F_s((-x)_-) = O(x^{-\alpha}) \qquad \text{as } x \to \infty.$$

If $X$ and $Y$ are i.i.d.r.v.'s with $F$ as d.f., and if $m$ is a median of $F$, then the left-hand side in the last relation

$$= P\{|X - Y| \geq x\}$$

$$\geq P\{X \geq x + m, Y \leq m\} + P\{X \leq -x + m, Y \geq m\}$$

$$\geq \tfrac{1}{2}P\{|X - m| \geq x\},$$

whence $1 - F(x) + F((-x)_-) = O(x^{-\alpha})$ as well, as $x \to \infty$. Writing $H(x) = F(x) - F((-x)_-)$, for $x > 0$, we have $1 - H(x) = O(x^{-\alpha})$, and we also have (considering only $t > 0$ in what follows, without loss of generality)

$$|1 - f(t)| \leq \int_{|x| \leq 1/t} |1 - e^{ixt}|\, dF(x) + 2\int_{|x| > 1/t} dF(x) = I_1 + I_2.$$

Assertion (i) then follows from

$$I_1 \leq t\int_{|x| \leq 1/t} |x|\, dF(x) = t\int_0^{1/t} x\, dH(x)$$

$$= t\int_0^{1/t} (1 - H(x)\, dx = O(t^{\alpha}) \tag{3.4.3}$$

and

$$I_2 = 2(1 - H(1/t)) = O(t^{\alpha}).$$

(ii) If $1 < \alpha < 2$, let $a = \int x\, dF(x)$. We may then argue as before, but now using the inequality $|1 + iu - e^{iu}| \leq \frac{1}{2}u^2$ for $u$ real, with $|u| \leq 1$.

(iii) If $\alpha = 1$, we have

$$\left|\frac{1 - f(t)}{t} - \frac{1 - f(\varepsilon t)}{\varepsilon t}\right| \leq \int_{|x| < 1/\varepsilon t} + \int_{|x| \geq 1/\varepsilon t} \left|\frac{1 - e^{ixt}}{t} - \frac{1 - e^{i\varepsilon xt}}{\varepsilon t}\right| dF(x)$$

$$= J_1 + J_2.$$

Noting that $x(1 - H(x))$ is bounded for $x > 0$, and using the same technique as in deriving (3.4.3), we have

$$J_1 = \int_{|x| < 1/\varepsilon t} \left|\frac{1 + ixt - e^{ixt}}{t} - \frac{1 + i\varepsilon xt - e^{i\varepsilon xt}}{\varepsilon t}\right| dF(x)$$

$$\leq \tfrac{1}{2}t\int_{|x| < 1/\varepsilon t} x^2\, dF(x) = O\left(t\int_0^{1/\varepsilon t} x^2\, dH(x)\right)$$

$$= O\left(t\int_0^{1/\varepsilon t} x(1 - H(x)\, dx\right) = O(t(\varepsilon t)^{-1}) = O(\varepsilon^{-1}),$$

and

$$J_2 \le (4/\varepsilon t)(1 - H(1/\varepsilon t)) = O(1).$$

Hence (iii), and the lemma, are proven.

We continue with the proof of Theorem 3.4.1 now (in the cases $0 < \alpha < 2$). From the representation for $\psi = -\log|f|$ already obtained, we have

$$1 - |f(t)| = O(|t|^\alpha) \qquad \text{as } t \to 0.$$

***The case*** $\alpha \neq 1$. By Lemma 3.4.4(i) and (ii), there exists a real $a(=0$ if $0 < \alpha < 1)$ such that, as $t \to 0$,

$$1 + iat - f(t) = O(|t|^\alpha).$$

Since $1 + iat - e^{iat} = O(t^2)$, we have, for $\alpha < 2$,

$$1 - e^{-iat}f(t) = e^{-iat}(e^{iat} - f(t)) = O(|t|^\alpha).$$

Now, for any complex $z$ with $|z| \le \frac{1}{2}$, we have

$$|z| \le |e^z - 1 - z| + |e^z - 1| \le |z|^2 + |e^z - 1| \le \tfrac{1}{2}|z| + |e^z - 1|,$$

and hence $|z| \le 2|e^z - 1|$. It follows that

$$-\psi(t) + iA(t) - iat = O(|t|^\alpha).$$

If $I(\cdot)$ is defined by $|t|^\alpha I(t) = A(t) - at$, then $I$ is an odd function, bounded on compact intervals, satisfying the functional equation

$$I(t) = \sum \gamma_{2j}\beta_{2j}^\alpha I(\beta_{2j}t) - \sum \gamma_{2j-1}\beta_{2j-1}^\alpha I(\beta_{2j-1}t), \qquad t \in \mathbb{R}. \quad (3.4.4)$$

It follows from Theorem 2.4.2 (on using an obvious change of variable) that $I(t)$ has the form $\Delta(\log t)$ on the interval $[0, T]$, for every $T > 0$, where $\Delta$ has the stated properties, and so for all $t > 0$. Since $I$ is an odd function, the stated representation then holds for $t \in \mathbb{R}$.

***The case*** $\alpha = 1$. Since $1 - |f(t)| = O(|t|)$ in this case, we have, from Lemma 3.4.4(i), that

$$\psi(t) = O(|t|) \qquad \text{and} \qquad 1 - f(t) = O(|t|^\gamma) \qquad \text{for every } \gamma < 1.$$

Choose and fix $\gamma \in (\frac{1}{2}, 1)$. Then, as before, from the inequality $|z| \le 2|e^z - 1|$ for $|z| < \frac{1}{2}$, we have

$$\phi(t) = -\psi(t) + iA(t) = O(|t|^\gamma),$$

so that $A(t) = O(|t|^\gamma)$. Now, since $e^z - 1 = z + O(|z|^2)$ as $z \to 0$, we have, for $t > 0$,

$$\frac{1 - f(t)}{t} - \frac{1 - f(\varepsilon t)}{\varepsilon t} = \frac{\phi(t) + O(t^{2\gamma})}{t} - \frac{\phi(\varepsilon t) + O((\varepsilon t)^{2\gamma})}{\varepsilon t}$$

$$= -\Gamma(\log t) + \Gamma(\log(\varepsilon t)) + I(t) - I(\varepsilon t) + O(t^{2\gamma - 1})$$

and so, $\Gamma$ being bounded, we see by Lemma 3.4.4(iii) that

$$\sup_{0 < t \le T} |I(t) - I(\varepsilon t)| < \infty, \tag{3.4.5}$$

for every $T > 0$ and every $0 < \varepsilon < 1$. It follows (on using an obvious change of variable) from Corollary 2.4.3 that $I(t) = \Delta(\log t)$ on $[0, T]$, with $\Delta$ having the stated properties, for every $T > 0$, and so for all $t > 0$. As before, since $I$ is an odd function, the stated representation holds for all $t \in \mathbb{R}$.

Finally, it is clear that if $f_0(t) = f(t)e^{-iat}$, then $f_0(t) = f_0(bt)^c$ with $b = e^{-2\rho}$, $c = e^{2\alpha\rho}$, $\rho > 0$ being as defined in case (b)(ii) of the enunciation of the theorem and arbitrary in other cases. Thus, $f$ differs from a semistable ch.f. only by a location parameter. Theorem 3.4.1 is now proven.

Theorem 3.4.1 may be recast into the following general form, the proof being essentially the same.

**Theorem 3.4.5.** *Let $f$ be a nondegenerate ch.f., nonvanishing on $\mathbb{R}$, and let $\phi = \log f$ satisfy the equation*

$$\phi(t) = \int_{(0,1]} \phi(tu)\, d\mu_1(u) + \int_{(0,1]} \phi(-tu)\, d\mu_2(u), \qquad t \in \mathbb{R},$$

*where $\mu_1$ and $\mu_2$ are $\sigma$-finite measures on the Borel subsets of $(0, 1]$. Then, there exists an $\alpha \in (0, 2]$ such that*

$$\int_{(0,1]} t^\alpha\, d(\mu_1 + \mu_2)(u) = 1.$$

*Then, the assertion of Theorem 3.4.1 holds with*

$$B_k = \{|\log u| : u \in \operatorname{supp} \mu_k\}, \; k = 1, 2,$$

*and with*

$$a\left(1 - \int_{(0,1]} u\, d\mu_1(u) + \int_{(0,1]} u\, d\mu_2(u)\right) = 0$$

in the case $\alpha > 1$.

## APPENDIX: SERIES EXPANSIONS FOR STABLE DENSITIES ($\alpha \neq 1, 2$)

For $0 < \alpha < 1$ or $1 < \alpha < 2$, consider the function

$$f(t) = e^{-|t|^\alpha \exp(i\pi\gamma \operatorname{sgn} t/2)},$$

where $|\gamma| < 1$, and the function $p(x; \alpha, \gamma)$ obtained by applying the inversion formula to $f$, namely,

$$\begin{aligned} p(x; \alpha, \gamma) &= (1/2\pi) \int_{-\infty}^{\infty} f(t) e^{-itx}\, dt \\ &= (1/\pi) \operatorname{Re} \int_0^\infty e^{-itx - t^\alpha \exp(i\pi\gamma/2)}\, dt. \end{aligned}$$

Obviously, $p(-x; \alpha, \gamma) = p(x; \alpha, -\gamma)$.

If $0 < \alpha < 1$, a routine application of Cauchy's theorem enables us to move the path of integration to the negative imaginary axis (i.e., to justify the formal substitution $t = (u/x)\exp(-i\pi/2)$), to obtain, for $x > 0$,

$$\begin{aligned} p(x; \alpha, \gamma) &= \operatorname{Re}\left[(-i/\pi x) \int_0^\infty e^{-u-(u/x)^\alpha \exp(i\pi(\gamma - a)/2)}\, du\right] \\ &= \operatorname{Re}\left[(-i/\pi x) \int_0^\infty e^{-u} \sum_{k=0}^\infty (-x^{-\alpha} e^{i\pi(\gamma-\alpha)/2})^k u^k/k!\right] \\ &= \operatorname{Re}\left[(-i/\pi x) \sum_{k=0}^\infty \frac{\Gamma(k\alpha + 1)}{k!} (-x^{-\alpha} e^{i\pi(\gamma-\alpha)/2})^k\right], \end{aligned}$$

which reduces to (3.1.6).

If $1 < \alpha < 2$, then the formal substitution $t = u^{1/\alpha} e^{-i\pi\gamma/2\alpha}$ can be similarly justified, and we have an integrand of the form

$$u^{(1/\alpha)-1} \exp(-u - cu^{1/\alpha}).$$

Expanding $\exp(-cu^{1/\alpha})$ in an exponential series, we have

$$\begin{aligned} p(x; \alpha, \gamma) &= (1/\alpha\pi) \operatorname{Re}\left((e^{-i\pi\gamma/2\alpha}) \sum_{k=0}^\infty \frac{\Gamma((k+1)/\alpha)}{k!} (-ixe^{-i\pi\gamma/2\alpha})^k\right) \\ &= (1/\pi x) \operatorname{Re}\left(i \sum_{k=1}^\infty \frac{\Gamma(1 + k/\alpha)}{k!} (-xe^{-i\pi(\gamma-\alpha)/2\alpha})^k\right), \end{aligned}$$

which reduces to (3.1.7). Stable densities correspond to: $|\gamma| \le \alpha$ if $0 < \alpha < 1$; $|\gamma| \le 2 - \alpha$ if $1 < \alpha < 2$.

Setting $\alpha = \frac{1}{2}$, $\gamma = -\frac{1}{2}$ in (3.1.6), we obtain, corresponding to the ch.f. $\exp(-|t|^{1/2}(1 - i\,\mathrm{sgn}\,t)/\sqrt{2})$, the p.d.f.

$$p(x; \tfrac{1}{2}, -\tfrac{1}{2}) = \begin{cases} \dfrac{1}{2\sqrt{\pi}} x^{-3/2} e^{-(1/4x)} & \text{for } x > 0, \\ 0 & \text{for } x < 0, \end{cases}$$

which belongs to a Lévy–Smirnov type distribution.

## NOTES AND REMARKS

The proofs given here for Theorem 3.1.2 and Lemma 3.1.4 and the alternative proof of the necessity part of Theorem 3.1.3 are from Ramachandran *et al.* (1976, 1980). The semistable laws (of Section 3.2) were defined by P. Lévy, who also studied their basic properties including the Lévy representation. The generalized stable laws (of Sections 3.3 and 3.4) were considered in Shimizu (1968), Ramachandran and Rao (1970), Davies and Shimizu (1976), and Shimizu (1978), and the final forms of the results thereof, as given in these sections, are from Ramachandran *et al.* (1988). The series expansions in the appendix are essentially as in Feller (1971); such expansions and interrelationships between stable laws with exponents $>1$ and those with exponents $<1$ were also studied by H. Bergström and V. M. Zolotarev.

CHAPTER

# 4

# Integrated Cauchy Functional Equations with Error Terms on $\mathbb{R}_+$

In Chapters 2 and 3, we have seen that certain properties of the exponential, geometric, stable (including normal), semistable, and other laws are characteristic of those laws, using results on the ICFE and related forms of functional equations. That some "perturbed" forms of the ICFE occur in investigations of what we shall call "stability" of such characterizations should come as no surprise. What is of even greater interest is the appearance of such perturbed forms also in the study of certain "exact" characterizations, as we shall see in Sections 4.2 and 4.3.

We shall consider two kinds of ICFE's with error terms. The first, simpler kind will be taken up in Section 4.1, and used in Section 4.2 to characterize the Weibull distributions, and in Section 4.3 to discuss an extended form of the functional equations (3.3.1)—also having the semistable laws as solutions. The second kind will be formulated in Section 4.4, where applications thereof to questions of the stability of certain characterizations will also be considered.

## 4.1. ICFE'S WITH ERROR TERMS ON $\mathbb{R}_+$: THE FIRST KIND

We shall use the notation: $\varepsilon$-ICFE $(\sigma, S)$ to denote an equation of the form

$$f(x) = \int_0^\infty f(x + y)\, d\sigma(y) + S(x), \qquad \forall\, x \geq 0, \tag{4.1.1}$$

where the "error term" $S$ is such that $|S(x)| \leq Ce^{-\varepsilon x}$ for all $x \geq 0$, for some $C, \varepsilon > 0$. In this section, we shall confine our attention to cases where $f \geq 0$ and $\sigma$ is a probability measure, or where $f$ is bounded while $\sigma$ is a subprobability measure or a signed measure with total variation $\leq 1$. The study of a somewhat different kind of ICFE with error term, also involving more general $\sigma$'s, will be taken up in Section 4.4.

**Theorem 4.1.1.** *Let $\sigma \neq \delta_0$ be a probability measure and $f \geq 0$ a locally integrable (with respect to Lebesgue measure) solution of the ε-ICFE $(\sigma, S)$. Then,*

$$f(x) = p(x) + A(x), \qquad \forall\, x \geq 0,$$

*where $p \geq 0$ has every element of* supp $\sigma$ *as a period and*

$$|A(x)| \leq Ce^{-\varepsilon x}/(1 - \gamma), \quad x \geq 0, \qquad \text{with } \gamma = \int_0^\infty e^{-\varepsilon y}\, d\sigma(y).$$

***Proof.*** Let, as before, $\sigma^0 = \delta_0$ and, for $n \geq 1$, let $\sigma^n$ denote the $n$-fold convolution of $\sigma$ with itself. It follows from (4.1.1) and Fubini's theorem that

$$\begin{aligned}
f(x) &= \int_0^\infty f(x + y)\, d\sigma(y) \\
&= \int_0^\infty \left( \int_0^\infty f(x + y + z)\, d\sigma(z) + S(x + y) \right) d\sigma(y) + S(x) \\
&= \int_0^\infty f(x + y)\, d\sigma^2(y) + \int_0^\infty S(x + y)\, d\sigma(y) + S(x) \\
&= \cdots \\
&= \int_0^\infty f(x + y)\, d\sigma^n(y) + \sum_{j=0}^{n-1} \int_0^\infty S(x + y)\, d\sigma^j(y). \qquad (4.1.2)
\end{aligned}$$

Let

$$A(x) = \sum_{j=0}^{\infty} \int_0^\infty S(x + y)\, d\sigma^j(y).$$

Then, with $\gamma$ defined as before, we have

$$|A(x)| \leq Ce^{-\varepsilon x} \sum_{j=0}^{\infty} \int_0^\infty e^{-\varepsilon y}\, d\sigma^j(y) = Ce^{-\varepsilon x} \sum_{j=0}^{\infty} \gamma^j = Ce^{-\varepsilon x}/(1 - \gamma).$$

It then follows from (4.1.2) that if $p := f - A$, then

$$p(x) = \lim_{n \to \infty} \int_0^\infty f(x + y)\, d\sigma^n(y) \qquad (x \geq 0)$$

exists and is nonnegative and locally integrable since $f$ and $A$ are; and further, for $x \geq 0$,

$$\int_0^\infty p(x + y)\, d\sigma(y) = \int_0^\infty f(x + y)\, d\sigma(y) - \int_0^\infty A(x + y)\, d\sigma(y)$$
$$= f(x) - S(x) - \sum_{j=0}^{\infty} \int_0^\infty S(x + y)\, d\sigma^{j+1}(y)$$
$$= f(x) - A(x) = p(x), \qquad (4.1.3)$$

so that, by Theorem 2.2.4, $p$ has every element of supp $\sigma$ as a period, and the theorem is proven.

**Theorem 4.1.2.** *Let $\sigma \neq \delta_0$ be a subprobability measure on $[0, \infty)$, and let $f$ be a bounded Borel measurable solution of the ε-ICFE $(\sigma, S)$. Then, $f$ admits the representation*

$$f(x) = p(x) + A(x), \qquad x \geq 0,$$

*where $A$ has the same properties as in the enunciation of Theorem* 4.1.1, *and $p \equiv 0$ if $\sigma[0, \infty) < 1$, and $p$ is a bounded function with every element of* supp $\sigma$ *as a period if $\sigma[0, \infty) = 1$.*

***Proof.*** The same proof as for the previous theorem applies. Further, if $\sigma[0, \infty) < 1$, then $\sigma^n(\mathbb{R}_+) \to 0$ as $n \to \infty$, whence $p \equiv 0$ in this case.

We pass to a discussion of the ε-ICFE $(\sigma, S)$ for suitable signed measures $\sigma$ (cf. also Theorem 2.4.1).

**Theorem 4.1.3.** *Let $\mu$ and $\nu$ (neither vanishing identically) be subprobability measures on $\mathbb{R}_+$ such that $\mu + \nu \neq \delta_0$ is also a subprobability measure. Let $f$ be a bounded Borel measurable solution of the* ε-ICFE *$(\mu - \nu, S)$. Then,*

$$f(x) = p(x) + A(x), \qquad x \geq 0,$$

*where $p \equiv 0$ if $(\mu + \nu)(\mathbb{R}_+) < 1$, and, if $(\mu + \nu)(\mathbb{R}_+) = 1$, then*

$$p(x + u) = \begin{cases} p(x) & \text{for } x \in \operatorname{supp}(\mu + \nu^2), \\ -p(x) & \text{for } x \in \operatorname{supp} \nu, \end{cases} \qquad (4.1.4)$$

*and, for some $C_1 > 0$, $|A(x)| \leq C_1 e^{-\varepsilon x}$ for $x \geq 0$.*

***Proof.*** Proceeding as in the proof of Theorem 2.4.1, we have on iteration that, for $n \in \mathbb{Z}_+$,

$$f = f \bullet \left(\mu + \nu^2 * \left(\sum_{j=0}^{n} \mu^j\right)\right) - f \bullet (\mu^{n+1} * \nu) + S \bullet \left(\delta_0 - \nu * \sum_{j=0}^{n} \mu^j\right).$$

Since $f$ is bounded and $\mu(\mathbb{R}_+) < 1$, the middle term on the right tends to zero pointwise on $\mathbb{R}_+$ as $n \to \infty$, and we have

$$f = f \bullet \sigma + S \bullet (\delta_0 - \nu * \tau) = f \bullet \sigma + T, \quad \text{say}, \tag{4.1.5}$$

where $\tau = \sum_{j=0}^{\infty} \mu^j$ and $\sigma = \mu + \nu^2 * \tau$. Also,

$$\begin{aligned} |T(x)| &\le |S(x)| + \int_0^\infty |S(x+y)|\, d\tau(y) \\ &\le Ce^{-\varepsilon x}\left(1 + \int_0^\infty e^{-\varepsilon y}\, d\tau(y)\right) = C_1 e^{-\varepsilon x} \end{aligned}$$

for some $C_1 > 0$. Applying Theorem 4.1.2 to (4.1.5), we have the assertion of the present theorem, with (4.1.4) following from (4.1.3)—$p = p \bullet \sigma$—and the fact that $\sigma$ is a probability measure when $\mu + \nu$ is one.

**Corollary 4.1.4.** *The assertion of Theorem* 4.1.3 *holds if (instead of assuming $f$ itself is bounded) $f(x + y) - f(x)$ is bounded for every fixed $y > 0$.*

***Proof.*** Fix $y > 0$, and let $f_y(x) = f(x + y) - f(x)$, $S_y(x) = S(x + y) - S(x)$. Then, $|S_y(x)| \le 2Ce^{-\varepsilon x}$ for $x \ge 0$, and, by Theorem 4.1.3, $f_y$, as a solution of the $\varepsilon$-ICFE $((\mu - \nu), S_y)$, is of the form

$$f_y(x) = p_y(x) + A_y(x), \qquad x \ge 0, \tag{4.1.6}$$

where $|A_y(x)| \le C_1 e^{-\varepsilon x}$ for $x \ge 0$, with $C_1$ independent of $y > 0$, and $p_y$ is as in Theorem 4.1.3.

If now $\mu + \nu$ is a strict subprobability measure, then $p_y \equiv 0$, and hence

$$|f(x + y) - f(x)| = |A_y(x)| \le C_1 e^{-\varepsilon x} \qquad \forall\, y > 0,\ x \ge 0.$$

This implies that $f$ is bounded, so that Theorem 4.1.3 applies, showing that $f$ has the stated form.

If $\mu + \nu$ is a probability measure, we first note that $p_{k\rho}(x) = 0$ for $x \ge 0$, for every $k \in \mathbb{Z}_+$ and $\rho \in \operatorname{supp}(\mu + \nu^2)$. In fact, for $x, y \ge 0$,

$$\begin{aligned} f(x + y + k\rho) - f(x + k\rho) &= p_y(x + k\rho) + A_y(x + k\rho) \\ &= p_y(x) + A_y(x + k\rho) \\ &= f(x + y) - f(x) + A_y(x + k\rho) - A_y(x). \end{aligned}$$

Hence,

$$\begin{aligned} f(x + y + k\rho) - f(x + y) &= \{f(x + k\rho) - f(x)\} + A_y(x + k\rho) - A_y(x) \\ &= p_{k\rho}(x) + A_{k\rho}(x) + A_y(x + k\rho) - A_y(x). \end{aligned}$$

Integrating over $\mathbb{R}_+$ with respect to $\mu - \nu$, we have (for fixed $x \geq 0$)

$$(1 - a)(p_{k\rho}(x) + A_{k\rho}(x)) = \int_0^\infty (A_y(x + k\rho) - A_y(x))\, d(\mu - \nu)(y) + S(x + k\rho) - S(x),$$

where $a = (\mu - \nu)(\mathbb{R}_+) < 1$, so that $|p_{k\rho}(x)| \leq C_2 e^{-\varepsilon x}$ for $x \geq 0$, for some $C_2 > 0$. Since $p_{k\rho}$ is periodic, it then follows that it vanishes identically, as claimed. We then have

$$f(x + k\rho) - f(x) = A_{k\rho}(x), \qquad x \geq 0, \tag{4.1.7}$$

for all $k \in \mathbb{Z}_+$ and $\rho \in \operatorname{supp}(\mu + \nu^2)$. For $m, n \in \mathbb{Z}_+$ with $m < n$,

$$\begin{aligned} |f(x + n\rho) - f(x + m\rho)| &\leq |A_\rho(x + (n - 1)\rho)| + \cdots + |A_\rho(x + m\rho)| \\ &\leq C_1 e^{-\varepsilon x}\left(\sum_{j=m}^{n-1} e^{-j\varepsilon\rho}\right) \to 0 \qquad \text{as } m, n \to \infty. \end{aligned}$$

Thus, $p(x) = \lim_{n\to\infty} f(x + n\rho)$ exists and, by its definition, is periodic with period $\rho$. Equation (4.1.7) then implies that $A(x) = \lim_{k\to\infty}\{-A_{k\rho}(x)\}$ exists, with $|A(x)| \leq C_1^{-\varepsilon x}$. Thus, $f = p + A$, with both $p$ (by its periodicity) and $A$ bounded, so that $f$ is itself bounded, Theorem 4.1.3 applies, and the corollary is proven.

## 4.2. CHARACTERIZATIONS OF THE WEIBULL DISTRIBUTION

In this section, we first solve the functional equation (4.2.1) using the results of the last section, and then apply the solution to characterize the Weibull distribution.

Let $(\Omega, \mathcal{B}, P)$ be a probability space and $U: \mathbb{R}_+ \times \Omega \mapsto \mathbb{R}$ a measurable map, where every $U(\cdot, \omega)$, $\omega \in \Omega$, is a d.f. on $[0, \infty)$ having no jump at the origin. Consider $G$, the $P$-mixture of the $U$'s, i.e., $G = \int_\Omega U(\cdot, \omega)\, dP(\omega)$. For convenience, we shall use the symbol $\mathcal{P}(G)$ to denote the class of all right-continuous functions with every element of supp $G$ as a period.

**Theorem 4.2.1.** *Let $g$ be a bounded, real-valued, right-continuous function on $[0, \infty)$, satisfying, for some $\alpha > 0$, the relation*

$$exp(e^{-\alpha x} g(x)) = \int_\Omega \exp\left[e^{-\alpha x}\int_0^\infty g(x + y)U(dy, \omega)\right] dP(\omega), \qquad \forall x \geq 0. \tag{4.2.1}$$

*Then, $g \in \mathcal{P}(G)$.*

***Proof.*** Let $B$ stand for a function bounded by a constant not depending on $\omega \in \Omega$, and not necessarily denoting the same function at each appearance. Then, we may rewrite the left and right sides of (4.2.1) in the forms

$$1 + e^{-\alpha x}g(x) + B(x)e^{-2\alpha x}, \qquad 1 + e^{-\alpha x}\int_0^\infty g(x+y)\,dG(y) + B(x)e^{-2\alpha x},$$

respectively, so that

$$g(x) = \int_0^\infty g(x+y)\,dG(y) + B(x)e^{-\alpha x}, \qquad x \geq 0.$$

Theorem 4.1.2 then implies that $g(x) = p(x) + \xi(x)e^{-\alpha x}$, where $p$ has every element of supp $G$ as a period and $\xi$ is bounded. The right-continuity of $g$ implies that of $p$ and $\xi$ as well; in particular, $p \in \mathcal{P}(G)$. The theorem will be proven if we show that $\xi$ vanishes identically. Substituting the preceding representation of $g$ in (4.2.1) and cancelling out $\exp(-e^{-\alpha x}p(x))$ on both sides of (4.2.1), we have

$$\exp(e^{-2\alpha x}\xi(x)) = \int_\Omega \exp\left[e^{-2\alpha x}\int_0^\infty \xi(x+y)e^{-\alpha y}U(dy,\omega)\right] dP(\omega). \qquad (4.2.2)$$

The right side of (4.2.2) is

$$= 1 + e^{-2\alpha x}(1 + B(x)e^{-2\alpha x})\int_0^\infty |\xi(x+y)|e^{-\alpha y}\,dG(y). \qquad (4.2.3)$$

Since $c := \int_0^\infty e^{-\alpha y}\,dG(y) < 1$, there exists $x_1 \geq 0$ such that

$$|c(1 + B(x)e^{-2\alpha x})| < d < 1 \qquad \text{for } x \geq x_1,$$

for some $d$ not depending on $x$. Taking the logarithms of both sides of (4.2.2) and using the previous equation, we see that there exists $x_2 \geq x_1$ such that, for all $x \geq x_2$,

$$|\xi(x)| \leq d\int_0^\infty |\xi(x+y)|\,dK(y),$$

where $K$ is the d.f. given by $dK(y) = c^{-1}e^{-\alpha y}\,dG(y)$. Hence,

$$\sup\{|\xi(x)| : x \geq x_2\} \leq d\sup\{|\xi(x)| : x \geq x_2\},$$

so that $\xi = 0$ on $[x_2, \infty)$.

Now, let

$$x_3 = \inf\{x \geq 0 : \xi(u) = 0 \text{ for } u \geq x\},$$

so that $x_3 \leq x_2$. Suppose that $x_3 > 0$. Then, for $\varepsilon > 0$ and $x \geq x_3 - \varepsilon\ (\geq 0)$, (4.2.3) becomes

$$1 + e^{-2\alpha x}(1 + B(x)e^{-2\alpha x})\int_0^\varepsilon |\xi(x+y)|e^{-\alpha y}\,dG(y)$$

and hence

$$|\xi(x)| \le 2(1 + B(x)e^{-2\alpha x}) \int_0^{\varepsilon} |\xi(x+y)| e^{-\alpha y}\, dG(y), \qquad x \ge x_3 - \varepsilon.$$

Since $G(0) = 0$, we may choose $\varepsilon > 0$ so small that, for $x \ge 0$,

$$2(1 + B(x)e^{-2\alpha x}) \int_0^{\varepsilon} e^{-\alpha y}\, dG(y) < d \quad (<1).$$

This implies that $\sup |\xi(x)| \le d \sup |\xi(x)|$, the supremum being taken over $[x_3 - \varepsilon, x_3]$, whence $\xi = 0$ there, contrary to the definition of $x_3$. Hence, we must have $x_3 = 0$.

The following proposition and corollary provide sufficient conditions for $g$ satisfying (4.2.1) to be bounded, so that Theorem 4.2.1 applies. We recall the definition of $G$ in the first paragraph of this section.

**Proposition 4.2.2.** *Suppose $G$ has finite, positive expectation:*

$$0 < \int x\, dG(x) = \int_{\Omega} \left[ \int_0^{\infty} xU(dx, \omega) \right] dP(\omega) < \infty. \tag{4.2.4}$$

*Suppose $g \ge 0$ satisfies, for some $\alpha > 0$, the inequality*

$$\exp(-e^{-\alpha x} g(x)) \le \int_{\Omega} \exp\left\{ -e^{-\alpha x} \int_0^{\infty} g(x+y) U(dy, \omega) \right\} dP(\omega), \qquad x \ge 0. \tag{4.2.5}$$

*Then, there exist $k_1, k_2 > 0$ such that*

$$\int_x^{x+k_1} e^{-g(y)}\, dy \ge k_2, \qquad \forall\, x \ge 0. \tag{4.2.6}$$

***Proof.*** Denote $\int_0^{\infty} g(x+y)U(dy, \omega)$ by $k(x, \omega)$. Then, by Jensen's inequality,

$$e^{-k(x,\omega)} \le \int_0^{\infty} e^{-g(x+y)} U(dy, \omega), \qquad \forall\, x \ge 0,\ \omega \in \Omega. \tag{4.2.7}$$

Also, since, for $0 < \delta \le 1$, $E(|X|^{\delta}) \le (E|X|)^{\delta}$ by the well-known Liapunov inequality, we have

$$\begin{aligned} \int_{\Omega} \exp(-e^{-\alpha x} k(x, \omega))\, dP(\omega) &= \int_{\Omega} [\exp(-k(x, \omega))]^{e^{-\alpha x}}\, dP(\omega) \\ &\le \left( \int_{\Omega} \exp(-k(x, \omega))\, dP(\omega) \right)^{e^{-\alpha x}}. \end{aligned}$$

This inequality and (4.2.5) imply that, for $x \geq 0$,

$$e^{-g(x)} \leq \int_\Omega \left( \int_0^\infty e^{-g(x+y)} U(dy, \omega) \right) dP(\omega) = \int_0^\infty e^{-g(x+y)} \, dG(y).$$

Hence, for any $x \geq 0$, we have, by Fubini's theorem,

$$\begin{aligned} \int_0^x e^{-g(t)} \, dt &\leq \int_0^x \left( \int_0^\infty e^{-g(t+y)} \, dG(y) \right) dt \\ &= \int_0^\infty \left( \int_y^{x+y} e^{-g(t)} \, dt \right) dG(y) \quad (\leq x \text{ obviously}). \end{aligned}$$

Since $\int dG(y) = 1$, this may be rewritten, after an obvious rearrangement, in the form

$$\begin{aligned} k := \int_0^\infty \left( \int_0^y e^{-g(t)} \, dt \right) dG(y) &\leq \int_0^\infty \left( \int_x^{x+y} e^{-g(t)} \, dt \right) dG(y) \\ &\leq \int y \, dG(y) < \infty. \end{aligned} \tag{4.2.7}$$

Obviously, $k > 0$. We fix $k_1$ such that $G(k_1) > 0$ *and* $\int_{k_1}^\infty y \, dG(y) < k/2$. Then, for any $x \geq 0$,

$$\begin{aligned} \int_0^{k_1} \left( \int_x^{x+k_1} e^{-g(t)} \, dt \right) dG(y) &\geq \int_0^{k_1} \left( \int_x^{x+y} e^{-g(t)} \, dt \right) dG(y) \\ &= \int_0^\infty - \int_{k_1}^\infty (\cdots) \, dG(y) \\ &\geq k - \int_{k_1}^\infty y \, dG(y) > k/2, \end{aligned}$$

so that

$$\int_x^{x+k_1} e^{-g(t)} \, dt \geq k/(2G(k_1)) = k_2, \quad \text{say}.$$

Hence the proposition is proven.

**Corollary 4.2.3.** *Let* $g: \mathbb{R}_+ \to \mathbb{R}_+$ *be right-continuous and bounded on every compact interval. Suppose g satisfies* (4.2.4) *and* (4.2.5), *and further suppose that there exist* $A > 0$, $\eta > 0$ *such that*

$$\sup\{g(x+y) : 0 \leq y \leq \eta\} \leq A g(x) \qquad \textit{for } x \geq 0. \tag{4.2.8}$$

*Then, g is bounded on* $\mathbb{R}_+$.

***Proof.*** Proposition 4.2.2 implies that (with $k_1, k_2$ as defined there)

$$\forall x \geq 0, \qquad \inf\{g(t): x \leq t \leq x + k_1\} \leq -\log(k_2/k_1) =: C.$$

Thus, there exists a sequence $\{x_n\} \to \infty$ such that $x_{n+1} \in [x_n + k_1, x_n + 2k_1]$ and $g(x_n) \leq C$. Iterating (4.2.8), we see that, for some $B > 0$ (for instance, $B = A^{2([k_1/\eta]+1)}$),

$$\sup\{g(x + y): 0 \leq y \leq 2k_1\} \leq Bg(x) \qquad \text{for } x \geq 0.$$

For $x \in [x_n, x_{n+1})$, we have $g(x) \leq Bg(x_n) \leq BC$, and the corollary is proven.

We proceed to obtain some characterizations of the Weibull distribution (whose d.f. is of the form $1 - \exp(-\lambda x^\alpha)$ for $x \geq 0$), which can be based on the following reformulation of the results just established.

**Proposition 4.2.4.** *Let $\{V(\cdot, \omega): \omega \in \Omega\}$ be a family of d.f.'s with support contained in* $[0, 1]$ *and such that $V(0, \omega) = 0$ and $V(1_-, \omega) = 1$ identically in $\omega$. Let further*

$$0 < \int_\Omega \left( \int_0^1 |\log y| V(dy, \omega) \right) dP(\omega) < \infty. \tag{4.2.9}$$

*Let $F$ be a d.f. on $\mathbb{R}_+$ such that $F(0) = 0$, $T(x) = 1 - F(x) > 0$ for all $x > 0$, and $T$ satisfies*

$$T(x) = \int_\Omega \exp\left( \int_0^1 y^{-\alpha} \log T(xy) V(dy, \omega) \right) dP(\omega), \qquad x \geq 0. \tag{4.2.10}$$

*Then,*

$$F(x) = 1 - \exp(-p(-\log x)x^\alpha), \qquad x \geq 0, \tag{4.2.11}$$

*where $p \in \mathcal{P}(G)$, with $G$ defined by*

$$G(x) = \int_\Omega (1 - V((e^{-x})_-, \omega))\, dP(\omega).$$

***Proof.*** Equation (4.2.9) is equivalent to (4.2.4) with

$$U(x, \omega) = 1 - V((e^{-x})_-, \omega), \qquad \omega \in \Omega;$$

i.e.,

$$0 < \int_\Omega \left( \int_0^\infty x U(dx, \omega) \right) dP(\omega) < \infty.$$

For every $x_0 < 0$, let $g$ be the nonnegative right-continuous function defined by

$$g(x) = -e^{\alpha x} \log(T(e^{-x})_-), \qquad x \geq x_0.$$

Then, the following inequality holds: $\sup\{g(x+y): 0 \le y \le 1\} \le e^{\alpha}g(x)$, for $x \ge x_0$, and the functional equation (4.2.10) reduces to the form (4.2.2) for $x \ge x_0$. Setting $\tilde{g}(x) = g(x + x_0)$, $x \ge 0$, we see that $\tilde{g}$ satisfies the conditions of Corollary 4.2.3 and Theorem 4.2.1. It follows that $\tilde{g}$, and hence $g$, is of the form $p$ in the statement of the proposition, and $T$ of the form (4.2.11) for $0 < x \le e^{-x_0}$. Since $x_0 < 0$ is arbitrary, it follows that (4.2.11) holds for all $x \ge 0$.

**Theorem 4.2.5.** *Let $\{X_n\}_{n=1}^{\infty}$ be a sequence of nonnegative i.i.d.r.v.'s with common d.f. $F \ne \delta_0$. Let $N$ be an integer-valued r.v. with $P\{N \ge 2\} = 1$, independent of the $X_n$ and such that* $\log N$ *has a nonlattice d.f. with finite expectation. If, for some $\alpha > 0$,*

$$N^{1/\alpha} \min\{X_j : 1 \le j \le N\} \sim X_1, \tag{4.2.12}$$

*then, for some $\lambda > 0$, $1 - F(x) = \exp(-\lambda x^{\alpha})$ for $x > 0$.*

***Proof.*** Let $T = 1 - F$, and let $p_n = P\{N = n\}$, $n \ge 2$. The equidistribution assumption (4.2.12) implies that

$$T(x) = \sum_{n=2}^{\infty} p_n T^n(n^{-1/\alpha}x), \qquad x > 0. \tag{4.2.13}$$

We begin by noting that $T(x) > 0$ for $x > 0$; for, if $T(x_0) = 0$ for some $x_0 > 0$, then $T(n^{-1/\alpha}x_0) = 0$ for some $n \ge 2$, and, repeating this argument, we see that $T(0) = 0$, contrary to our assumption that $F \ne \delta_0$. Hence, $\log T$ is defined on $\mathbb{R}_+$, and (4.2.13) may be recast in the form (4.2.10) with

$$\Omega = \{2, 3, \ldots\}, \qquad P\{n\} = p_n \quad \text{for } n \in \Omega, \qquad V(\cdot, n) = \delta_{n^{-1/\alpha}}.$$

The assumption that $E(\log N) < \infty$ implies that (4.2.9) is satisfied; $G$ is nonlattice since $\log N$ has a nonlattice d.f. by assumption, and consequently $p(\cdot)$ appearing in (4.2.11) is a constant. The present theorem is then an immediate consequence of Proposition 4.2.4.

**Theorem 4.2.6.** *Let $m \ge 2$ be a fixed positive integer. Let $\{X_j\}_{j=1}^{m}$ be nonnegative i.i.d.r.v.'s, with common d.f. $F \ne \delta_0$. Let $\{a_j\}_{j=1}^{m}$ be positive r.v.'s independent of the $X_j$, satisfying the conditions:*

$$\left.\begin{array}{l} \text{For some } \alpha > 0,\ P\{a_1^{\alpha} + \cdots + a_m^{\alpha} = 1\} = 1. \\ P\{\log a_j / \log a_k \text{ is irrational for some } j, k\} > 0. \end{array}\right\} \tag{4.2.14}$$

*If*

$$Y := \min\{X_j / a_j : 1 \le j \le m\} \sim X_1,$$

*then, for some $\lambda > 0$, $1 - F(x) = \exp(-\lambda x^{\alpha})$ for $x > 0$.*

***Proof.*** Let, as before, $T = 1 - F$. If $(\Omega, \mathcal{B}, P)$ is the probability space on which the r.v.'s $X_j$, $a_j$ are defined, we have, as a consequence of our equidistribution assumption, that, for every $x > 0$,

$$T(x) = P\{Y > x\} \equiv \int_\Omega \prod_{j=1}^{m} P\{X_j(\omega) > xa_j(\omega)\}\, dP(\omega). \qquad (4.2.15)$$

We check again as in the previous proof that $T(x) > 0$ for all $x > 0$. If we define

$$V(y, \omega) = \sum_{j=1}^{m} a_j^\alpha(\omega)\, \delta_{y - a_j(\omega)}(\omega),$$

then (4.2.15) reduces to (4.2.10), and Proposition 4.2.4 applies. The second of the conditions (4.2.14) implies that supp $G$ is nonlattice; it then follows as before that the function $p(\cdot)$ appearing in (4.2.11) reduces to a constant, say $\lambda$, and the theorem is proven.

## 4.3. A CHARACTERIZATION OF SEMISTABLE LAWS

Let $(\Omega, \mathcal{B}, P)$ be a probability space. Suppose

$$\left.\begin{array}{l} V: \Omega \times [-1, 1] \to \mathbb{R} \text{ is a measurable map such} \\ \text{that, for every } \omega \in \Omega,\ V(\cdot, \omega) \text{ is a d.f. on} \\ [-1, 1] \text{ with } V(-1, \omega) = 0 \text{ and } V(1_-, \omega) = 1. \end{array}\right\} \qquad (4.3.1)$$

In this section, we consider ch.f.'s satisfying, for such a $V$, the functional equation

$$f(t) = \int_\Omega \exp\left\{\int_{-1}^{1} |u|^{-\alpha}\phi(ut)V(du, \omega)\right\} dP(\omega), \qquad |t| < t_0, \quad (4.3.2)$$

for some $\alpha > 0$, where $\phi = \log f$ and $t_0 = \sup\{u: f(t) \neq 0 \text{ for } |t| < u\}$. The functional equation studied in Sections 3.3 and 3.4, namely,

$$f(t) = \prod_{j=1}^{\infty} f(\beta_j t)^{\gamma_j}, \qquad \text{for } |t| < t_0,$$

with $0 < |\beta_j| < 1$ and $\gamma_j > 0$ for all $j$, is a special case with $\Omega = \mathbb{N}$, $V(u, j) = \delta_{\beta_j}(u)$, $P = \sum_{j=1}^{\infty} \gamma_j |\beta_j|^\alpha \delta_j$, $\alpha$ satisfying $\sum_{j=1}^{\infty} \gamma_j |\beta_j|^\alpha = 1$.

We define, for $x > 0$,

$$U_1(x, \omega) = 1 - V(e_-^{-x}, \omega), \qquad U_2(x, \omega) = V(-e^{-x}, \omega);$$

$$G_k(x) = \int U_k(x, \omega)\, dP(\omega), \qquad k = 1, 2;$$

$$U = U_1 + U_2, \quad \tilde{U} = U_1 - U_2, \quad G = G_1 + G_2 \quad \text{and} \quad \tilde{G} = G_1 - G_2.$$

By assumption, $\phi(e^{-x})$ is defined for $x > -\log t_0$; we set

$$g(x) = \operatorname{Re}(-e^{-\alpha x}\phi(e^{-x})) = -e^{-\alpha x}\log f(e^{-x}),$$

$$h(x) = \operatorname{Im}(-e^{-\alpha x}\phi(e^{-x})) = -e^{-\alpha x}\arg f(e^{-x}).$$

Then, (4.3.2) may be written as

$$\exp[-e^{-\alpha x}(g(x) + ih(x))]$$

$$= \int_\Omega \exp\left\{-e^{-\alpha x}\left[\int_0^\infty g(x+y)U(dy,\omega) + i\int_0^\infty h(x+y)\tilde{U}(dy,\omega)\right]\right\} dP(\omega). \tag{4.3.3}$$

We introduce the following additional notation:

For any sub-d.f.'s $F_1, F_2, G$, $\mathcal{P}(F_1, F_2)$ is the class of all right-continuous functions $p$ such that

$$p(x+\rho) = \begin{cases} p(x) & \text{for } \rho \in \operatorname{supp} F_1, \\ -p(x) & \text{for } \rho \in \operatorname{supp} F_2. \end{cases}$$

Also, $\mathcal{P}(G) := \mathcal{P}(G, 0)$.

Then, we have the following.

**Theorem 4.3.1.** *Let* $\{V(\cdot,\omega), \omega \in \Omega\}$ *be a family of d.f.'s on* $[-1, 1]$ *satisfying the conditions* (4.3.1). *We assume further that*

$$0 < \int x\, dG(x) < \infty \qquad \text{in the case } \alpha \neq 1, \tag{4.3.4}$$

*and that there exists* $\delta > 0$ *such that*

$$0 < \int x^{1+\delta}\, dG(x) < \infty \qquad \text{in the case } \alpha = 1. \tag{4.3.4'}$$

*If a ch.f.* $f$ *satisfies* (4.3.2) *for such a* $V$, *then* $t_0 = \infty$, *and* $\phi$ *admits the representation*

$$\phi(t) = i\mu t - |t|^\alpha\{g(-\log|t|) + i(\operatorname{sgn} t)h(-\log|t|)\}, \qquad t \in \mathbb{R}, \tag{4.3.5}$$

*where* $\mu \in \mathbb{R}$, *and*

(I) $g \equiv h \equiv 0$ *if* $\alpha > 2$;
(II) $g \in \mathcal{P}(G)$ *and* $h \in \mathcal{P}(G_1, G_2)$ *if*

(a) $0 < \alpha \leq 1$; *or*

(b) $1 < \alpha \leq 2$, *and if* any one *of the following two conditions is satisfied: either*

(i) $\int_\Omega \left\{ \int_{-1}^{1} |y|^{1-\alpha} V(dy, \omega) \right\} dP(\omega) < \infty,$ *and*

$$\int_\Omega \int_{-1}^{1} y|y|^{-\alpha} V(dy, \omega)\, dP(\omega) \neq 1;$$

*or*

(ii) $$\int_{-1}^{1} y|y|^{-\alpha} V(dy, \omega) = 1 \qquad a.a.\ [P]\omega.$$

*Moreover, we have* $\mu = 0$ *in case* (a) *and case* (b)(i), *and* $\mu$ *is arbitrary in case* (b)(ii); *in all cases,* $f(t)e^{-i\mu t}$ *is the ch.f. of a semistable law with exponent* $\alpha$ *and, in particular,* $f$ *is a normal ch.f. if* $\alpha = 2$.

Later in this section, we consider two applications of this theorem—Theorems 4.3.5 and 4.3.6. We begin the proof of the above result with two auxiliary results.

**Lemma 4.3.2.** *If* $p \in \mathcal{P}(G_1, G_2)$ *and is U-integrable for every* $\omega \in \Omega$, *then there exists an* $\Omega_0 \subset \Omega$ *with* $P(\Omega_0) = 1$ *such that*

$$p(x) = \int p(x + y)U(dy, \omega) \qquad \text{for } x \in \mathbb{R},\ \omega \in \Omega_0.$$

***Proof.*** If supp $G$ is not contained in some lattice with the origin as a lattice point, then $p$ is a constant and the lemma holds trivially. If supp $G \subseteq \{n\rho : n \in \mathbb{Z}_+\} =: L$, consider rationals $u$, $v$ such that $u < v$ and $[u, v] \cap L$ is empty. Then, for $k = 1, 2$,

$$0 = G_k(v) - G_k(u) = \int_\Omega \{U_k(v, \omega) - U_k(u, \omega)\}\, dP(\omega).$$

Since $U_k(\cdot, \omega)$ is nondecreasing, it follows that $U_k(v, \omega) - U_k(u, \omega) = 0$ for a.a. $[P]$ $\omega$; i.e., there exists an $\Omega_{u,v}$ such that $P(\Omega_{u,v}) = 1$, and $U_k(v, \omega) = U_k(u, \omega)$ for $\omega \in \Omega_{u,v}$. Define $\Omega_0 = \bigcap_{u<v} \Omega_{u,v}$ (with $u$, $v$ rational). Then, $P(\Omega_0) = 1$, and $U_k(\cdot, \omega)$ is constant on each of the open intervals $(n\rho, (n + 1)\rho)$ if $\omega \in \Omega_0$. The assertion of the lemma then follows from the definition of the family $\mathcal{P}(G_1, G_2)$.

**Lemma 4.3.3.** *Let, for some* $x_0 \in \mathbb{R}$, *g and h be real-valued right- (or left-) continuous functions defined on* $\{x_0, \infty)$, *satisfying* (4.3.3) *for* $x \geq x_0$. *Suppose that*

(a) *if* $\alpha \neq 1$, *both g and h are bounded, and*

(b) *if* $\alpha = 1$, *g is bounded, while* $h(x + y) - h(x)$ *is bounded for every* $y > 0$, *and, further,*

$$\left.\begin{aligned} |h(x)| &\le c(1 + x - x_0) \qquad \text{for } x \ge x_0, \quad \text{for some } c > 0, \\ \int y^{1+\delta}\, dG(y) &< \infty \qquad \text{for some } \delta > 0. \end{aligned}\right\} \tag{4.3.6}$$

*Then,* $g \in \mathcal{P}(G)$ *and* $h \in \mathcal{P}(G_1, G_2)$ *on* $[x_0, \infty)$.

***Proof.*** As in the last section, let $B$ denote a constant or a function (of $x$ and $\omega$ or of $x$ alone) admitting a uniform bound not depending on $x$ and $\omega$, and again not necessarily the same at each appearance. The arguments below are provided for the case $\alpha = 1$; the modifications needed for the simpler cases $\alpha \ne 1$ ($h$ then being assumed to be bounded) should be obvious. Using the assumptions (4.3.6), the imaginary part of the right side of (4.3.3) may be written as

$$-\int_\Omega (1 + B(x, \omega)e^{-\alpha x}) \sin\left(e^{-\alpha x} \int h(x + y)\tilde{U}(dy, \omega)\right) dP(\omega).$$

Noting that $\sin x = x + B(x)|x|^{1+\delta}$ for $x \in \mathbb{R}$, and that

$$\begin{aligned} \int_\Omega \left\{\int |h(x + y| U(dy, \omega)\right\}^{1+\delta} dP(\omega) &\le B \int (1 + x - x_0 + y)^{1+\delta}\, dG(y) \\ &\le B\left\{(1 + x - x_0)^{1+\delta} + \int y^{1+\delta}\, dG(y)\right\}, \end{aligned}$$

we see that this quantity is

$$= -e^{-\alpha x} \int h(x + y)\, d\tilde{G}(y) + B(x)e^{-(\alpha+\varepsilon)x}$$

for any fixed $\varepsilon \in (0, \alpha\delta)$. On the other hand, the imaginary part of the left side of (4.3.3) is

$$\begin{aligned} &= -e^{-\alpha x}h(x) + Be^{-\alpha(1+\delta)x}(1 + x - x_0)^{1+\delta} \\ &= -e^{-\alpha x}h(x) + B(x)e^{-(\alpha+\varepsilon)x}. \end{aligned}$$

Similar arguments in the case $\alpha \ne 1$ ($h$ being assumed bounded) lead to similar expressions, with the "remainder terms" for both the left and right sides of (4.3.3) then taking the form $B(x)e^{-2\alpha x}$. Thus, in all cases (for a fixed $\varepsilon \in (0, \alpha\delta)$ in the case $\alpha = 1$, and with $\varepsilon = \alpha$ in the case $\alpha \ne 1$), we have

$$h(x) = \int h(x + y)\, d\tilde{G}(y) + B(x)e^{-\varepsilon x}.$$

It follows then from Corollary 4.1.4 (from Theorem 4.1.3 itself for $\alpha \neq 1$) that there exist bounded functions $p \in \mathcal{P}(G_1, G_2)$ and $\eta$ such that

$$h(x) = p(x) + \eta(x)e^{-\varepsilon x}, \qquad x \geq x_0. \tag{4.3.7}$$

By Lemma 4.3.2,

$$p(x) = \int p(x+y)U(dy, \omega) \qquad \text{for a.a. } [P]\omega. \tag{4.3.8}$$

We claim that $\eta \equiv 0$, whence $h \in \mathcal{P}(G_1, G_2)$. Then, substituting from (4.3.7) into (4.3.2), we have, for $x \geq x_0$,

$$\exp\{-e^{-\alpha x}g(x)\} = \int_\Omega \exp\left\{-e^{-\alpha x}\int_0^\infty g(x+y)U(dy, \omega)\right\} dP(\omega),$$

and it follows from Theorem 4.2.1 that $g \in \mathcal{P}(G)$.

It therefore remains only to show that $\eta \equiv 0$. We adopt the same method as for proving that $\xi \equiv 0$ in the proof of Theorem 4.2.1. Substituting from (4.3.7) into (4.3.2), we have, for $x \geq x_0$,

$$\begin{aligned} &\exp[-\{g(x) + i\eta(x)e^{-\varepsilon x}\}e^{-\alpha x}] \\ &= \int_\Omega \exp\left[-e^{-\alpha x}\left\{\int_0^\infty g(x+y)U(dy, \omega)\right.\right. \\ &\qquad \left.\left. + ie^{-\varepsilon x}\int_0^\infty \eta(x+y)e^{-\varepsilon y}\tilde{U}(dy, \omega)\right\}\right] dP(\omega). \end{aligned}$$

If $M = \sup |g|$, then the imaginary part of the right side is, in absolute value,

$$\begin{aligned} &\leq \exp(Me^{-\alpha x})e^{-(\alpha+\varepsilon)x}\int_\Omega \left\{\int |\eta(x+y)|e^{-\varepsilon y}U(dy, \omega)\right\} dP(\omega) \\ &= \exp(Me^{-\alpha x})e^{-(\alpha+\varepsilon)x}\int |\eta(x+y)|e^{-\varepsilon y}\, dG(y) \\ &= c\exp(Me^{-\alpha x})e^{-(\alpha+\varepsilon)x}\int |\eta(x+y)|\, dK(y), \end{aligned}$$

where (as before) $c = \int e^{-\varepsilon y}\, dG(y)$ $(<1)$ and $K$ is the d.f. given by $dK(y) = c^{-1}e^{-\varepsilon y}\, dG(y)$. On the other hand, the absolute value of the imaginary part of the left side is

$$\begin{aligned} &\exp(-e^{-\alpha x}g(x))|\sin(e^{-(\alpha+\varepsilon)x}\eta(x))| \\ &\qquad \geq \exp(-Me^{-\alpha x})\{1 - (\tfrac{1}{6})e^{-2(\alpha+\varepsilon)x}(\eta(x))^2\}e^{-(\alpha+\varepsilon)x}|\eta(x)|. \end{aligned}$$

Denote the second factor on the right-hand side by $L(x)$. Then, we have, from the two inequalities just established, that

$$L(x)|\eta(x)| \leq c \exp(2Me^{-\alpha x}) \int |\eta(x + y)| \, dK(y), \qquad x \geq x_0.$$

Hence, as before, there exist $x_1 \geq x_0$ and $0 < c' < 1$ such that

$$|\eta(x)| \leq c' \int |\eta(x + y)| \, dK(y) \qquad \text{for } x \geq x_1,$$

so that $\eta$ vanishes on $[x_1, \infty)$. Then, we conclude as before that $\eta$ vanishes on $[x_0, \infty)$ itself, and the lemma is proved.

We are now in a position to continue with the proof of Theorem 4.3.1.

***Proof of Theorem 4.3.1.*** Considering the absolute values of both sides of (4.3.2), we obtain the inequality (4.2.5) for $x \geq x_0$, for any $x_0 > -\log t_0$. Since the condition (4.3.4) is equivalent to (4.2.4), we conclude from Proposition 4.2.2 that there exist positive $k_1$ and $k_2$ such that

$$\int_x^{x+k_1} e^{-g(u)} \, du < k_2 \qquad \text{for } x \geq x_0.$$

Then, as in the proof of Corollary 4.2.3, there exists a sequence $\{x_n\}$ with $x_{n+1} \in [x_n + k_1, x_n + 2k_1]$ such that $g(x_n) \leq -\log(k_2/k_1)$. Setting $t_n = e^{-x_n}$, we see that $-\log|f(t_n)|/t_n^\alpha$ is bounded. Theorem 1.3.8(b) at once implies that, if $\alpha > 2$, then $|f| \equiv 1$, proving assertion (I). Let $0 < \alpha \leq 2$. Lemma 3.4.3 applies with $a = e^{-2k_1}$ and $b = e^{-k_1}$, enabling us to conclude that $\log|f(t)|/|t|^\alpha$ is bounded as $t \to 0$. This means that $g$ is bounded, and in order to be able to apply Lemma 4.3.3, we have to check on $h$. For this purpose, we now divide the argument into three cases.

***Case 1.*** $0 < \alpha < 1$. The boundedness of $\{1 - |f(t)|\}/|t|^\alpha$ implies that of $|1 - f(t)|/|t|^\alpha$ in this case, according to Lemma 3.4.4(i). This means that $h$ also is bounded on $[x_0, \infty)$, for any $x_0 > -\log t_0$. Lemma 4.3.3 implies then that $g \in \mathcal{P}(G)$ and $h \in \mathcal{P}(G_1, G_2)$ on $[x_0, \infty)$, and so on $(-\log t_0, \infty)$. If $M = \sup|g|$ (on the latter interval), then $|f(t)| \geq \exp(-M|t|^\alpha)$ for $|t| < t_0$, and this holds also for $t = t_0$ by the continuity of $f$, so that $f$ cannot vanish at $t_0$ (if $t_0 \in \mathbb{R}$). Since the basic functional equation has been assumed to hold in the largest interval around the origin in which $f$ does not vanish, we must have $t_0 = \infty$. Hence, $g \in \mathcal{P}(G)$ and $h \in \mathcal{P}(G_1, G_2)$ on $\mathbb{R}$ itself, and assertion (IIa) is proved in this case.

***Case 2.*** $\alpha = 1$. Lemma 3.4.3 implies that $\log |f(t)|/|t|$ is bounded as $t \to 0$, and $xT(x)$ is bounded as $x \to \infty$, where $T(x) = 1 - F(x) + F(-x_-)$. It follows in turn that, for some $C > 0$, $|1 - f(t)| \le C(|t| - |t| \log |t|)$ as $t \to 0$: For, taking $t > 0$,

$$|1 - f(t)| \le \int |1 - e^{itx}| \, dF(x) \le t \int_{|x| \le 1/t} |x| \, dF(x) + 2T(1/t)$$

$$= -t \int_0^{1/t} x \, dT(x) + 2T(1/t) = S(t) + 2T(1/t), \quad \text{say.}$$

For $t \ge 1$, $0 \le S(t) \le 1$, and, for $0 < t < 1$,

$$0 \le S(t) \le t - t \int_1^{1/t} x \, dT(x) \le t + t \int_1^{1/t} T(x) \, dx = t + O(-t \log t).$$

This fact and the assumption (4.3.4′) imply that the conditions (4.3.6) are satisfied, and Lemma 3.4.4(iii) implies that $h(x + y) - h(x)$ is bounded for every $y > 0$. Hence, Lemma 4.3.3 applies, and then we may conclude as in Case 1 that $t_0 = \infty$ and that (4.3.5) holds for all $t \in \mathbb{R}$, with $g \in \mathcal{P}(G)$ and $h \in \mathcal{P}(G_1, G_2)$ on $\mathbb{R}$. This establishes assertion (IIa) in the case $\alpha = 1$.

***Case 3.*** $1 < \alpha \le 2$. Lemma 3.4.3 implies that $\{1 - |f(t)|\}/|t|^\alpha$ is bounded, and then Lemma 3.4.4 implies that, for some $\mu \in \mathbb{R}$, $|1 + i\mu t - f(t)|/|t|^\alpha$ is bounded. Hence, there exists a bounded function $h_0$ such that

$$h(x) = -\mu e^{(\alpha-1)x} + h_0(x), \qquad x \ge x_0. \tag{4.3.9}$$

We show separately in Lemma 4.3.4 that, under condition (i) of Part (IIb) of Theorem 4.3.1, $\mu = 0$, so that $h$ is bounded, whereas, under condition (ii), (4.3.3) may be rewritten with $h_0$ in place of $h$ (the factors on either side that involve $\mu$ become identical, and being nonzero, may be cancelled out). Thus, under either of these conditions, Lemma 4.3.3 applies and the argument may be concluded as before.

In particular, $\alpha = 2$ corresponds to normal laws.

Thus, the proof of Theorem 4.3.1 will be complete once we establish the following.

**Lemma 4.3.4.** *Under condition* (i) *of Part II(b) of Theorem* 4.3.1, $\mu = 0$ *in the representation* (4.3.9) *for h* (*so h is bounded*).

***Proof.*** As already seen, in our discussion of Case 3,

$$\phi(t) = i\mu t + |t|^\alpha \{A(t) + iB(t)\} \qquad \text{for } |t| < t_0,$$

where $A$ and $B$ are some real-valued continuous functions. Writing

$$\xi(\omega) = \int_{-1}^{1} u|u|^{-\alpha} V(du, \omega), \qquad \xi^*(\omega) = \int_{-1}^{1} |u|^{1-\alpha} V(du, \omega),$$

and equating the imaginary parts of both sides of the basic functional equation, we have

$$\begin{aligned} &\exp\{|t|^\alpha A(t)\} \sin\{\mu t + |t|^\alpha B(t)\} \\ &= \int_\Omega \exp\{|t|^\alpha \int_{-1}^{1} A(ut) V(du, w)\} \\ &\qquad \times \sin\{\mu t \xi(w) + |t|^\alpha \int_{-1}^{1} B(ut) V(du, \omega)\} \, dP(\omega). \end{aligned}$$

The boundedness of the functions $A$ and $B$ on a fixed compact interval contained in $(-t_0, t_0)$, the fact that $|\sin \theta| \leq |\theta|$ for real $\theta$, and the assumption that $\int \xi^*(\omega) \, dP(\omega) < \infty$ enable us to take the limit as $t \to 0$ under the integral sign on the right side of the relation obtained after dividing through by $t$. We arrive at the relation

$$\mu = \mu \int \xi(\omega) \, dP(\omega).$$

Since, by assumption, the integral on the right $\neq 1$, it follows that $\mu = 0$. This proves the lemma and thus completes the proof of theorem 4.3.1.

We conclude this section with two applications of Theorem 4.3.1.

**Theorem 4.3.5.** *Let $\{X_n\}_{n=1}^{\infty}$ be a sequence of i.i.d.r.v.'s with common d.f. F. Let N be a positive integer-valued r.v. independent of the $X_n$, such that the d.f. of* $\log N$ *is not concentrated on a lattice with the origin as a lattice point. Let $0 < \alpha \leq 2$, and suppose that one of the following conditions on N is satisfied:*

(i) $E(\log N) < \infty$ *if* $0 < \alpha < 1$;
(ii) $E(\log N)^{1+\delta} < \infty$ *for some* $\delta > 0$, *if* $\alpha = 1$;
(iii) $E(N) < \infty$ *if* $\alpha > 1$ *(so* $E(\log N) < \infty$ *as well).*

*Then, the identical-distribution assumption*

$$N^{-1/\alpha}(X_1 + \cdots + X_N) \sim X_1$$

*implies that F is a stable d.f. with exponent $\alpha$; more precisely,* (4.3.5) *holds with g and h constants; and $\mu = 0$ in the cases $\alpha \neq 1, 2$.*

***Remark.*** Stable ch.f.'s of the form (4.3.5) with $g$ and $h$ constants are sometimes referred to as "strictly stable." $F$ in the above set-up is thus "strictly stable." Strictly stable laws with exponent $\alpha = 1$ form a proper subclass of the class of stable laws with that exponent; for $\alpha \neq 1$, the two classes coincide.

***Proof.*** Let $p_n = P\{N = n\}$. The equidistribution assumption implies that, if $(-t_0, t_0)$ is the largest open interval around the origin on which $f$, the ch.f. of $F$, does not vanish, then ($\phi = \log f$ as usual)

$$f(t) = \sum_{n=1}^{\infty} p_n \exp\{n\phi(n^{-1/\alpha}t)\} \qquad \text{for } |t| < t_0.$$

This is of the form (4.3.1) with $\Omega = \mathbb{N}$, $P\{n\} = p_n$, and $V(\cdot, n) = \delta_{n^{-1/\alpha}}$ for $n \in \Omega$. Also, $G = \tilde{G} =$ the d.f. of $\log N$ here. Conditions (i)–(iii) correspond respectively to the conditions imposed in Theorem 4.3.1 for the three ranges of values of $\alpha$ (in particular, for $1 < \alpha \leq 2$, $1 = \sum p_n < \sum p_n n^{1-1/\alpha} < \infty$ under (iii)). Hence, Theorem 4.3.1 applies, and $g$ and $h$ in the representation (4.3.5) are constants since $G$ is not a lattice d.f. with the origin as a lattice point. The same theorem also tells us that $\mu = 0$ (under the conditions assumed) if $\alpha \neq 1$ or 2.

**Theorem 4.3.6.** *Let $2 \leq m \in \mathbb{N}$, and let $\{X_j\}_{j=1}^m$ be i.i.d.r.v.'s with common d.f. $F$. Suppose $\{a_j\}_{j=1}^m$ are r.v.'s, each nonzero with probability one and independent of the $X_j$, and satisfying the conditions:*

$$\textit{for some } \alpha \in (0, 2], \qquad P\{|a_1|^\alpha + \cdots + |a_m|^\alpha = 1\} = 1; \tag{4.3.10a}$$

*and*

$$P\{\log |a_j|/\log |a_k| \textit{ is irrational for some } j, k\} > 0. \tag{4.3.10b}$$

*We shall also assume that one of the following conditions is satisfied:*

(i) $\sum E(|a_j|^\alpha |\log |a_j||) < \infty$ *if* $0 < \alpha < 1$;
(ii) $\sum E(|a_j| |\log |a_j||)^{1+\delta} < \infty$ *if* $\alpha = 1$, *for some* $\delta > 0$;
(iii) $\sum E(|a_j|) < \infty$ and $\sum Ea_j \neq 1$ *if* $1 < \alpha \leq 2$.

*Then, the equidistribution assumption $\sum_{j=1}^m a_j X_j \sim X_1$ implies that $F$ is a "strictly stable" d.f. with exponent $\alpha$, and that $\mu = 0$ if $\alpha \neq 1$ or 2.*

***Remarks.*** (1) The first condition in (iii) in the cases $1 < \alpha \leq 2$ also implies that the condition (i) holds as well for such $\alpha$, since $x^{\alpha-1} \log x$ is bounded on (0, 1) for such $\alpha$.

(2) If the conditions in both (ii) and (iii) are assumed for $\alpha = 1$, then $F$ is a Cauchy-type d.f. with median zero (i.e., has ch.f. of the form $\exp(-c|t|)$ for some $c > 0$.)

***Proof.*** The equidistribution assumption implies that

$$f(t) = \int_{\Omega} \prod_{j=1}^{m} f(ta_j(\omega)), \qquad \forall\, t \in \mathbb{R}, \tag{4.3.11}$$

where $f$ is the ch.f. of $F$.

If $(-t_0, t_0)$ is the largest open interval around the origin on which $f$ does not vanish, then (4.3.11) can be recast in the form (4.3.1) with (recalling (4.3.10a))

$$V(y, \omega) = \sum_{j=1}^{m} |a_j(\omega)|^{\alpha}\, \delta_{a_j(\omega)}(y).$$

As in the previous theorem, conditions (i), (ii), and (iii) here also correspond respectively to the conditions imposed in Theorem 4.3.1 for the relevant ranges of values of $\alpha$; we recall Remark 1 in this context. Hence, Theorem 4.3.1 applies, and $f$ has the representation (4.3.5) with $g$ and $h$ constants, in view of (4.3.10b), which implies that $G$ is not a lattice d.f. with the origin as a lattice point. Hence, the theorem follows. As for Remark 2, if we substitute $\phi(t) = i\mu t - c|t|$ in the basic equation (4.3.11) and recall (4.3.10a) for $\alpha = 1$, we obtain

$$Ee^{it\mu(\sum a_j(\omega))} = e^{it\mu}, \qquad \forall\, t \in \mathbb{R},$$

which implies that either $\mu = 0$, or $\sum a_j(\omega) = 1$ with probability one. The latter possibility is ruled out since then we must have $\sum Ea_j(\omega) = 1$ as well, contrary to assumption.

## 4.4. ICFE'S WITH ERROR TERMS ON $\mathbb{R}_+$: THE SECOND KIND

In this section, we shall be concerned with equations of the form

$$f(x)\{1 - S(x)\} = \int_0^{\infty} f(x + y)\, d\sigma(y), \qquad x \geq 0, \tag{4.4.1}$$

where $f$ and $S$ are real-valued functions on $\mathbb{R}_+$, $\sigma$ is a $\sigma$-finite Borel measure on $\mathbb{R}_+$ with $\sigma\{0\} < 1$, and $S(\cdot)$, the "error term," satisfies the condition

$$|S(x)| \leq Ce^{-\varepsilon x} \quad \forall\, x \geq 0, \qquad \text{for some } C > 0,\ \varepsilon > 0.$$

We shall use the notation $\varepsilon$-ICFE$'(\sigma, S)$ to denote such an equation. Since we have assumed that $\sigma\{0\} < 1$, it is clear that we may suitably redefine $f$, $S$, and $\sigma$ to assume without loss of generality that $\sigma\{0\} = 0$. In what follows, we will do this.

**Theorem 4.4.1.** *Let $f$ be a nonnegative real-valued function on $\mathbb{R}_+$, locally integrable with respect to the Lebesgue measure $\omega$ there, not vanishing identically on any interval $(a, \infty)$, and satisfying an $\varepsilon$-ICFE$'(\sigma, S)$ with $\sigma\{0\} = 0$ for almost all $[\omega]$ $x \geq 0$. Then, there exists a unique $\alpha$ such that*

$$\int_0^\infty e^{\alpha y}\, d\sigma(y) = 1,$$

*and $f$ admits the representation*

$$f(x) = p(x)e^{\alpha x}\{1 + k(x)\} \qquad a.e.\ [\omega]x \geq 0, \tag{4.4.2}$$

*where $p$ has every element of* supp $\sigma$ *as a period and $k$ satisfies*

$$|k(x)| \leq C_0 e^{-\varepsilon x} \qquad \text{for } x \geq x_0, \qquad \text{for some } x_0 \geq 0,\ C_0 \geq 0.$$

***Remarks.*** (1) If we take an (arbitrary) $f(\geq 0)$ vanishing on some interval $(a, \infty)$, $a > 0$, then (4.4.1) obviously holds with a suitable $S$—namely, any $S$ given on $[0, a]$ by

$$f(x)\{1 - S(x)\} = \int_0^{a-x} f(x + y)\, d\sigma(y).$$

Thus, nothing in general can be said of an $f$ vanishing outside a compact interval and satisfying (4.4.1).

(2) Theorem 4.1.2 reduces to a special case of the present theorem. The idea of proof remains the same, but a more elaborate argument is needed to estimate the growth of $f$. Compare in particular the $A(\cdot)$ of Theorem 4.1.1 with the $A(\cdot)$ defined by (4.4.6).

***Proof.*** Let $x_1 > 0$ be such that $|S(x)| < \frac{1}{2}$ for $x \geq x_1$. We recall that $\sigma\{0\} = 0$ by assumption; then, by an obvious modification of the proof of Lemma 2.2.2, using the local integrability of $f$ and the $\sigma$-finiteness of $\sigma$, we arrive at the inequality

$$\int_{x_1}^x \{1 - S(u)\}f(u)\, du \geq \int_A \left(\int_{y+x_1}^{y+x} f(u)\, du\right) d\sigma(y), \qquad x \geq x_1,$$

for a suitable Borel set $A$; we conclude that, for a suitable real $\beta$,

$$\int_{x_1}^\infty \{1 - S(u)\}f(u)e^{-\beta u}\, du < \infty.$$

It follows that $\int_0^\infty f(u)e^{-\beta u}\, du < \infty$, and hence $\int_0^\infty \int_y^\infty f(u)e^{-2\beta u}\, du\, dy < \infty$. Let

$$\tilde{f}(x) = \int_x^\infty \left[\int_y^\infty f(u)e^{-2\beta u}\, du\right] dy, \qquad x \geq 0.$$

We claim that $\tilde{f}(x) \leq C_1 e^{-(\tau-(\varepsilon/2))}$, $x \geq 0$, for suitable constants $C_1 > 0$ and $\tau$. Indeed, $\tilde{f}$ is strictly positive, since $f$ does not (by assumption) vanish identically on any interval of the form $(a, \infty)$. $\tilde{f}$ is also decreasing and convex, in particular continuous, on $(0, \infty)$. Further, $\tilde{f}$ satisfies

$$\tilde{f}(x)\{1 - T(x)\} = \int_0^\infty \tilde{f}(x + u)\, d\nu(u), \tag{4.4.3}$$

where

$$T(x) = \left\{\int_x^\infty \left[\int_y^\infty f(u)e^{-2\beta u} S(u)\, du\right] dy\right\} \Big/ \tilde{f}(x); \qquad d\nu(u) = e^{2\beta u}\, d\sigma(u).$$

We note that

$$|T(x)| \leq Ce^{-\varepsilon x}, \qquad \forall\, x \geq 0$$

(with the same $C$ as for $S$). We define

$$\tilde{f}_T(x) = \{1 - T(x)\}\tilde{f}(x), \qquad \tilde{g}(y) = \overline{\lim_{x\to\infty}}\, \{\tilde{f}(x + y)/\tilde{f}(x)\};$$

then, $\tilde{g}(y)$ equals $\overline{\lim}_{x\to\infty}\, \tilde{f}_T(x + y)/\tilde{f}_T(x)$ also. Following the same proof as in Theorem 2.2.4, with some obvious modifications, we have $\tilde{g}(y) = e^{-\tau y}$ for all $y \in \operatorname{supp} \sigma$, and that $\int_0^\infty e^{-\tau u}\, d\nu(u) = 1$. (We need only mention here the form of the basic inequality: For fixed $y \in \operatorname{supp} \sigma$, denote $\tilde{g}(y)$ by $c$. Then, for arbitrary $\delta > 0$, $\eta > 0$, using the fact that $f$ is decreasing, we have, for all $x \geq x_0(\delta)$,

$$\begin{aligned} c + \delta &- \frac{\tilde{f}_T(x + y)}{\tilde{f}_T(x)} \\ &\geq \frac{\tilde{f}_T(x + y + \eta)}{\tilde{f}_T(x)} \left(c + \delta - \frac{\tilde{f}(x + 2y - \eta)}{\tilde{f}(x + y + \eta)}\right) \sigma(y - \eta, y + \eta). \end{aligned}$$

We then use, as before, the convexity of $\tilde{f}$ and the fact that $\tilde{f}_T(x)/\tilde{f}(x) \to 1$ as $x \to \infty$.) $\tau$ is unique and the $\alpha$ of the theorem is equal to $2\beta - \tau$. Now,

$$\tilde{f}(x + 1)/\tilde{f}(x) < e^{-\tau+\varepsilon/2} \qquad \text{for all large } x.$$

Since $\tilde{f}$ decreases, a routine argument using the preceding fact enables us to conclude that, for some $C_1 > 0$,

$$\tilde{f}(x) \leq C_1 e^{-x(\tau-\varepsilon/2)}, \qquad \forall\, x \geq 0, \tag{4.4.4}$$

as claimed.

Iterating (4.4.1) as in the proof of Theorem 4.1.1, we have, for every $n \in \mathbb{N}$,

$$f(x) = \int_0^\infty f(x + u)\, d\sigma^n(u) + \sum_{j=0}^{n-1} \int_0^\infty f(x + u)S(x + u)\, d\sigma^j(u). \tag{4.4.5}$$

Let (formally at first)

$$A(x) = \sum_{j=0}^{\infty} \int_0^{\infty} f(x + u)S(x + u)\, d\sigma^j(u). \tag{4.4.6}$$

Then,

$$\int_x^{\infty} \int_y^{\infty} e^{-2\beta z}|A(z)|\, dz\, dy$$

$$\le C \sum \int_0^{\infty} \left\{ \int_x^{\infty} \int_y^{\infty} e^{-2\beta z} f(z + u) e^{-\varepsilon(z+u)}\, dz\, dy \right\} d\sigma^j(u)$$

$$\le C \sum \int_0^{\infty} e^{-\varepsilon(x+u)} \left\{ \int_x^{\infty} \int_y^{\infty} e^{-2\beta z} f(z + u)\, dz\, dy \right\} d\sigma^j(u)$$

$$\le C \sum \int_0^{\infty} e^{-\varepsilon(x+u)} e^{2\beta u} \tilde{f}(x + u)\, d\sigma^j(u) \tag{4.4.7}$$

$$\le CC_1 \sum \int_0^{\infty} e^{-(\tau+\varepsilon/2)(x+u)} e^{2\beta u}\, d\sigma^j(u)$$

$$\le CC_1 e^{-(\tau+\varepsilon/2)x}/(1 - \delta), \tag{4.4.8}$$

with

$$\delta = \int_0^{\infty} e^{(\alpha-\varepsilon/2)u}\, d\sigma(u) \quad (<1),$$

recalling that $\alpha = 2\beta - \tau$ and

$$\int_0^{\infty} e^{\alpha y}\, d\sigma(y) = 1,$$

and (4.4.4). In particular,

$$\int_0^{\infty} \int_y^{\infty} e^{-2\beta z}|A(z)|\, dz\, dy < \infty,$$

so that $A(z)$ exists finitely for $[\omega]$ almost all $z$. We define $g = f - A$, so that

$$g(x) = \lim_{n\to\infty} \int_0^{\infty} f(x + y)\, d\sigma^n(y) \qquad \text{exists a.e. } [\omega].$$

Then (as in the proof of Theorem 4.1.1), we have ($g \ge 0$) and

$$g(x) = \int_0^{\infty} g(x + y)\, d\sigma(y) \qquad \text{a.e.},$$

so that $g(x) = p(x)e^{\alpha x}$ a.e., where $p$ has every element of supp $\sigma$ as a period, and we have $f = g + A$.

We now claim that $g$ cannot vanish a.e. For, we must then have $f = A$ a.e., and (4.4.8) then implies that

$$(0<)\tilde{f}(x) \leq CC_1 e^{-(\tau+\varepsilon/2)x}/(1-\delta).$$

Substituting this back in (4.4.7), we obtain

$$\tilde{f}(x) \leq C_1 e^{-(\tau+\varepsilon/2)x}\left(\frac{Ce^{-\varepsilon x/2}}{1-\delta}\right)^2.$$

Repeating this argument, we have

$$\tilde{f}(x) \leq C_1 e^{-(\tau+\varepsilon/2)x}\left(\frac{Ce^{-\varepsilon x/2}}{1-\delta}\right)^n, \qquad \forall n \in \mathbb{N}.$$

This implies that $\tilde{f}$ vanishes on some interval $(a, \infty)$, whereas $\tilde{f}$ is strictly positive by our construction. Hence, $g = 0$ a.e. is impossible. We thus have the representation

$$f(x) = p(x)e^{\alpha x} + A(x),$$

with $\int_0^\infty e^{\alpha y}\, d\sigma(y) = 1$, and have to show now that $A(x) = p(x)k(x)e^{\alpha x}$ a.e. to obtain (4.4.2). For convenience of notation, we shall take $\alpha = 0$ (so $\tau = 2\beta$), the changes otherwise necessary in what follows being obvious. Substituting the representation $f(x) = p(x) + A(x)$ in the definition of $A(\cdot)$ given by (4.4.6), we have

$$\begin{aligned} A(x) &= \sum \int_0^\infty f(x+u)S(x+u)\, d\sigma^j(u) \\ &= \sum \int_0^\infty \{p(x) + A(x+u)\}S(x+u)\, d\sigma^j(u) \\ &= p(x)A_1(x) + B_1(x), \end{aligned} \tag{4.4.9}$$

where

$$A_1(x) = \sum \int_0^\infty S(x+u)\, d\sigma^j(u),$$

$$B_1(x) = \sum \int_0^\infty A(x+u)S(x+u)\, d\sigma^j(u).$$

We have ($\sigma$ being a probability measure, in view of our "adjustment" above), with $\eta = \int_0^\infty e^{-\varepsilon u}\, d\sigma(u) < 1$,

$$|A_1(x)| \leq C \sum \int_0^\infty e^{-\varepsilon(x+u)}\, d\sigma^j(u) = Ce^{-\varepsilon x}/(1-\eta).$$

Substituting (4.4.9) in the expression for $B_1$, we have

$$B_1(x) = p(x)A_2(x) + B_2(x)$$

where

$$A_2(x) = \sum \int_0^\infty A_1(x + u)S(x + u)\, d\sigma^j(u),$$

$$B_2(x) = \sum \int_0^\infty B_1(x + u)S(x + u)\, d\sigma^j(u).$$

An argument similar to that for estimating $A_1(\cdot)$ shows that

$$|A_2(x)| \leq [Ce^{-\varepsilon x}/(1 - \eta)]^2.$$

Thus proceeding, we have

$$A(x) = p(x)\left(\sum_{k=1}^{n} A_k(x)\right) + B_n(x), \tag{4.4.10}$$

where, inductively,

$$|A_k(x)| \leq [Ce^{-\varepsilon x}/(1 - \eta)]^k \tag{4.4.11}$$

and

$$B_n(x) = \sum_{j=0}^{\infty} \int_0^\infty B_{n-1}(x + u)S(x + u)\, d\sigma^j(u), \qquad n \in \mathbb{N};$$

$$B_0 = A. \tag{4.4.12}$$

It is clear from (4.4.11) that, for a suitable $x_0 \geq 0$, $\sum_{k=1}^{\infty} A_k(x)$ converges for all $x \geq x_0$, and then from (4.4.10) that $\lim_{n\to\infty} B_n(x)$ exists for such $x$. We show in what follows that this limit is zero a.e. $[\omega]$.

Recalling that $B_0$ denotes $A$, and the estimate (4.4.8), we have (here, we make no explicit use of $\alpha = 0$, $\tau = 2\beta$)

$$\int_x^\infty \int_y^\infty e^{-2\beta z}|B_1(z)|\, dz\, dy$$

$$\leq C \sum \int_0^\infty e^{-\varepsilon(x+u)} \left\{\int_x^\infty \int_y^\infty e^{-2\beta z}|B_0(z + u)|\, dz\, dy\right\} d\sigma^j(u)$$

$$\leq C \sum \int_0^\infty e^{-\varepsilon(x+u)+\beta u} \left\{\int_{x+u}^\infty \int_y^\infty e^{-2\beta z}|B_0(z)|\, dz\, dy\right\} d\sigma^j(u)$$

$$\leq C^2 C_1 \sum \int_0^\infty e^{-\varepsilon(x+u)+\beta u-(\tau+\varepsilon/2)(x+u)}\, d\sigma^j(u)/(1 - \delta)$$

$$\leq C^2 C_1 e^{-(\tau+3\varepsilon/2)x} \sum \left[\int_0^\infty e^{-(\gamma+3\varepsilon/2)u}\, d\sigma(u)\right]^j \Big/ (1 - \delta)$$

$$\leq CC_1[e^{-(\tau+\varepsilon/2)x}/(1 - \delta)]\left(\frac{Ce^{-\varepsilon x}}{1 - \eta}\right).$$

Arguing similarly and inductively, using (4.4.12), we have

$$\int_x^\infty \int_y^\infty e^{-2\beta z}|B_n(z)|\,dz\,dy \le CC_1(e^{-(\tau+\varepsilon/2)x}/(1-\delta))\left(\frac{Ce^{-\varepsilon x}}{1-\eta}\right)^n, \qquad n \in \mathbb{N},$$

whence, for $x \ge x_0$, $B_n(\cdot)$ tends to zero in measure, so that its pointwise limit (existing for such $x$) is also zero a.e. $[\omega]$. Hence, we have (4.4.2), with $k(x) = \sum_{j=1}^\infty A_j(x)$ for $x \ge x_0$, the estimate for $k(\cdot)$ given in the statement of the theorem following at once from (4.4.11). Hence, the theorem is proven.

***Remark.*** Recalling that $\sigma\{0\} = 0$, we see that if $x_0$ is taken to satisfy $(Ce^{-\varepsilon x_0})/(1-\eta) \le \frac{1}{2}$, then we may take $C_0 = 2C/(1-\eta)$ in the estimate for $k(\cdot)$ given in the theorem, where $\eta = \int_0^\infty e^{(\alpha-\varepsilon)}\,d\sigma(y)$ $(<1)$. In particular, we may take $x_0 = 0$ itself if $2C \le 1 - \eta$. These remarks may be specialized in the contexts dealt with by Theorems 4.4.2 and 4.4.3; we omit the details there.

We conclude this section with two sample results, which can be obtained using Theorem 4.4.1, concerning the "stability" of the characterizations of certain probability distributions that were derived in Section 2.5. The other results given in Section 2.5 can be dealt with similarly. We have not aimed at formulating the most general conditions possible; in particular, the "lattice cases" have been omitted.

The "error terms" $S(\cdot)$ in what follows will be taken to satisfy the conditions stated in Theorem 4.4.1.

**Theorem 4.4.2.** *Let $X$ and $Y$ be independent nonnegative r.v.'s with $F$ and $G$ as their d.f.'s. Suppose that $F \ne \delta_0$, $G(0) < P\{X > Y\} < 1$, and that $G$ is not a lattice d.f. with the origin as a lattice point. If*

$$P\{X > Y + x \mid X > Y\} = P\{X > x\}(1 - S(x)), \qquad \forall\, x \ge 0,$$

*then, for $\alpha > 0$ defined by $\int_0^\infty e^{-\alpha y}\,dG(y) = P\{X > Y\}$, we have*

$$1 - F(x) = pe^{-\alpha x}(1 + k(x)) \qquad \text{for } x \ge 0,$$

*where $p$ is a constant and $k(x) = O(e^{-\varepsilon x})$ as $x \to \infty$.*

**Theorem 4.4.3.** *Let $X$ be a nonnegative r.v. with $F$ as d.f., and suppose that $F \ne \delta_0$ is not a lattice d.f. with the origin as a lattice point. Suppose further that $E(X^\lambda) < \infty$ for some $\lambda > 0$, and that*

$$E[(X - x)^\lambda \mid X > x] = E(X^\lambda)(1 - S(x)), \qquad \forall\, x \ge 0.$$

*Then, for* $\alpha$ *satisfying the equation*

$$E(X^{\lambda}) = \lambda \int_0^{\infty} e^{-\alpha y} y^{\lambda - 1} \, dy,$$

*the same conclusion as in Theorem* 4.4.2 *holds.*

## NOTES AND REMARKS

The results given here are (suitably amended versions in some cases) from Shimizu (1980), Shimizu and Davies (1981), Gu and Lau (1984), and Ramachandran and Lau (1990).

CHAPTER

# 5

# Independent/Identically Distributed Linear Forms, and the Normal Laws

In Section 3.3, we have already met with a characterization of the normal laws through the hypothesis that $X_1$ and $\sum a_j X_j$ are identically distributed, where $\{X_j\}$ is a sequence of i.i.d.r.v.'s and $\{a_j\}$ is a sequence of real constants. In Sections 5.1–5.4, we shall consider a similar but more general problem, namely, the characterization of normal laws through the hypothesis that two linear forms $\sum a_j X_j$ and $\sum b_j X_j$ are identically distributed, where $\{X_j\}$ is a sequence of i.i.d.r.v.'s, and $\{a_j\}$, $\{b_j\}$ are sequences of real constants. The assumption of identical distribution is equivalent to the functional equation

$$\prod_{j=1}^{\infty} f(a_j t) = \prod_{j=1}^{\infty} f(b_j t), \qquad t \in \mathbb{R},$$

where $f$ is the common ch.f. of the $X_j$. The simplest situation is where the $X_j$ are assumed to have moments of all orders; in this case, the $X_j$ are normally distributed (without any conditions other than that the sequences $\{|a_j|\}$, $\{|b_j|\}$ are not permutations of each other being imposed on the coefficients $a_j, b_j$). This result, due to J. Marcinkiewicz, is given as Theorem 5.1.5. If the moment assumption is dropped, the problem naturally becomes more complicated. The basic results (imposing conditions on the $a_j, b_j$ in order that the only "admissible" distributions be normal), due to Yu. V. Linnik and A. A. Zinger, are presented as Theorems 5.1.8 and 5.4.1, respectively. In the proof of the sufficiency parts of these results, we follow the approach (suitably amended) due to M. Riedel (1985).

In Section 5.5, we consider a characterization of normal laws through the assumption that two linear forms in independent r.v.'s are themselves independent.

## 5.1. IDENTICALLY DISTRIBUTED LINEAR FORMS

We begin with a few auxiliary results.

**Lemma 5.1.1.** *If a sequence $\{g_j\}$ of ch.f.'s converges to a ch.f. $g$ on $\mathbb{R}$, then the convergence is uniform on every compact set.*

*Also, if $I$ is any compact interval around the origin on which $g$ is nonvanishing, and if $\phi_j$, $\phi$ are the (distinguished versions of the) logarithms of $g_j$, $g$, respectively, then $\phi_j$ is defined on $I$ for all large $j$ and $\phi_j \to \phi$ uniformly on $I$.*

**Corollary 5.1.2.** *If $t_j \to t_0 \in \mathbb{R}$, then $g_j(t_j) \to g(t_0)$ in the preceding set-up.*

***Proof.*** For proofs of the first assertion of the lemma and of the corollary (which follows at once from it), see for instance, Loève (1977), Theorem 12.2.C and Corollary 1 thereof. The second assertion of the lemma is almost immediate from the first: Let $|g(t)| \geq c > 0$ for $t \in I$. Then, $|g_j(t)| \geq c/2$ for all large $j$, uniformly for all $t \in I$, $\phi_j$ is defined on $I$ for such $j$, and we have, from the logarithmic series expansion, that

$$|\phi_j(t) - \phi(t)| \leq \sum_{j=1}^{\infty} |g_j(t)/g(t) - 1|^n/n \leq \sum_{j=1}^{\infty} |g_j(t) - g(t)|^n/(nc^n),$$

whence the required assertion.

**Lemma 5.1.3.** *If a ch.f. $g$ is the infinite product of a sequence $\{f_j\}$ of ch.f.'s, then $g$ vanishes at a point on $\mathbb{R}$ if and only if, for some $j$, $f_j$ does so (i.e., a product of nonvanishing ch.f.'s converging on $\mathbb{R}$ to a ch.f. cannot "diverge to 0" at any point).*

***Proof.*** Since $g = \prod f_j$, we may assume through the usual symmetrization ($|g|^2 = \prod |f_j|^2$) that $g$ and the $f_j$ are nonnegative and real-valued. On an interval $I$ around the origin where $g$ does not vanish, we have that

$$g_{m,n}(t) = \prod_{m+1}^{n} f_j(t) \to 1 \quad \text{as } m, n \to \infty, \qquad \text{for } t \in I.$$

By the easily verified inequality $1 - \operatorname{Re} f(2t) \le 4(1 - \operatorname{Re} f(t))$, satisfied by every ch.f. $f$, applied to the ch.f.'s $g_{m,n}$, we see that

$$g_{m,n}(t) \to 1 \quad \text{as } m, n \to \infty, \qquad \text{for } t \in 2I.$$

Repeating the argument, it follows that this convergence holds for all real $t$. Thus, for any fixed $t_0 \in \mathbb{R}$, there exists $N$ such that $|\prod_{m+1}^{n} f_j(t_0) - 1| < \frac{1}{2}$ for $m, n \ge N$, so that, in particular, $\prod_{N+1}^{\infty} f_j(t_0)$ converges to a nonzero limit. If now $t_0$ is an arbitrary zero of $g$, then, from

$$0 = g(t_0) = \prod_{1}^{N} f_j(t_0) \times \prod_{N+1}^{\infty} f_j(t_0),$$

it follows that $\prod_1^N f_j(t_0) = 0$, and some $f_j$ necessarily vanishes at $t_0$. Hence, the lemma is proven.

**Lemma 5.1.4.** *Let $\{X_j\}$ be a sequence of nondegenerate i.i.d.r.v.'s, and $\{a_j\}$ a sequence of real numbers. If $\sum a_j X_j$ converges a.s., then $\sum a_j^2$ converges.*

***Proof.*** This assertion is an immediate consequence of Kolmogorov's three-series theorem (see for instance Chung (1974), pp. 118–119): If $\sum Y_j$ is a series whose summands $Y_j$ are independent r.v.'s, then one of the necessary conditions for $\sum Y_j$ to converge a.s. is that, for every $c > 0$, $\sum \operatorname{var} Y_j^c < \infty$, where $Y_j^c$ is obtained from $Y_j$ by "truncation" at $c$, i.e., $Y_j^c = Y_j$ if $|Y_j| \le c$ and zero otherwise. Fixing $c$ such that $\operatorname{var} X_j^c = \operatorname{var} X_1^c > 0$, the lemma follows at once upon noting that $\operatorname{var}(a_j X_j)^c = a_j^2 \operatorname{var} X_j^c$.

For an analytical proof, we first note that $a_j \to 0$. (This may be seen probabilistically as follows: Since $a_j X_j \to 0$ a.s., and so converges to 0 in probability as well, we have $P\{|a_j X_j| > \varepsilon\} = P\{|a_j X_1| > \varepsilon\} \to 0$ as $j \to \infty$, for every $\varepsilon > 0$. If there is a subsequence of $\{a_j\}$ that does not converge to zero, it would follow that $X_1 = 0$ a.s., contrary to hypothesis.) Since $a_j X_j \to 0$ a.s., we have $f(a_j t) \to 1$ as $j \to \infty$, for every $t \in \mathbb{R}$, where $f$ is the ch.f. of the $X_j$. If, for some subsequence $\{j'\}$ of the positive integers, $a_{j'} \to \alpha \ne 0$, then by Corollary 5.1.2 with $g_j = f(a_j \cdot)$ and $g \equiv 1$, we have, for every $t \in \mathbb{R}$ (since $t/a_{j'} \to t/\alpha$),

$$f(t) = g_{j'}(t/a_{j'}) \to 1 \qquad \text{as } j' \to \infty,$$

so that $f \equiv 1$ and $X_1 = 0$ a.s. The same contradiction is arrived at if it is assumed that $a_{j'} \to \pm\infty$ for some subsequence $\{j'\}$. Hence, $a_j \to 0$ as $j \to \infty$. If $c = \max |a_j|$, then by considering $\{a_j/c\}$ in place of $\{a_j\}$, we may assume without loss of generality that $|a_j| \le 1$. Let now $G = F * \tilde{F}$, so that $G$ has $|f|^2$ as its ch.f. If $h$ is the ch.f. of $\sum a_j X_j$, then

$$h(t) = \prod_{j=1}^{\infty} f(a_j t), \qquad |h(t)|^2 = \prod_{j=1}^{\infty} g(|a_j| t).$$

From the elementary inequality $\sin \theta/\theta \geq 2/\pi$ for $0 \leq \theta \leq \pi/2$,

$$\begin{aligned} g(t_0) &= 1 - 2 \int \sin^2(t_0 x/2)\, dG(x) \\ &\leq \exp\left\{-2 \int \sin^2(t_0 x/2)\, dG(x)\right\} \\ &\leq \exp\left\{-2 \int_{|x| \leq \pi/t_0} (t_0 x/\pi)^2\, dG(x)\right\}. \end{aligned}$$

We have, on choosing $t_0$ such that $h(t_0) \neq 0$, $\int_{|x| \leq \pi/t_0} x^2\, dG(x) = A > 0$, and, noting that $|a_j| \leq 1$ for all $j$,

$$|h(t_0)|^2 \leq \exp\{-2t_0^2/\pi^2(\textstyle\sum a_j^2)A\}$$

whence the convergence of $\sum a_j^2$.

Proceeding to the main results, let $\{X_j\}$ be a sequence of i.i.d.r.v.'s, with $X_1 \sim N(\mu, \sigma^2)$, and let $\{a_j\}$ be a real sequence. If $\sum a_j X_j$ converges a.s., it follows (from Lemma 5.1.4 also) that $\sum a_j^2 < \infty$; and $\sum a_j$ converges also if $\mu \neq 0$ (note that if for $\alpha_j$ real, $\{\exp(i\alpha_j t)\}$ converges to a limit as $j \to \infty$, for every $t \in I$, an interval around the origin, then $\{\alpha_j\}$ converges). Furthermore, $\sum a_j X_j$ has an $N(0, (\sum a_j^2)\sigma^2)$ distribution if $\mu = 0$ and an $N(\mu(\sum a_j), (\sum a_j^2)\sigma^2)$ distribution if $\mu \neq 0$.

If $\sum a_j X_j$ and $\sum b_j X_j$ converge a.s., for two real sequences $\{a_j\}$ and $\{b_j\}$, and are identically distributed, with the $X_j$ as before, then we conclude that $\sum a_j^2 = \sum b_j^2$ and that, if $\mu \neq 0$, then both $\sum a_j$ and $\sum b_j$ converge and are equal.

We are concerned below with the converse: If $\sum a_j X_j$ and $\sum b_j X_j$ are identically distributed, where the $X_j$ are i.i.d.r.v.'s and the $a_j$, $b_j$ are real numbers, under what conditions on these coefficients can we assert that $X_1$ is necessarily normally distributed? We begin with a fundamental result (due to J. Marcinkiewicz) on this question.

**Theorem 5.1.5.** *Let $\{X_j\}$ be a sequence of i.i.d.r.v.'s, having moments of all orders. Suppose that for two real sequences $\{a_j\}$ and $\{b_j\}$, the two infinite linear forms $L_1 = \sum a_j X_j$ and $L_2 = \sum b_j X_j$ are defined a.s. and are identically distributed. Then, either $\{|a_j|\}$ is a permutation of $\{|b_j|\}$, or the $X_j$'s are normally distributed. (In the latter case, the $a_j$, $b_j$ obey the equality relations stated above.)*

***Proof.*** Let $F$ be the d.f. of the $X_j$, and $f$ the ch.f. of $F$. The identical distribution of $L_1$ and $L_2$ is equivalent to the functional equation

$$\prod f(a_j t) = \prod f(b_j t), \qquad t \in \mathbb{R}. \tag{5.1.1}$$

In view of the Lévy–Cramér theorem (Theorem 1.3.13), we may assume (considering, as usual, $|f|^2$ instead of $f$) that $f$ is nonnegative, real-valued, and even, and that $a_j, b_j > 0$. By Lemma 5.1.4, $\sum a_j^2$ and $\sum b_j^2$ converge, $a_j, b_j \to 0$ as $j \to \infty$, and we may also assume without loss of generality that $\max\{a_j, b_j : j \in \mathbb{N}\} = 1$. Then, choosing a neighborhood $I$ of the origin where $f \geq \frac{1}{2}$, we see that $f(a_j t), f(b_j t) \geq \frac{1}{2}$ also, for all $j$ and $t \in I$. Let $\phi(t) = -\log f(t)$, $t \in I$. Then,

$$\sum \phi(a_j t) = \sum \phi(b_j t), \qquad t \in I. \tag{5.1.2}$$

By Lemma 5.1.1, the convergence of both series is uniform on $I$. We claim that, in view of the assumption that the $X_j$'s have moments of all orders, we may differentiate term by term as often as we please in the above equality to conclude that

$$\sum a_j^n \phi^{(n)}(a_j t) = \sum b_j^n \phi^{(n)}(b_j t), \qquad n \in \mathbb{Z}_+,\ t \in I. \tag{5.1.3}$$

Take $n = 1$ first, and let $\pm t_0$ be the endpoints of $I$. Since $f(t) = \int \cos tx\, dF(x)$, we have, using $f(a_n t) \geq \frac{1}{2}$ for $t \in I$,

$$\begin{aligned} |a_n \phi'(a_n t)| &= a_n |f'(a_n t)/f(a_n t)| \\ &\leq 2a_n \left| \int x \sin(a_n tx)\, dF(x) \right| \\ &\leq 2a_n^2 t_0 \int x^2\, dF(x). \end{aligned}$$

Since $\sum a_n^2$ converges, it follows that, for $n = 1$, the series on the left-hand side and similarly on the right-hand side of (5.1.3) converge (absolutely and) uniformly on $I$. Term-by-term differentiation of both sides of (5.1.2) is therefore valid, and (5.1.3) holds for $n = 1$. For $n \geq 2$, we make use of the form of $\phi^{(n)}$ in the preamble to Theorem 1.3.9 and obtain constants $c_n$—depending only on $t_0$—such that for $t \in I$, $|\phi^{(n)}(a_j t)|, |\phi^{(n)}(b_j t)| \leq c_n$. Since $\sum a_j^n < \infty$, $\sum b_j^n < \infty$ for $n \geq 2$ also, a similar argument holds as for $n = 1$, and (5.1.3) holds for all $n$. Setting $t = 0$ there, we have

$$\phi(^{(n)}(0)(\textstyle\sum a_j^n - \sum b_j^n) = 0, \qquad n \in \mathbb{Z}_+.$$

If $\{a_j\}$ is not a permutation of $\{b_j\}$, it follows from Corollary 1.2.5 that the relation $\sum a_j^n = \sum b_j^n$ can hold only for finitely many $n \in \mathbb{N}$. Hence, $\phi^{(n)}(0) = 0$ for all large $n \in \mathbb{N}$, and $\phi$ is therefore a polynomial (Proposition 1.3.18). Theorem 1.3.16 and the Lévy–Cramér theorem then imply that the $X_j$ are normally distributed.

The above result can be extended as follows:

**Theorem 5.1.6.** *Let $F$ be a nondegenerate d.f. having moments of all orders. Suppose $f$, its ch.f., satisfies, in an interval $I$ around the origin, the relation*

$$\prod f(a_j t)^{\gamma_j} = \prod f(b_j t)^{\delta_j}, \qquad t \in I,$$

*where $\{a_j\}$, $\{b_j\}$ are bounded, and $\{\gamma_j\}$, $\{\delta_j\}$ are nonnegative. Then, either $\{|a_j|\}$ is a permutation of $\{|b_j|\}$, or $f$ is a normal ch.f.*

***Remark.*** If $c = \max\{|a_j|, b_j|: j \in \mathbb{N}\}$, $I$ is taken to be such that $f(ct) \neq 0$ there in order for the above relation to make sense.

***Proof.*** With the notation of the previous remark, we may take $c = 1$ without loss of generality. A proof identical to that for the previous theorem applies, suitably using Corollary 1.2.5.

In what follows, we shall consider two identically distributed linear forms in a sequence of i.i.d.r.v.'s, dropping the assumption that the r.v.'s concerned have moments of all orders. We introduce some notation and conditions: $\{a_j\}$ and $\{b_j\}$ are real sequences such that

$$\{|a_j|\} \text{ is not a permutation of } \{|b_j|\}, \tag{5.1.4a}$$

and

$$\sum (a_j^2 + b_j^2) < \infty. \tag{5.1.4b}$$

We define

$$\rho = \sup\{x \in \mathbb{R}: \sum (|a_j|^{-x} + |b_j|^{-x}) < \infty\} \tag{5.1.4c}$$

and

$$G(z) = \sum |a_j|^{-z} - \sum |b_j|^{-z}, \qquad \text{for } \operatorname{Re} z < \rho,\ z \in \mathbb{C}. \tag{5.1.4d}$$

Then, $G \not\equiv 0$ by (5.1.4a), and $\rho \geq -2$ by (5.1.4b). If only finitely many of the $a_j$, $b_j$ are nonzero, then $\rho = \infty$; otherwise, $\rho \leq 0$. Thus, in all cases,

$$\rho \in [-2, 0] \cup \{\infty\}. \tag{5.1.5}$$

In view of (5.1.5), we define

$$\mathfrak{F}_\rho = \begin{cases} \text{the set of all d.f.'s on } \mathbb{R} & \text{if } \rho = \infty; \\ \text{the set of all d.f.'s on } \mathbb{R} \text{ with moment} & \\ \text{of } \textit{some} \text{ order} > -\rho & \text{if } -2 < \rho \leq 0. \end{cases} \tag{5.1.6}$$

We shall not consider the case $\rho = -2$; our analysis fails then (see the conditions of Lemma 5.2.3).

Since (5.1.4b) implies that $a_j$, $b_j \to 0$ as $j \to \infty$, let

$$c = \max\{|a_j|, |b_j| : j \in \mathbb{N}\}, \tag{5.1.7a}$$

$$l_1 = \#\{j : |a_j| = c\}, \qquad l_2 = \#\{j : |b_j| = c\}. \tag{5.1.7b}$$

Consider the relation (for $x$ real)

$$G(x) = c^{-x}\{(l_1 - l_2) + \textstyle\sum^* (|a_j|/c)^{-x} - \sum^* (|b_j|/c)^{-x}\},$$

where $\sum^*$ runs over those indices $j$ for which $|a_j| < c$, $|b_j| < c$. If $l_1 \neq l_2$, it follows from the above and from (5.1.4b) that $G(x) \neq 0$ for all large negative $x$. If $l_1 = l_2$, then we consider the largest among the $|a_j|$, $|b_j|$ occurring with unequal cardinality in the two forms (use (5.1.4a)) and come to the same conclusion: that $G(x) \neq 0$ for all negatively large $x$. It is therefore meaningful to speak of the smallest negative zero (the negative zero of maximum absolute value) of $G$ if $G$ has negative zeros.

We are now in a position to introduce a set of conditions postulated in our present context by Yu. V. Linnik. Note that, condition (a) being satisfied, it makes sense to speak of the smallest negative zero of $G$, as we do in conditions (b) and (c).

**Definition 5.1.7.** *G is said to satisfy the* Linnik conditions *if*:

(a) $G(-2) = 0$ (*i.e.*, $\sum a_j^2 = \sum b_j^2$);

(b) *all the negative zeros of G, except possibly the smallest, are even integers, and are simple zeros as well*;

(c) *the smallest negative zero $x_0$ of G satisfies one of the following mutually exclusive conditions*:

(i) $x_0 \equiv 0 \pmod 4$ *and is a simple zero*;
(ii) $x_0 \equiv 2 \pmod 4$ *and is at most a double zero*;
(iii) $x_0$ *is not an even integer, it is a simple zero, and* $[|x_0|/2]$ *is odd.*

We then have the following.

**Theorem 5.1.8.** *Let* $L_1 = \sum_{j=1}^{\infty} a_j X_j$ *and* $L_2 = \sum_{j=1}^{\infty} b_j X_j$ *be two linear forms in a sequence* $\{X_j\}$ *of nondegenerate i.i.d.r.v.'s with F as d.f.*, $L_1$, $L_2$ *being assumed to be defined a.s. Suppose* $l_1 \neq l_2$. *Then, in order that the assumption*

(A) $F \in \mathfrak{F}_p$, $L_1$ *and* $L_2$ *are identically distributed,*

*should imply that*

(B) *F is a normal d.f.*,

*it is necessary and sufficient that the Linnik conditions hold.*

We remark that the case where $L_1$ and $L_2$ are both finite linear forms is the subject matter of the classic study of Linnik (1953a, b). It was pointed out in Zinger (1975, 1977) that the condition $l_1 \neq l_2$ is the appropriate condition here, rather than the stringent one that $\max_j |a_j| \neq \max_j |b_j|$ in Linnik's original formulation.

The discussion of the complementary case $l_1 = l_2$ is due to Zinger (1975, 1977). It will be studied in Section 5.4.

## 5.2. PROOF OF THE SUFFICIENCY PART OF LINNIK'S THEOREM

We first consider the basic implications of the assumption that $L_1$ and $L_2$ are identically distributed. As in (5.1.1), we have, under this equidistribution assumption,

$$\prod_{j=1}^{\infty} f(a_j t) = \prod_{j=1}^{\infty} f(b_j t), \qquad t \in \mathbb{R}. \tag{5.2.1}$$

We may assume that (as usual, considering $|f|^2$ in place of $f$ and appealing to the Lévy–Cramér theorem to establish that $f$ is a normal ch.f.) $f \geq 0$ and that $a_j > 0$, $b_j > 0$ for all $j$. By Lemma 5.1.4, $\sum (a_j^2 + b_j^2) < \infty$; $a_j, b_j \to 0$ as $j \to \infty$; and if $c = \max\{a_j, b_j : j \in \mathbb{N}\}$, we may assume without loss of generality that $c = 1$.

We begin by noting that if $l_1 \neq l_2$, then (5.2.1) implies that $f$ is nonvanishing on $\mathbb{R}$. Suppose not, and let $\pm t_0$ be the zeros of $f$ nearest to the origin. Assuming without loss of generality that $l_1 > l_2$, consider the two ch.f.'s obtained by omitting the $l_2$ factors corresponding to $a_j = b_j = 1$, on either side of (5.2.1); these are the ch.f.'s of the r.v.'s obtained by omitting the corresponding $l_2$ terms respectively from $L_1$ and $L_2$. These two ch.f.'s agree on $(-t_0, t_0)$, and so, by continuity, at $\pm t_0$ as well. But the "new" ch.f. on the left vanishes at $t_0$ (since $l_1 > l_2$), while the other does not, by Lemma 5.1.3. Thus, $f$ cannot have zeros on $\mathbb{R}$.

Let $\phi = -\log f$ on $\mathbb{R}$; we have $\phi \geq 0$ and

$$\sum_{j=1}^{\infty} \phi(a_j t) = \sum_{j=1}^{\infty} \phi(b_j t), \qquad t \in \mathbb{R}. \tag{5.2.2}$$

By Lemma 5.1.1, the convergence of both these series is uniform on compact sets.

We now consider the Mellin transform of $\phi$:

$$\hat{\phi}(z; A, B) = \int_A^B t^{z-1} \phi(t)\, dt, \qquad z \in \mathbb{C},$$

where $0 \le A < B \le \infty$. If $0 < A < B < \infty$, then $\hat{\phi}(z; A, B)$ is an entire function of $z$. For $0 < B < \infty$, $\hat{\phi}(z; 0, B)$ is defined and analytic in the strip

$$\operatorname{Re} z > \sigma_\phi := \inf\{x: \hat{\phi}(x; 0, B) < \infty\}.$$

Dually, for $0 < A < \infty$, $\hat{\phi}(z; A, \infty)$ is defined and analytic in the strip

$$\operatorname{Re} z < \gamma_\phi := \sup\{x: \hat{\phi}(x; A, \infty) < \infty\} \qquad \text{if } \gamma_\phi > -\infty.$$

We note that $\sigma_\phi$, $\gamma_\phi$ are, respectively, independent of the choice of $B$, $A$ in $(0, \infty)$.

**Lemma 5.2.1.**

$$\phi(t) = O(t^r) \text{ as } t \to \infty, \qquad \text{for some } r > 0. \tag{5.2.3}$$

***Proof.*** Let $\rho$ be defined as in (5.1.4c).

***Case 1.*** $\rho = \infty$. In this case, both $L_1$ and $L_2$ are finite linear forms (since $G(0)$ exists finitely). Since $\phi$ is $\ge 0$, (5.2.2) implies that

$$\phi(t) \le (l_1 - l_2)\phi(t) \le \textstyle\sum^* \phi(b_j t) \qquad \text{for } t \ge 0,$$

where $\sum^*$ runs over only those $j$ for which $b_j < 1$; since $\phi$ is continuous, we have

$$\phi(t) \le n\phi(tb(t)),$$

for some $0 \le b(t) \le b = \max\{b_j: b_j < 1\}$, and $n$ is the number of terms in $\sum^*$. If $B = 1/b$ and $C = \max\{\phi(t): 0 \le t \le 1\}$, then, for $k \in \mathbb{Z}_+$,

$$\max\{\phi(t): B^{k-1} \le t \le B^k\} \le \max\{\phi(t): 0 \le t \le B^k\} \le Cn^k,$$

so that, for $t \ge 1$,

$$(0 \le)\phi(t) \le Cn^{1 + (\log t/\log B)} = Cnt^{\log n/\log B},$$

whence the lemma by taking any $r > \log n/\log B$.

***Case 2.*** $\rho \in (-2, 0]$ (recall that we do not consider the case $\rho = -2$). In this case, $F \in \mathfrak{F}_\rho$ is equivalent to $\sigma_\phi < \rho$, so that, for any $u \in (\sigma_\phi, \rho)$, $\sum (a_j^{-u} + b_j^{-u}) < \infty$ and $\int_0^1 t^{u-1}\phi(t)\,dt < \infty$. Writing $\lambda = -u$, we have $0 < \lambda < 2$, and it follows from Theorems 1.3.5 and 1.3.6 that $\phi(t)/t^\lambda \to 0$ as $t \to 0+$. Writing $\phi_1(t) = \phi(t)/t^\lambda$ for $t > 0$, we therefore see that $\phi_1$ is bounded in $(0, 1]$ and

$$\sum_{j=1}^{\infty} a_j^\lambda \phi_1(a_j t) = \sum_{j=1}^{\infty} b_j^\lambda \phi_1(b_j t).$$

It follows that

$$\phi_1(t) \le (l_1 - l_2)\phi_1(t) \le \left(\sum_{j=1}^{\infty} b_j^{\lambda}\right) \max\{\phi_1(x): 0 < x \le bt\}.$$

As in Case 1, we conclude that $\phi_1(t) = O(t^s)$ as $t \to \infty$, for some $s > 0$, and the assertion of the lemma follows.

**Lemma 5.2.2.** *If $\phi \ge 0$ is continuous, and satisfies* (5.2.2), *then $\sigma_\phi \le 0$ and $\gamma_\phi > -\infty$.*

***Proof.*** The fact that $\phi$ is bounded on $[0, B]$ yields the first assertion, and Lemma 5.2.1 yields the second.

Recalling that $G$ is analytic on $\operatorname{Re} z < \rho$, we have the following fundamental factorization result.

**Lemma 5.2.3.** *Suppose $\phi$ ($\ge 0$) satisfies* (5.2.2) *and that $\sigma_\phi < \rho$. For $D > 0$, let*

$$K_\phi(z; D) = \sum a_j^{-z}\hat{\phi}(z; a_j D, D) - \sum b_j^{-z}\hat{\phi}(z; b_j D, D).$$

*Then, for any $u$ with $\sigma_\phi < u < \rho$, $K_\phi(z; D)$ is defined for* $\operatorname{Re} z \le u$, *and*

$$G(z)\hat{\phi}(z; 0, 1) = K_\phi(z; 1) \qquad \textit{for } \operatorname{Re} z = u, \tag{5.2.4a}$$

$$G(z)\hat{\phi}(z; 1, \infty) = -K_\phi(z; 1) \qquad \textit{for } \operatorname{Re} z < \min(u, \gamma_\phi). \tag{5.2.4b}$$

***Proof.*** We recall the assumption that $c = 1$, where $c$ is defined by (5.1.7a). Let $0 < A < B$; for $0 < a < 1$, a simple computation gives

$$\int_A^B t^{z-1}\phi(at)\,dt = a^{-z}\{\hat{\phi}(z; A, B) + \hat{\phi}(z; aA, A) - \hat{\phi}(z; aB, B)\}. \tag{5.2.5}$$

Since the convergence of $\sum \phi(a_j t)$ is uniform on compact sets (Lemma 5.1.1), we may multiply the series by $t^{z-1}$, integrate term by term over $[A, B]$, and invoke (5.2.5) to get

$$G(z)\hat{\phi}(z; A, B) + K_\phi(z; A) - K_\phi(z; B) = 0 \qquad \text{for } \operatorname{Re} z \le u. \tag{5.2.6}$$

Now, by assumption, $\hat{\phi}(u; 0, A)$ and $G(u)$ exist finitely for some $u > -2$; it follows that

$$|K_\phi(z; A)| \le \sum (a_j^{-u} + b_j^{-u})\hat{\phi}(u; 0, A)A^{\operatorname{Re} z - u} \qquad \text{for } \operatorname{Re} z \le u. \tag{5.2.7}$$

Equation (5.2.4a) follows from (5.2.6) and (5.2.7) upon letting $A \to 0$. To prove (5.2.4b), we need only show that, with $u$ as before,

$$K_\phi(z; B) \to 0 \quad \text{as } B \to \infty, \qquad \text{if } \operatorname{Re} z < \min(u, \gamma_\phi), \tag{5.2.8}$$

and apply (5.2.6). It obviously suffices to show that, for $\operatorname{Re} z < \min(u, \gamma_\phi)$,

$$L(z; B) = \sum a_j^{-z}\hat{\phi}(z; a_j B, B) \to 0 \qquad \text{as } B \to \infty. \tag{5.2.9}$$

Let $\varepsilon > 0$ be given. For $x = \operatorname{Re} z < \min(u, \gamma_\phi)$, fix $B_0$ such that $\hat{\phi}(x, B_0, \infty) < \varepsilon$, and consider $B > B_0$. Let $\sum_{\mathrm{I}}$ and $\sum_{\mathrm{II}}$ denote summation over $j$ such that $a_j B > B_0$ and $a_j B \le B_0$, respectively. Let $S_{\mathrm{I}}$, $S_{\mathrm{II}}$ denote the sums of the terms $a_j^{-z}\hat{\phi}(z; a_j B, B)$ corresponding respectively to these ranges of values of $j$. Then, noting that, for $aB \le B_0$, $\hat{\phi}(x; aB, B_0) \le (aB)^{x-u}\hat{\phi}(u; 0, B_0)$ since $x < u$, we have

$$\begin{aligned}
|S_{\mathrm{I}}| &\le \varepsilon(\textstyle\sum_{\mathrm{I}} a_j^{-x}) \le \varepsilon(\sum a_j^{-x}),\\
|S_{\mathrm{II}}| &\le \textstyle\sum_{\mathrm{II}} a_j^{-x}\hat{\phi}(x; a_j B, B_0) + \sum_{\mathrm{II}} a_j^{-x}\hat{\phi}(x; B_0, B)\\
&\le B^{x-u} \textstyle\sum_{\mathrm{II}} a_j^{-u}\hat{\phi}(u; 0, B_0) + \sum_{\mathrm{II}} a_j^{-x}\hat{\phi}(x; B_0, \infty)\\
&\le \varepsilon(\textstyle\sum a_j^{-u} + \sum a_j^{-x})
\end{aligned}$$

for all large $B$. Thus,

$$|L(z, B)| \le \varepsilon(2 \textstyle\sum a_j^{-x} + \sum a_j^{-u})$$

for all large $B$, and (5.2.9), and therefore (5.2.8), have been proven.

**Corollary 5.2.4.** *Under the assumptions of Lemma* 5.2.3, $\hat{\phi}(z; 0, 1)$ *is meromorphic on* $\mathbb{C}$, *and* $\sigma_\phi < 0$.

***Proof.*** Since $\phi$ satisfies (5.2.2), and since $G(u)$ and $\hat{\phi}(u; 0, 1)$ exist finitely for some $u > -2$, it follows that (a) $\hat{\phi}(z; 0, 1)$ is analytic on $\operatorname{Re} z > u$ and continuous on its closure, while (b) $G(z)$ and $K_\phi(z; 1)$ are analytic on $\operatorname{Re} z < u$ and continuous on its closure. Equation (5.2.4a) and the Schwarz reflection principle (Theorem 1.2.9) then imply that $\hat{\phi}(z; 0, 1)$ is meromorphic on $\mathbb{C}$.

Also, since $\phi$ is continuous and $\phi(0) = 0$, we have

$$x\hat{\phi}(x; 0, 1) = \int_0^\infty e^{-t}\phi(e^{-t/x})\, dt \to 0 \qquad \text{as } x \to 0+,$$

by the dominated convergence theorem. Thus, the origin cannot be a pole for $\hat{\phi}(z; 0, 1)$. But, since $\phi \ge 0$, $\sigma_\phi$ must be a singularity—hence necessarily a pole—for the function, by Theorem 1.2.7 (with a change of variable). Hence, $\sigma_\phi \ne 0$, and so $<0$, by Lemma 5.2.2.

We now proceed to the proof of the main result (Theorem 5.1.8: sufficiency of the Linnik conditions for (A) to imply (B)). The assumption that $F \in \mathfrak{F}_\rho$ is equivalent to $\sigma_\phi < \rho$. Lemma 5.2.3 and Corollary 5.2.4 are therefore applicable. Since $\sigma_\phi < 0$, and since $G$ has no zeros in the interval $(-2, 0)$ under the Linnik conditions, we have

$$\sigma_\phi \le -2. \tag{5.2.10}$$

The arguments that follow (proof of Lemma 5.2.5, (i), case (b)), will show in particular that $\sigma_\phi = -2$ (and that $F$ has finite second moment).

Recall that $\{a_j\}$ and $\{b_j\}$ have been assumed positive without loss of generality, and are not permutations of each other. By Theorem 5.1.5, we need only show that the $X_j$ have moments of all orders. Suppose that, for some integer $k \in \mathbb{Z}_+$, the $X_j$ have the moment of order $2k$ but not of order $2k + 2$. For the case $k = 0$, the following arguments down to (5.2.15) are irrelevant, and we may proceed straight to (5.2.15)—which is then the same as (5.2.2). For $k \ge 1$, we proceed as follows: $\phi^{(2k)}(t)$ exists for all real $t$, and a finite Taylor expansion for the even function $\phi$ is given (as readily seen by integrating by parts repeatedly in (5.2.12)) by

$$\phi(t) = \sum_{m=1}^{k} d_m t^{2m} + (-1)^k \phi_k(t), \tag{5.2.11}$$

where the $d_m$ are real constants, and

$$\phi_k(t) = \frac{(-1)^k}{(2k-1)!} \int_0^t (t - x)^{2k-1} \{\phi^{(2k)}(x) - \phi^{(2k)}(0)\}\, dx. \tag{5.2.12}$$

Theorem 1.3.10 then implies that $\phi_k(t) > 0$ for all $t$ in a suitable neighborhood of the origin, if the $(2k + 2)$th moment of $F$ does not exist finitely. We may assume without loss of generality that this neighborhood contains $[-1, 1]$. Equations (5.2.2) and (5.2.11) imply that, for $t \in \mathbb{R}$,

$$\{\sum \phi_k(a_j t) - \sum \phi_k(b_j t)\} + (-1)^k \sum_{m=1}^{k} d_m G(-2m) t^{2m} = 0. \tag{5.2.13}$$

By a well-known property of the Taylor expansion for ch.f.'s of d.f.'s with finite moment of order $2k$, the remainder term in (5.2.11) is $o(t^{2k})$, i.e.,

$$t^{-2k} \phi_k(t) \to 0 \qquad \text{as } t \to 0. \tag{5.2.14}$$

Hence, the expression in the curly brackets in the left-hand side of (5.2.13) is $o(t^{2k})$ as $t \to 0$, so that we must have $d_m G(-2m) = 0$ for $m = 1, \ldots, k$. Thus, $\phi_k$ also satisfies (5.2.2), i.e.,

$$\sum \phi_k(a_j t) = \sum \phi_k(b_j t), \qquad t \in \mathbb{R}. \tag{5.2.15}$$

As remarked earlier, to discuss the case $k = 0$, we start from this stage, with $\phi_0 = \phi$—in other words, from (5.2.2) itself—and proceed directly to the proof of Lemma 5.2.5(i), case (b), to conclude that equality holds in

(5.2.10). For $k \geq 1$, we first note that, unlike $\phi$, $\phi_k$ cannot be claimed to be nonnegative throughout $\mathbb{R}$. However, its definition (5.2.11) implies:

(i) that the convergence of $\sum \phi_k(a_j t)$ is uniform on compact sets since $\sum \phi(a_j t)$ has that property; similarly for $\sum \phi_k(b_j t)$.

(ii) that an analog of Lemma 5.2.1 holds for $\phi_k$ also, namely,

$$\phi_k(t) = O(t^{r_k}) \quad \text{as } t \to \infty, \qquad \text{for } r_k = \max(r, 2k), \tag{5.2.16}$$

with $r$ as in Lemma 5.2.1; and

(iii) $K_{\phi_k}(z; 1)$—denoted in the following by $K_k(z)$—is defined for $\operatorname{Re} z \leq u$, for all $u \in (\sigma_\phi, \rho)$, taking (5.2.7) into account.

We may now introduce $\hat{\phi}_k(z; A, B)$ for $0 \leq A < B \leq \infty$, $\sigma_{\phi_k}$, and $\gamma_{\phi_k}$—to be denoted by $\sigma_k$ and $\gamma_k$, respectively—in an obvious manner, with (5.2.16) ensuring that $\gamma_k > -\infty$. Equation (5.2.11) also implies that $\hat{\phi}_k(u; 0, 1)$ exists finitely for $u \in (\sigma_\phi, \rho)$, since $\hat{\phi}(u; 0, 1)$ does. Then, in view of (i), (ii), and (5.2.15), the analogs of (5.2.4a, b) apply, with $\phi_k$ in place of $\phi$—and for the same $u$'s, namely, $u \in (\sigma_\phi, \rho)$. It follows that $\hat{\phi}_k(z; 0, 1)$, $K_k(z)/G(z)$, and $-\hat{\phi}_k(z; 1, \infty)$ represent in various half-planes the same meromorphic function on $\mathbb{C}$. We also obtain, from (5.2.11) directly, explicit relations between the meromorphic functions $\hat{\phi}(z; 0, 1)$ and $\hat{\phi}_k(z; 0, 1)$, equivalently, between $\hat{\phi}(z; 1, \infty)$ and $\hat{\phi}_k(z; 1, \infty)$:

$$\hat{\phi}(z; 0, 1) = (-1)^k \hat{\phi}_k(z; 0, 1) + \sum_{m=1}^{k} d_m/(z + 2m), \tag{5.2.17a}$$

$$\hat{\phi}(z; 1, \infty) = (-1)^k \hat{\phi}_k(z; 1, \infty) - \sum_{m=1}^{k} d_m/(z + 2m). \tag{5.2.17b}$$

It follows from (5.2.14) that

$$\sigma_k \leq -2k. \tag{5.2.18}$$

We then have the following.

**Lemma 5.2.5.** *If the moment of order $2k$ exists but not that of order $2k + 2$, then, for $\phi_k$ as defined:*

(i) $\sigma_k = -(2k + 2)$;

(ii) *$\sigma_k$ is a simple pole for $\hat{\phi}_k(z; 0, 1)$.*

***Proof.*** (i) In view of (5.2.18), we only have to rule out the three other possibilities: (a) $\sigma_k = -2k$; (b) $\sigma_k < -(2k + 2)$; (c) $-(2k + 2) < \sigma_k < -2k$.

(a) Suppose that $\sigma_k = -2k$; then, as argued in the proof of Corollary 5.2.4, we have, for $x \in \mathbb{R}$,

$$(x - \sigma_k)\hat{\phi}_k(x; 0, 1) = (x - \sigma_k)\int_0^\infty e^{-ux}\phi_k(e^{-u})\,du \to 0 \qquad \text{as } x \to \sigma_k+$$

in view of (5.2.14). On the other hand, $\phi_k \geq 0$ on $[0, 1]$ implies that $\hat{\phi}_k(z; 0, 1)$ should have a singularity at $\sigma_k$ (Theorem 1.2.8), which is necessarily a simple pole. This implies that $\lim_{z \to \sigma_k}(z - \sigma_k)\hat{\phi}_k(z; 0, 1)$ exists and is nonzero, a contradiction.

(b) Suppose next that $\sigma_k < -(2k + 2)$, so that

$$\int_0^1 \phi_k(t)t^{-(2k+3+\delta)}\,dt < \infty \qquad \text{for some } \delta > 0.$$

This implies that $\liminf_{t \to 0} \phi_k(t)/t^{2k+2} = 0$ (for otherwise, the integral cannot exist finitely since $\int_0^a t^{-(1+\delta)}\,dt = \infty$ for every $a > 0$). Hence, there exists a null sequence $\{t_n\}$ such that $\phi_k(t_n)/t_n^{2k+2} \to 0$ as $n \to \infty$. The formula (5.2.12) for $\phi_k(t)$ and the mean value theorem then imply that there exist $u_n \in (0, t_n)$ such that $\{\phi^{(2k)}(u_n) - \phi^{(2k)}(0)\}/u_n^2 \to 0$ as $n \to \infty$. This in turn means that the $X_j$ have moment of order $2k + 2$ (by Theorem 1.3.10), and we have a contradiction to our assumption.

(c) Suppose finally that $-(2k + 2) < \sigma_k < -2k$; $k = 0$ is then ruled out, by (5.2.10). If we fix $u \in (-2, \rho)$ such that the relations (5.2.4a, b) hold, then $\sigma_k < u$, and then analogs of the relations (5.2.4a, b) hold for $\phi_k$. Now, since $\hat{\phi}_k(z; 0, 1)$ is analytic on $\operatorname{Re} z > \sigma_k$, while $K_k(z)$ and $G(z)$ are analytic on $\operatorname{Re} z < u$ and continuous on $\operatorname{Re} z \leq u$, and since the analog of (5.2.4a) holds for $\phi_k$, we have

$$G(z)\hat{\phi}_k(z; 0, 1) = K_k(z) \qquad \text{for } u \geq \operatorname{Re} z > \sigma_k \tag{5.2.19a}$$

Also, since $\hat{\phi}_k(z; 0, 1)$ necessarily has a pole at $\sigma_k$ (since $\phi_k \geq 0$ on $[0, 1]$), we then see that $G(z)$ must have a zero at $\sigma_k$. Then, under the Linnik conditions, $\sigma_k$ must be a simple zero and the smallest negative zero of $G(z)$, and, further, $k = [|\sigma_k|/2]$ must be odd. In particular, $K_k(z)/G(z)$ is analytic on $\operatorname{Re} z < \sigma_k$. Then, the analog of (5.2.4b) for $\phi_k$ implies that $\hat{\phi}_k(z; 1, \infty)$ is also analytic there, and we have

$$G(z)\hat{\phi}_k(z; 1, \infty) = -K_k(z) \qquad \text{for } \operatorname{Re} z < \sigma_k. \tag{5.2.19b}$$

Since the zero of $G$ at $\sigma_k$ is simple by assumption, $G'(\sigma_k) \neq 0$ and we have, from the relations (5.2.19), that

$$\begin{aligned} 0 \leq \lim_{x \to \sigma_k+}(x - \sigma_k)\hat{\phi}_k(x; 0, 1) &= -\lim_{x \to \sigma_k-}(x - \sigma_k)\hat{\phi}_k(x; 1, \infty) \\ &= K_k(\sigma_k)/G'(\sigma_k). \end{aligned} \tag{5.2.20}$$

But (5.2.17b) implies, in view of $\varphi \geq 0$, that ($k$ being odd)

$$\lim_{x \to \sigma_k -} (x - \sigma_k)\hat{\phi}_k(x; 1, \infty) \geq 0. \tag{5.2.21}$$

Equations (5.2.20) and (5.2.21) imply that $K_k(\sigma_k) = 0$, contradicting the fact that $\hat{\phi}_k(z; 0, 1)$ has a (simple) pole at $\sigma_k$.

Thus, statement (i) of the lemma is proven.

(ii) Suppose that $\sigma_k$ is not a simple pole for $\hat{\phi}_k(z; 0, 1)$. Then, under the Linnik conditions, the zero of $G$ at $\sigma_k$ must be a double zero and also the smallest negative zero; further, $k$ must be even. As before, then, $\hat{\phi}_k(z; 0, 1)$ is analytic on $\operatorname{Re} z > \sigma_k$, and $\hat{\phi}_k(z; 1, \infty)$ on $\operatorname{Re} z < \sigma_k$, and we have from the relations (5.2.19a, b) that

$$\begin{aligned} 0 \leq \lim_{x \to \sigma_k +} (x - \sigma_k)^2 \hat{\phi}_k(x; 0, 1) &= -\lim_{x \to \sigma_k -} (x - \sigma_k)^2 \hat{\phi}_k(x; 1, \infty) \\ &= -K_k(\sigma_k)/G''(\sigma_k). \end{aligned} \tag{5.2.22}$$

On the other hand, (5.2.17b) implies that

$$\lim_{x \to \sigma_k -} (x - \sigma_k)^2 \hat{\phi}_k(x; 1, \infty) \geq 0. \tag{5.2.23}$$

It follows from (5.2.22) and (5.2.23) that $K_k(\sigma_k) = 0$, so that $\hat{\phi}_k(z; 0, 1)$ has a simple pole at $\sigma_k$, a contradiction to our assumption. Hence, (ii) is proven.

We now conclude our proof. By Lemma 5.2.5(ii),

$$\lim_{z \to \sigma_k} (z - \sigma_k)\hat{\phi}_k(z; 0, 1)$$

exists finitely. This implies that $\liminf_{t \to 0} \phi_k(t)/t^{2k+2} < \infty$. (For, if not, there exists, for every $C > 0$, a $u > 0$ such that

$$\phi_k(t)/t^{2k+2} \geq C \qquad \text{for } 0 < t < u.$$

It follows that, for $x \in \mathbb{R}$,

$$\lim_{x \to \sigma_k +} (x - \sigma_k)\hat{\phi}_k(x; 0, 1) \geq \lim_{x \to \sigma_k +} C u^{x - \sigma_k} = C.$$

Since $C > 0$ is arbitrary, we obtain a contradiction.) Applying the mean value theorem to $\phi_k(t)$ as given by the integral formula (5.2.12), we conclude that there exists a null sequence $\{u_n\}$ such that

$$\lim_{n \to \infty} (\phi^{(2k)}(u_n) - \phi^{2k}(0))/u_n^2$$

exists finitely. This implies that $F$ has finite $(2k + 2)$th moment, a contradiction with our assumption.

Thus, $F$ has moments of all orders, Theorem 5.1.5 applies, and the proof of the fact that assumption (A) and the Linnik conditions imply the normality of the $X_j$ is thus completed.

## 5.3. PROOF OF THE NECESSITY PART OF LINNIK'S THEOREM

We first set up the essentials of the proof, and prove the needed auxiliary results later as Lemmas 5.3.1–5.3.3.

It follows from Lemma 5.1.4 that, if the given condition holds, then $\sum a_j^2$, $\sum b_j^2$ converge and are equal, and hence $G(-2) = 0$. In the following, we list all the possibilities and show that if the conditions on $G$ in (b) or (c) do not hold, then there exists a nonnormal ch.f. $f$ for which $\phi = \log f$, which satisfies

$$\sum \phi(|a_j|t) = \sum \phi(|b_j|t), \qquad \forall\, t \in \mathbb{R}.$$

The $\gamma_1$, $\gamma_2$ involved in the following will be taken to be positive.

(1) If $-\gamma_1$ is a zero of $G$ such that $\gamma_1 \equiv 0 \pmod 4$, and is not a simple zero of $G$, then take $\phi(t) = -At^2 - |t|^{\gamma_1} \log |t|$ (Lemma 5.3.3).

This implies conditions (b) and (c)(i) on $G$ must hold.

(2) If $-\gamma_1$ is a zero of $G$ such that $\gamma_1 \equiv 2 \pmod 4$, and has multiplicity $>2$, then take $\phi(t) = -At^2 - |t|^{\gamma_1}(\log |t|)^2$ (Lemma 5.3.3).

(3) If $-\gamma_1$ is a zero of $G$ such that $\gamma_1 \equiv 2 \pmod 4$ and there exists a zero $-\gamma_2$ of $G$ with $-\gamma_2 < -\gamma_1$, then take

$$\phi(t) = -At^2 + |t|^{\gamma_1} \log |t| - |t|^{\gamma_2}$$

(Lemma 5.3.3).

Statements (2) and (3) imply that condition (c)(ii) must hold.

(4) If a zero $-\gamma_1$ of $G$ is not an even integer and if $[\gamma_1/2]$ is even, then take $\phi(t) = -At^2 - |t|^{\gamma_1}$ (Lemma 5.3.1).

(5) If a zero $-\gamma_1$ of $G$ is not an even integer, if $[\gamma_1/2]$ is odd, and if there exists a zero $-\gamma_2$ of $G$ such that $-\gamma_2 < -\gamma_1$, then take $\phi(t) = -At^2 + |t|^{\gamma_1} - |t|^{\gamma_2}$ (Lemma 5.3.2).

(6) If a zero $-\gamma_1$ of $G$ is not an even integer, if $[\gamma_1/2]$ is odd, and if $-\gamma_1$ is not a simple zero, then take $\phi(t) = -At^2 - |t|^{\gamma_1} \log |t|$ (Lemma 5.3.2).

Statements (4), (5), and (6) imply that condition (c)(iii) must hold.

We proceed to prove these three lemmas.

**Lemma 5.3.1.** *If $\gamma_1 > 2$ is not an even integer and if $[\gamma_1/2]$ is even, then $\exp(-At^2 - |t|^{\gamma_1})$ is a ch.f. for all large $A > 0$.*

***Proof.*** Let $f(z) = \exp(-Az^2 - z^{\gamma_1})$, for $z \in \mathbb{C}$, $z \neq 0$, $\operatorname{Re} z \geq 0$. For $x \in \mathbb{R}$, define

$$h(x) = \frac{1}{2\pi}\int_{\mathbb{R}} e^{itx} f(t)\,dt = \frac{1}{\pi}\operatorname{Re}\int_0^\infty e^{itx} f(t)\,dt. \tag{5.3.1}$$

We will show that $h \geq 0$ for sufficiently large $A$; hence, $f(|t|)$ will be a ch.f.

For any fixed $0 < \delta < 1$, let $2\gamma_1 < \lambda < 4\gamma_1/(1 + \delta)$, and let

$$\theta(A) = (\lambda(1 + \delta)A \log A)^{1/2}, \qquad \Lambda(x) = \lambda \log x/x.$$

It is clear that $\Lambda(x)$ is a decreasing function on $(e, \infty)$, and $0 \leq \Lambda(x) < \lambda$ for $x > 1$. We will divide the estimation of the integral in (5.3.1) into two cases: (a) $x \geq \theta(A)$; (b) $0 < x < \theta(A)$, where $A \geq A_0$ and $A_0$ is chosen and fixed in what follows. For convenience, we use $\Lambda$ to denote $\Lambda(x)$; also, we denote the different constants arising in the estimates that follow by one and the same symbol $C$ unless otherwise specified.

(a) $x \geq \theta(A)$. The Cauchy theorem (Theorem 1.2.8) implies that

$$\int_0^\infty e^{itx} f(t)\,dt = \int_{L_1} + \int_{L_2} e^{izx} f(z)\,dz = I_1 + I_2,$$

where $L_1 = \{z = t + i\Lambda : 0 \leq t < \infty\}$, $L_2 = \{z = it : 0 \leq t \leq \Lambda\}$. In view of

$$A(\Lambda(x))^2 \leq A(\Lambda(\theta(A)))^2 \leq \lambda \log A/4,$$

we have

$$\begin{aligned}
|I_1| &\leq e^{-x\Lambda + A\Lambda^2}\int_0^\infty e^{-At^2}|e^{-(t+i\Lambda)^{\gamma_1}}|\,dt \\
&\leq e^{-\lambda\log x + A\Lambda^2}(C/\sqrt{A}) \leq CA^{(\lambda/4)-(1/2)}x^{-\lambda} \\
&\leq Cx^{-(1+\lambda/2)} = o(x^{-(\gamma_1+1)}) \qquad \text{for large } x.
\end{aligned} \tag{5.3.2}$$

The last equality holds since $x \geq \sqrt{A}$. To estimate

$$I_2 = i\int_0^\Lambda e^{-tx + At^2 - (it)^{\gamma_1}}\,dt,$$

let

$$I_3 = i\int_0^\Lambda e^{-tx + At^2}(1 - (it)^{\gamma_1})\,dt;$$

using

$$|e^z - 1 - z| \le |z|^2 e^{|z|},$$

we have (note that $x \ge (2 + \delta)A\Lambda(x)$ for $x \ge \theta(A)$, for all large $A$)

$$|\operatorname{Re} I_2 - \operatorname{Re} I_3| \le C \int_0^{\Lambda} e^{-tx + At^2} t^{2\gamma_1} e^{t^{\gamma_1}}\, dt$$

$$\le C \int_0^{\infty} e^{-tx/2} t^{2\gamma_1}\, dt \le C x^{-(2\gamma_1 + 1)}, \tag{5.3.3}$$

provided that $\Lambda$ is sufficiently small, which is the case if $x > \theta(A)$ and $A$ is large. It follows from the assumption on $\gamma_1$ that

$$\operatorname{Re}(i^{\gamma_1 + 1}) = \cos\frac{\pi}{2}(\gamma_1 + 1) < 0,$$

and hence

$$\operatorname{Re} I_3 = -\operatorname{Re}\left(i^{\gamma_1+1} \int_0^{\Lambda} e^{-tx + At^2} t^{\gamma_1}\, dt\right)$$

$$= C \int_0^{\Lambda} e^{-tx + At^2} t^{\gamma_1}\, dt$$

$$\ge C \int_0^{\Lambda} e^{-tx} t^{\gamma_1}\, dt$$

$$= C\left(\int_0^{\infty} - \int_{\Lambda}^{\infty} e^{-tx} t^{\gamma_1}\, dt\right). \tag{5.3.4}$$

The second integral is

$$\le e^{-\Lambda x/2} \int_0^{\infty} e^{-tx/2} t^{\gamma_1}\, dt \le C x^{-(\gamma_1 + 1 + \lambda/2)};$$

hence, we see that the expression in (5.3.4) is $> Cx^{-(\gamma_1+1)}$ for $x \ge \theta(A)$, for all large $A$. The same estimate holds for $\operatorname{Re} I_2$ in view of (5.3.3). It then follows from (5.3.2) that, for $x \ge \theta(A)$, for large $A$,

$$h(x) = (1/\pi) \operatorname{Re}(I_1 + I_2) \ge 0.$$

(b) $0 < x \le \theta(A)$. Let

$$I_4 = \int_0^{\infty} e^{itx - At^2}\, dt, \qquad I_5 = \int_0^{\infty} e^{itx - At^2}\, t^{\gamma_1}\, dt.$$

In view of $0 \le e^{-x} - 1 + x \le x^2/2$ for $x > 0$, we have

$$\left| \operatorname{Re} \int_0^\infty e^{itx - At^2 - t^{\gamma_1}}\, dt - \operatorname{Re}(I_4 - I_5) \right| \le \int_0^\infty e^{-At^2} t^{2\gamma_1}\, dt$$
$$= CA^{-(\gamma_1 + 1/2)}. \tag{5.3.5}$$

Also note that $\operatorname{Re} I_4 = Ce^{-x^2/4A}$. Setting $\xi = x/2A$, and applying Cauchy's theorem to replace the line of integration in the first of the following integrals by the segment $[0, i\xi]$ of the imaginary axis followed by the half-line $\{t + i\xi, t \ge 0\}$, we have

$$\operatorname{Re} I_5 = e^{-x^2/4A} \operatorname{Re} \int_0^\infty e^{-A(t - i\xi)^2} t^{\gamma_1}\, dt$$
$$= e^{-x^2/4A} \operatorname{Re} \left\{ i^{\gamma_1 + 1} \int_0^\xi e^{A(t-\xi)^2} t^{\gamma_1}\, dt + \int_0^\infty e^{-At^2} (t + i\xi)^{\gamma_1}\, dt \right\}.$$

Denoting the expressions within the curly brackets by $I_6$ and $I_7$, respectively, we have $\operatorname{Re} I_6 < 0$ in view of $\operatorname{Re}(i^{\gamma_1 + 1}) < 0$—under our assumption on $\gamma_1$. Also, in view of

$$(t + \xi)^{\gamma_1} \le C(t^{\gamma_1} + \xi^{\gamma_1}), \qquad \text{for } t \ge 0,$$

we have

$$|\operatorname{Re} I_7| \le \int_0^\infty e^{-At^2} (t^{\gamma_1} + \xi^{\gamma_1})\, dt$$
$$\le \frac{C}{\sqrt{A}} \left( \frac{1}{A^{\gamma_1/2}} + \xi^{\gamma_1} \right) = o(A^{-1/2}),$$

since $\xi^{\gamma_1} = (x/2A)^{\gamma_1} \le C(\log A/A)^{\gamma_1/2} = o(1)$ as $A \to \infty$. Hence, for $A$ large,

$$\operatorname{Re}(I_4 - I_5) \ge Ce^{-x^2/4A}$$
$$= CA^{-\lambda(1+\delta)/4} \ge A^{-\gamma_1}. \tag{5.3.6}$$

Then, (5.3.5) and (5.3.6) imply that $h(x) > 0$ (for such $x$ and $A$).

**Lemma 5.3.2.** *If $\gamma_2 > \gamma_1 > 2$, if $\gamma_1$ is not an even integer, and if $[\gamma_1/2]$ is odd, then there exists $A_0$ such that, for all $A \ge A_0$, the functions*

$$\text{(i)}\ \exp(-At^2 - |t|^{\gamma_1} \log |t|), \qquad \text{(ii)}\ \exp(-At^2 + |t|^{\gamma_1} - |t|^{\gamma_2})$$

*are ch.f.'s.*

***Proof.*** The procedure of proof is the same as for Lemma 5.3.1; we only give a proof of (i). Using the same notation as before, we consider the two cases:

(a) $x \geq \theta(A)$. $I_1$ is estimated as before by (5.3.2). Let $C_1$ be such that

$$|I_1| \leq C_1 x^{-(\gamma_1+1)}.$$

Let $C_2$ be chosen as in (5.3.7) following, and $A_0 > 0$ such that, for $x > \theta(A)$, $A > A_0$,

$$|\log t| > \max(C_2, \pi) \qquad \text{for } 0 < t < \Lambda(x).$$

Let

$$I_3 = i \int_0^{\Lambda} \exp(-tx + At^2)(1 - (it)^{\gamma_1}(\log t + i\pi/2))\, dt.$$

Then, as in (5.3.3),

$$\begin{aligned}
|\mathrm{Re}\, I_2 - \mathrm{Re}\, I_3| &\leq C \int_0^{\Lambda} e^{-tx+At^2} t^{2\gamma_1} e^{t^{\gamma_1}(|\log t| + \pi/2)} (|\log t| + \pi/2)^2\, dt \\
&\leq C \int_0^{\Lambda} e^{-tx/2} t^{2\gamma_1 - \alpha}\, dt \\
&\leq C x^{-(2\gamma_1 + 1 - \alpha)},
\end{aligned}$$

for any fixed $\alpha > 0$. Under the assumptions on $\gamma_1$, $\mathrm{Re}(i^{\gamma_1+1}) > 0$, and as in (5.3.4) we have

$$\begin{aligned}
\mathrm{Re}\, I_3 &= -\mathrm{Re}\left( i^{\gamma_1+1} \int_0^{\Lambda} e^{-tx+At^2} t^{\gamma_1} \left( \log t + i\frac{\pi}{2} \right) dt \right) \\
&\geq C \int_0^{\Lambda} e^{-tx+At^2} t^{\gamma_1} |\log t|\, dt - \frac{\pi}{2} \int_0^{\Lambda} e^{-tx+At^2} t^{\gamma_1}\, dt \\
&\geq \frac{C}{2} \int_0^{\Lambda} e^{-tx+At^2} t^{\gamma_1} |\log t|\, dt \\
&\qquad \text{(by assumption, } |\log t| > \pi \text{ on } (0, \Lambda)) \\
&\geq \frac{CC_2}{2} \int_0^{\Lambda} e^{-tx} t^{\gamma_1}\, dt \qquad (5.3.7) \\
&\geq C_3 x^{-(\gamma_1+1)}.
\end{aligned}$$

(The proof of the last inequality is the same as in the part following (5.3.4).) It follows that, for $x \geq \theta(A)$, $A$ large, $h(x) = (1/\pi)\, \mathrm{Re}(I_1 + I_2) \geq 0$.

(b) $0 \leq x \leq \theta(A)$. Let

$$I_4 = \int_0^{\infty} e^{itx - At^2}\, dt, \qquad I_5 = \int_0^{\infty} e^{itx - At^2} t^{\gamma_1} \log t\, dt.$$

Then, for any fixed $\alpha > 0$,

$$\left|\operatorname{Re}\int_0^\infty e^{itx-At^2-t^{\gamma_1}\log t}\,dt - \operatorname{Re}(I_4 - I_5)\right|$$

$$\le C\int_0^\infty e^{-At^2}t^{2\gamma_1}(\log t)^2\,dt$$

$$\le C\left(\int_0^1 e^{-At^2}t^{2\gamma_1-\alpha}\,dt + \int_0^\infty e^{-At^2}t^{2\gamma_1+\alpha}\,dt\right)$$

$$\le C(A^{-(\gamma_1+(1-\alpha)/2)} + A^{-(\gamma_1+(1+\alpha)/2)})$$

$$\le CA^{-(\gamma_1+1/2)+\alpha/2}.$$

The estimate for $\operatorname{Re} I_4$ is the same as in the last lemma. Set $\xi = x/2A$. Since $0 < x \le \theta(A)$, we can assume without loss of generality that $\xi < e^{-\pi/2}$. By the Cauchy theorem,

$$\operatorname{Re} I_5 = e^{-x^2/4A}\left\{\operatorname{Re}\left(i^{\gamma_1+1}\int_0^\xi e^{A(t-\xi)^2}t^{\gamma_1}\left(\log t + i\frac{\pi}{2}\right)dt\right)\right.$$

$$\left. + \operatorname{Re}\int_0^\infty e^{-At^2}(t+i\xi)^{\gamma_1}\log(t+i\xi)\,dt\right\}.$$

By the conditions on $\gamma_1$, $\operatorname{Re}(i^{\gamma_1+1}) > 0$. Also, note that on $(0, \xi)$, $\log t < 0$, and $|\log t| > |\log \xi| > \pi/2$. The first term inside the bracket is hence negative. The absolute value of the second integral is less than

$$C_\alpha\left(\int_0^\infty e^{-At^2}(t^{\gamma_1-\alpha} + \xi^{\gamma_1-\alpha})\,dt + \int_0^\infty e^{-At^2}(t^{\gamma_1+\alpha} + \xi^{\gamma_1+\alpha})\,dt\right),$$

for any $\alpha > 0$. By applying the same argument as in the last part of the proof of Lemma 5.3.1, we conclude that $h(x) \ge 0$ for $0 \le x < \theta(A)$.

The proof of the following lemma is essentially the same, and hence is omitted.

**Lemma 5.3.3.** *Let* $2 < \gamma_1 < \gamma_2$, *and* $\gamma_1$ *be an even integer. Then, for sufficiently large* $A > 0$, *we have*

(i) $\exp(-At^2 - |t|^{\gamma_1}\log|t|)$ *is a ch.f. if* $\gamma_1 \equiv 0 \pmod 4$,
(ii) $\exp(-At^2 - |t|^{\gamma_1}(\log|t|)^2)$ *is a ch.f. if* $\gamma_1 \equiv 2 \pmod 4$,
(iii) $\exp(-At^2 + |t|^{\gamma_1}\log|t| - |t|^{\gamma_2})$ *is a ch.f. if* $\gamma_2 \equiv 2 \pmod 4$.

## 5.4. ZINGER'S THEOREM

In this section, we consider Zinger's result concerning the case complementary to that studied by Linnik, namely, the case $l_1 = l_2$ in the set-up of Theorem 5.1.8. Recalling the definitions (5.1.4)–(5.1.7), we have the following.

**Theorem 5.4.1.** *Let* $L_1 = \sum a_j X_j$ *and* $L_2 = \sum b_j X_j$ *be two linear forms in the sequence* $\{X_j\}$ *of nondegenerate i.i.d.r.v.'s, with F as common d.f., both forms being taken to converge a.s. Suppose* $l_1 = l_2$. *Then, in order that the assumption*

(A) $F \in \mathfrak{F}_p$, $L_1$ *and* $L_2$ *are identically distributed,*

*should imply that*

(B) *F is a normal d.f.,*

*it is necessary and sufficient that the following hold:*

(i) $G(-2) = 0$ (*i.e.*, $\sum a_j^2 = \sum b_j^2$);
(ii) *all the negative zeros of G are even integers and are simple.*

***Proof.*** Sufficiency. Suppose (A), (i), and (ii) hold. To prove (B), we may as usual assume (in view of the Lévy–Cramér theorem) that $f$, the ch.f. of $F$, is nonnegative real-valued, and even, and that $a_j, b_j > 0$ for all $j$. Also, as before, (A) implies that $\sum (a_j^2 + b_j^2) < \infty$, and so in particular $a_j, b_j \to 0$ as $j \to \infty$. We may then assume without loss of generality (noting that $l_1 = l_2$) that $\max_j a_j = \max_j b_j = 1$. Let $I$ be an interval around the origin where $f$ is nonvanishing, and let $\phi = -\log f$ there; again, assume without loss of generality that $[-1, 1] \subset I$. By Lemma 5.1.1, the convergence of $\sum a_j X_j$ and $\sum b_j X_j$ a.s. then implies the uniform convergence on $I$ of the series $\sum \phi(a_j t)$ and $\sum \phi(b_j t)$, and (A) implies their equality on $I$. We may then proceed as in Section 5.2, noticing the following points of difference. We do not have or need an analog of Lemma 5.2.1. We do not have to consider $\gamma_\phi$ and $\hat{\phi}(z; 1, \infty)$. The analog of (5.2.4a) holds, but (5.2.4b) has no analog here. In the argument by contradiction that follows—where it is assumed that $F$ has moment of order $2k$ but not of order $2k + 2$, for some $k \in \mathbb{Z}_+$—(5.2.15) and (5.2.17a) hold. As for Lemma 5.2.5(i), assumption (ii) of the theorem rules out the possibility (c), and we need only rule out as before the possibilities (a) and (b). Assumption (ii) also rules out the necessity for proving Part (ii) of this lemma. The rest of the proof applies and the sufficiency part of the theorem is established.

Necessity. (i) is immediate if (A) and (B) hold, as pointed out immediately before Theorem 5.1.5. That condition (ii) is necessary follows almost at once from the following lemma, which we shall prove subsequently.

**Lemma 5.4.2.** *Let $Q(\cdot)$ be a real polynomial of degree $\leq m$ with $Q(0) = 0$. Let $0 < \delta \leq 2$, $\alpha > 0$, and $\beta > 1$ be given constants. Then, there exist positive $A$, $t_0$ and a ch.f. $f$ such that*

$$f(t) = \begin{cases} \exp\{-At^2 + Q(t^2) + (-1)^{m-1}t^{2m}w(t, \delta)\} & \text{for } |t| \leq t_0, \\ 0 & \text{for } |t| \geq \beta t_0, \end{cases}$$

*where*

$$w(t, \delta) = \begin{cases} |t|^\delta & \text{if } 0 < \delta < 2, \\ -t^2 \log |t| & \text{if } \delta = 2. \end{cases}$$

Assuming the truth of this lemma, suppose that $G$ has a negative zero $-\gamma$, where $\gamma$ either is not an even integer or is an even integer with the order of the zero of $G$ at $-\gamma$ greater than one. Write $\gamma = 2m + \delta$, $m \in \mathbb{Z}_+$, $0 < \delta \leq 2$. Recalling that $a_j$, $b_j > 0$ and $\max a_j = \max b_j = 1$, let $\max^*(a_j, b_j) = 1/\beta$ where $\max^*$ is taken over $N^* = \{j \in \mathbb{N} : a_j, b_j < 1\}$. For such a choice of $m$, $\delta$, and $\beta$ $(>1)$, take $f$ as per the lemma, with $Q \equiv 0$. Then we obtain a nonnormal ch.f. satisfying

$$\prod f(a_j t) = \prod f(b_j t), \qquad t \in \mathbb{R}.$$

To see this, we note that if $|t| \geq \beta t_0$, then $f(a_j t) = 0$, $f(b_j t) = 0$ for those $j$ for which $a_j = 1$, $b_j = 1$, such $j$ existing by assumption. For $|t| < \beta t_0$, the products of $l(=l_1 = l_2)$ factors on either side, each factor equal to $f(t)$, agree separately. It is therefore only left for us to check that

$$\prod_{j \in N^*} f(a_j t) = \prod_{j \in N^*} f(b_j t), \qquad \text{for } |t| < \beta t_0.$$

But, for $j \in N^*$, $|a_j t| \leq t_0$, $|b_j t| \leq t_0$, for such $t$, the "exponential form" for $f$ in $|t| \leq t_0$ applies, and the preceding equality follows from our assumption on $\gamma$, which implies that

(a) $\sum_{j \in N^*} a_j^\gamma = \sum_{j \in N^*} b_j^\gamma$ in all cases $0 < \delta \leq 2$, and further
(b) $\sum_{j \in N^*} a_j^\gamma \log a_j = \sum_{j \in N^*} b_j^\gamma \log b_j$ in the case $\delta = 2$.

The necessity part of the proof is therefore established once Lemma 5.4.2 is proved. We proceed to do so through two auxiliary results, namely, Lemmas 5.4.3 and 5.4.4. Lemma 5.4.3 is of some independent interest and we use the method of its proof rather than the lemma itself in our proof of Lemma 5.4.2.

**Lemma 5.4.3.** *Let $k \in \mathbb{Z}_+$, $\alpha > 0$, $0 < \delta < 1$ be given. Let $\theta$ be a real-valued function on $[0, \tau]$ for some $\tau > 0$, having (a possibly piecewise-defined) bounded derivative of order $(2k + 1)$ in $(0, \tau)$. Then, we can specify a polynomial $P$ of degree $k$, two positive constants $t_0 \in (0, \tau]$ and $T_0 > t_0$, and a ch.f. $\Psi$ such that*

$$\Psi(t) = \begin{cases} \exp\{P(t^2) + (-1)^{k-1}\alpha|t|^{2k+\delta} + \theta(|t|)\} & \text{for } |t| \le t_0, \\ 0 & \text{for } |t| \ge T_0. \end{cases}$$

*Further, $t_0$ may be chosen arbitrarily small.*

***Proof.*** Our assumption implies that $\theta^{(j)}(t) \to$ a finite limit as $t \to 0$, for $j = 0, 1, \ldots, 2k$. Making the necessary adjustments in the final choice of $P$, we may assume without loss of generality that

$$\theta^{(j)}(t) \to 0 \quad \text{as } t \to 0, \qquad \text{for } j = 0, 1, \ldots, 2k. \tag{5.4.1}$$

Let $h_\lambda(t) = \lambda - |t|$ for $|t| \le \lambda$, and zero otherwise. Let $\psi$ be a continuous even function on $\mathbb{R}$ vanishing for $|t| \ge \Delta$, and let $\psi_\lambda = h_\lambda * \psi$, where $*$ stands for the usual convolution operation. Then, $\psi_\lambda$ is also continuous and even, and straightforward computations show that its first two derivatives are given by

$$\psi'_\lambda(t) = \int_t^{t+\lambda} \psi(u)\,du - \int_{t-\lambda}^{t} \psi(u)\,du,$$

$$\psi''_\lambda(t) = \psi(t + \lambda) - 2\psi(t) + \psi(t - \lambda);$$

if $\lambda > 2\Delta$, we therefore have

$$\psi''_\lambda(t) = -2\psi(t) \quad \text{for } |t| < \Delta, \qquad \psi''_\lambda(0) = -2\psi(0), \qquad \psi'_\lambda(0) = 0. \tag{5.4.2}$$

Let $\lambda_j > 0$ for all $j$, $0 \le j \le k$, and $\Lambda_j = (\lambda_1, \lambda_2, \ldots, \lambda_j)$ for $j = 1, \ldots, k$. Let $H_{\Lambda_j} = h_{\lambda_1} * \cdots * h_{\lambda_j}$. Consider the function $g$ defined by

$$g = h_k * \psi. \tag{5.4.3}$$

If we impose the restrictions on $\Lambda_k$ given by

$$\lambda_1 > 2\lambda_0, \ldots, \lambda_j > 2(\lambda_0 + \cdots + \lambda_{j-1}), \qquad \text{for } j = 2, \ldots, k, \tag{5.4.4}$$

then for any continuous, even $\psi$ vanishing for $|t| \ge \lambda_0$, we have, from (5.4.2) and (5.4.3), that

$$g^{(2k)}(t) = (-2)^k \psi(t) \qquad \text{for } |t| < \lambda_0, \tag{5.4.5a}$$

$$g^{(2m)}(0) = (-2)^m (H_{\Lambda_{k-m}} * \psi)(0), \tag{5.4.5b}$$

$$g^{(2m+1)}(0) = 0, \tag{5.4.5c}$$

for $m = 0, 1, \ldots, k-1$. Set $\xi = cH_{\Lambda_k} * h_{\lambda_0}$, with $c$ a normalizing constant such that $\xi(0) = 1$. Then, (5.4.5a) with $\psi = h_{\lambda_0}$ implies by straightforward computation that, for appropriate real constants $c_j$, $0 \le j \le k$,

$$\xi(t) = 1 + c_1 t^2 + \cdots + c_k t^{2k} + c_0 |t|^{2k+1} \qquad \text{for } |t| \le \lambda_0.$$

Let $\gamma_1, \ldots, \gamma_k$ be such that

$$\exp(\gamma_1 t^2 + \cdots + \gamma_k t^{2k}) - \xi(t) = O(|t|^{2k+1}) \qquad \text{as } t \to 0.$$

We define $f$ according to

$$f(t) = \exp\{\gamma_1 t^2 + \cdots + \gamma_k t^{2k} + \alpha_0 t^{2k} + (-1)^{k-1}\alpha |t|^{2k+\delta} + \theta(|t|)\}, \qquad t \in I,$$

for a real constant $\alpha_0$ and an interval $I \subset (-\tau, \tau)$ around the origin, both to be chosen in the following such that $f$ coincides with a ch.f. on an interval around the origin. We may rewrite $f$ as

$$f(t) = \xi(t) + \alpha_0 t^{2k} + (-1)^{k-1}\alpha |t|^{2k+\delta} + \theta_1(|t|), \qquad t \in I, \quad (5.4.6)$$

where $\theta_1$ satisfies (5.4.1) and generally is of the same nature as $\theta$. Differentiating (5.4.6) $2k$ times and writing $\rho = f - \xi$, we have

$$\rho^{(2k)}(t) = (-1)^k\{\tilde{\alpha}_0 - \tilde{\alpha}_1 t^\delta + \theta_2(t)\} \qquad \text{for } 0 < t \in I,$$

where $\theta_2$ has a (possibly piecewise-defined) bounded derivative in the interval $I \cap (0, \tau)$, and $\tilde{\alpha}_1 > 0$ (since $\alpha > 0$). We choose $t_0 \in (0, \tau]$ so small that $\theta_2'$ is defined on $(0, t_0)$ and

$$\sup\{\theta_2'(t)\colon t > 0, t \in I\} \le \min\{\tilde{\alpha}_1 \delta t_0^{\delta-1}, \tfrac{1}{2}\tilde{\alpha}_1 \delta(1-\delta) t_0^{\delta-2}\}.$$

Now, we define $\tilde{\alpha}_0 = \tilde{\alpha}_1 t_0^\delta - \theta_2(t_0)$, $I = [-t_0, t_0]$,

$$\tilde{\rho}(t) = \begin{cases} \tilde{\alpha}_0 - \tilde{\alpha}_1 |t|^\delta + \theta_2(|t|) & \text{for } t \in I, \\ 0 & \text{otherwise.} \end{cases} \qquad (5.4.7)$$

Our choice of $t_0$ and $\tilde{\rho}$ implies that $\tilde{\rho}'$ is negative and decreases on $(0, t_0)$. Thus, $\tilde{\rho}$ is continuous and nonnegative on $\mathbb{R}$, and (decreasing and) convex on $[0, \infty)$. The sufficient conditions for a ch.f. given by the well-known theorem of Polya (see for instance Chung, 1974, p. 82, or Lukacs, 1970, p. 83) are satisfied by $\tilde{\rho}(\cdot)/\tilde{\alpha}_0$. Since $0 \le \tilde{\rho}(t) \le \tilde{\alpha}$ for $t \in \mathbb{R}$, we see that (since $\tilde{\alpha}_0 \to 0$ as $t_0 \to 0$) $\tilde{\rho} \to 0$ uniformly on $\mathbb{R}$ as $t_0 \to 0$.

In the following, we will exhibit the preceding $f$ as the restriction to $I$ of a ch.f. (namely, $\Psi$) that vanishes outside a compact interval; we can find $\tilde{\Lambda}_k = (\tilde{\lambda}_1, \ldots, \tilde{\lambda}_k)$ satisfying the relations

$$(H_{\tilde{\Lambda}_m} * (ch_{\lambda_0} + \tilde{\rho}))(0) = (H_{\Lambda_m} * ch_{\lambda_0})(0) \qquad \text{for } m = 1, \ldots, k. \quad (5.4.8)$$

To see that such a $\tilde{\Lambda}_k$ exists, suppose $\tilde{\lambda}_1, \tilde{\lambda}_2, \ldots, \tilde{\lambda}_{m-1}$ have been chosen, where $2 \le m \le k$, satisfying (5.4.8), and consider the continuous function $g_m$ given by

$$g_m(\lambda) = (H_{\tilde{\Lambda}_{m-1}} * h_\lambda * (ch_{\lambda_0} + \tilde{\rho}))(0), \qquad 0 < \lambda < \infty.$$

As $\lambda \to 0$, $g_m(\lambda) \to 0$, and as $\lambda \to \infty$, $g_m(\lambda) \to \infty$. Hence, there exists a $\tilde{\lambda}_m$ in $(0, \infty)$ for which (5.4.8) holds.

We claim that as $t_0 \to 0$, $\tilde{\Lambda}_k \to \Lambda_k$. As noted already, $\tilde{\rho}(t) \to 0$ uniformly on $\mathbb{R}$ as $t_0 \to 0$. Hence, it follows from (5.4.8) that

$$(H_{\tilde{\Lambda}_m} * h_{\lambda_0})(0) \to (H_{\Lambda_m} * h_{\lambda_0})(0) \quad \text{as } t_0 \to 0, \qquad m = 1, \ldots, k. \tag{5.4.9}$$

Again we proceed by induction. Suppose it has been proven that $\tilde{\Lambda}_{m-1} \to \Lambda_{m-1}$ as $t_0 \to 0$. Then, (5.4.9) implies that

$$\int_0^{\tilde{\lambda}_{m-1}} (\tilde{\lambda}_m - t)\psi_{\tilde{\Lambda}_{m-1}}(t)\,dt \to \int_0^{\lambda_{m-1}} (\lambda_m - t)\psi_{\Lambda_{m-1}}(t)\,dt$$

for a suitable nonnegative $\psi_{\Lambda_{m-1}}(t)$, continuous in $\lambda_1, \ldots, \lambda_{m-1}$, $t$, and positive on a set of positive measure for each fixed $\Lambda_{m-1}$. Since $\tilde{\Lambda}_{m-1} \to \Lambda_{m-1}$ as $t_0 \to 0$, the preceding relation implies that $\tilde{\lambda}_m \to \lambda_m$ as well.

Thus, for a small enough $t_0 > 0$, $\tilde{\Lambda}_k$ also satisfies the restrictions (5.4.4). If, for this choice of $t_0$ and the corresponding $\tilde{\rho}$, we define

$$\Psi = \{H_{\tilde{\Lambda}_m} * (ch_{\lambda_0} + \tilde{\rho})\},$$

then (5.4.8) is equivalent to

$$\Psi^{(2m)}(0) = \xi^{(m)}(0) \qquad \text{for } 0 \le m \le k-1,$$

so that

$$\Psi^{(r)}(0) = f^{(r)}(0) \qquad \text{for } 0 \le r \le 2k-1.$$

Also, (5.4.5a) implies that $\Psi^{(2k)}$ agrees with $f^{(2k)}$ on $(0, t_0)$, so that $f$ agrees with $\Psi$ on $[0, t_0]$. But, by definition, $\Psi$ is a ch.f. (note that $\Psi(0) = \xi(0) = 1$ yields the normalizing condition). Thus, $\Psi$ is a ch.f. of the stated form.

Our next lemma is close in spirit and in method of proof to the lemmas of Section 4.3, that were formulated in the context of the necessity part of Linnik's theorem.

**Lemma 5.4.4.** *Let $P(u) = p_1 u + \cdots + p_r u^r$ be a polynomial with real coefficients, of degree $r$ (vanishing at the origin). For arbitrary $l \in \mathbb{Z}_+$, $0 < \delta \le 2$, $\alpha > 0$, and $\gamma > 2\max(l+1, r)$, there exists $A_0 > 0$ such that*

$$f(t) = \exp\{-A_0 t^2 + P(t^2) + (-1)^{l-1} t^{2l} w(t, \delta) - |t|^\gamma\}$$

*is a ch.f., where $w(t, \delta)$ is as in the statement of Lemma* 5.4.2.

***Proof.*** Let

$$h(x) = (1/\pi)\,\mathrm{Re}\int_0^\infty f(t)e^{itx}\,dt, \qquad x \geq 0.$$

(The term $-|t|^\gamma$ provides the necessary convergence.) For the case $0 < \delta < 2$, we proceed essentially (with necessary minor changes) as in the proof of Lemma 5.3.1, with $2k + \delta$ in place of $\gamma_1$, and obtain the fact that, for large $A$, and $x \geq \theta(A)$,

$$h(x) \geq \Gamma(2k + \delta + 1)\sin(\pi\delta/2)/(2x^{2k+\delta+1}),$$

and that, for $0 \leq x \leq \theta(A)$,

$$h(x) \geq (4\sqrt{2A})^{-1}\exp(-x^2/4A).$$

For the case $\delta = 2$, we proceed as in the proof of Lemma 5.3.2(i), with the difference that the even integer $2k + 2$ takes the place of $\gamma_1$; in fact, this simplifies the computations somewhat in view of $\exp(i\gamma_1/2)$ being $= \pm 1$. We arrive at estimates of the form $h(x) \geq cx^{-(\gamma_1+1)}$ for $x \geq \theta(A)$, $h(x) \geq (c/\sqrt{A})\exp(-x^2/4A)$ for $0 \leq x \leq \theta(A)$, and for some constant $c > 0$, independent of $x$, both estimates holding for all sufficiently large $A > 0$.

We now proceed to prove Lemma 5.4.2. If $h_1$ is the triangular function already defined, its $m$-fold convolution is given by

$$h_1^{*m}(x) = \frac{1}{(2m-1)!}\left\{(x+m)_+^{2m-1} - \binom{2m}{1}(x+m-1)_+^{2m-1} + \cdots + (x-m)_+^{2m-1}\right\}$$

(being zero for $|x| > m$), where $u_+ = \max(u, 0)$, for real $u$. This formula is easily verified by induction (*cf.* Cramér (1946), p. 245). Thus, $h_1^{*m}$ has a bounded derivative of order $2m - 1$ in each of the intervals $(j, j+1)$, $-m \leq j < m$, and continuous derivatives of all lower orders on $(-m, m)$. Let $\tilde{\psi}_m$ denote the normalized $h_1^{*m}$, so that $\tilde{\psi}_m(0) = 1$, and let $\psi_m = \tilde{\psi}_m(m\,\cdot)$. Then,

$$\psi_m(t)\begin{cases} = 0 & \text{for } |t| \geq 1, \\ > 0 & \text{for } |t| < 1. \end{cases}$$

Now, let $n \in \mathbb{N}$ be fixed such that $2n > \max(k, \deg P)$, where $k$, $P$ are as in the statement of Lemma 5.4.3. Then, $4n + 3 > 2k + 2$. Denote $\psi_{2n+2}$ by $\psi$ for simplicity; then,

$$\psi(t) = \exp\{\tilde{Q}(t^2) + R(|t|)\} \qquad \text{for } |t| < 1,$$

where $\tilde{Q}$ is a polynomial of degree $2n + 1$ with $\tilde{Q}(0) = 0$, and $R$ is such that

$$R^{(j)}(0) = 0 \qquad \text{for } j = 0, 1, \ldots, 4n + 2,$$

and has a piecewise-defined derivative of order $4n + 3$ on $(0, 1)$, which is bounded on every interval $(0, \tau)$ with $\tau < 1$. We define, for $t_0 > 0$ (and for the given $\beta > 1$),

$$R_{t_0}(t) = t_0^{4n+3} R(|t|/\beta t_0) \qquad \text{for } |t| < \beta t_0.$$

Then,

$$R_{t_0}^{(j)}(0) = 0 \qquad \text{for } j = 0, 1, \ldots, 4n + 2,$$

and $R_{t_0}^{(4n+3)}$ is defined piecewise on $(-\beta t_0, \beta t_0)$ and is bounded on $(-t_0, t_0)$, admitting a bound which is *uniform* with respect to $t_0$—namely,

$$\sup\{|R^{(4n+3)}(t)|\beta^{-(4n+3)} : |t| \leq 1/\beta\}.$$

We may therefore apply the argument used to prove Lemma 5.4.3, with $\tau = t_0$, $I = (-t_0, t_0)$, and $\theta(t) = -\alpha_1 R_{t_0}(t)$, and conclude (in view of the uniformity of the bound) that $t_0$ may *also* be so chosen that, for given $\alpha_0, \alpha_1 > 0$, for a suitable polynomial $P$ of degree $2n + 1$ (not depending on $t_0$),

$$f_{t_0}(t) = \exp\{P(t^2) + \alpha_0 |t|^{4n+5/2} - \alpha_1 R_{t_0}(t)\}, \qquad \text{for } |t| \leq t_0,$$

extends to a ch.f. on $\mathbb{R}$; call this ch.f. $\Psi_1$. Since $t_0 > 0$ can be taken to be arbitrarily small, let $t_0$ be such that $\alpha_1 t_0^{4n+3} = 1/N$, $N \in \mathbb{N}$. Now, we note the fact that $\Psi_2(t) = \psi(t/\beta t_0)$ is a ch.f., having the form

$$\Psi_2(t) = \begin{cases} \exp\{\tilde{P}(t^2) + R(|t|/\beta t_0)\} & \text{for } |t| \leq t_0, \\ 0 & \text{for } |t| \geq \beta t_0. \end{cases}$$

Finally, by Lemma 5.4.4, there exists a ch.f. $\Psi_3$ of the form

$$\Psi_3(t) = \exp\{-A_0 t^2 - NP(t^2) - \tilde{P}(t^2) + Q(t^2) - N\alpha_0 |t|^{4n+5/2} + (-1)^{k-1} t^{2k} w(t, \delta)\}.$$

Taking $\Psi = \Psi_1^N \Psi_2 \Psi_3$, we obtain a ch.f. of the form specified in the statement of the lemma.

This completes the proof of Zinger's theorem in its extended form.

## 5.5. INDEPENDENCE OF LINEAR FORMS IN INDEPENDENT R.V.'S

Let $\{X_j\}$ be a sequence of independent r.v.'s, and let $L_1 = \sum a_j X_j$, $L_2 = \sum b_j X_j$ be two (real) linear forms in them, assumed to converge a.s. (equivalently, in probability or in distribution). It is immediate that if the $X_j$ are normally distributed, and if $\sum a_j b_j \sigma_j^2 = 0$, where $\sigma_j^2 = \operatorname{var} X_j$, then

$L_1$ and $L_2$ are independent (as a straightforward computation using ch.f.'s readily shows). In this section, we show that the independence of $L_1$ and $L_2$ essentially characterizes the normality of the $X_j$. We begin with an auxiliary result.

**Lemma 5.5.1.** *Let $\psi_j$, $j \in \mathbb{N}$, $A$ and $B$ be nonnegative real-valued Borel measurable functions, defined on some neighborhood of the origin, and let $I$ be a compact neighborhood of the origin such that, for some real sequence $\{c_j\}$,*

$$\sum_{j=1}^{\infty} \psi_j(u + c_j v) = A(u) + B(v), \qquad \text{for } u, v \in I, \tag{5.5.1}$$

*holds. If $\{c_j\}$ is a bounded sequence and if the series on the left-hand side converges uniformly for $u, v \in I$, then $\sum c_j^2 \psi_j$ is a polynomial of degree two at most, on $I$.*

***Proof.*** In view of the assumption of uniform convergence, we may multiply both sides of the relation (5.5.1) by $(x - u)$ and integrate over $[0, x]$, $0 < x \in I$; suitable changes for $0 > x \in I$ are obvious. We obtain

$$\sum \int_0^x (x - u)\psi_j(u + c_j v)\, du = C(x) + x^2 B(v)/2 \qquad \text{for } x, v \in I, \tag{5.5.2}$$

where $C$ is a suitable continuous function on $I$. The series on the left-hand side may be rewritten as

$$\sum \int_{c_j v}^{x + c_j v} (x + c_j v - t)\psi_j(t)\, dt,$$

so that we have

$$\begin{aligned}
\sum \int_0^{x + c_j v} (x + c_j v - t)\psi_j(t)\, dt
&= \sum \int_0^{c_j v} (x + c_j v - t)\psi_j(t)\, dt + C(x) + x^2 B(v)/2 \\
&= xD(v) + E(v) + C(x) + x^2 B(v)/2,
\end{aligned}$$

for suitable continuous functions $D$ and $E$ on $I$. The two series obtained by differentiating (for every fixed $x \in I$) the left-hand side term by term with respect to $v$ twice in succession are seen to be uniformly convergent on $I$, so that such termwise differentiation is justified. In particular, the second derivative of the left-hand side is given by the series

$$\sum c_j^2 \psi_j(x + c_j v).$$

By considering three distinct values of $x$ in $I$, we see that the functions $B$, $D$, $E$ are individually twice-differentiable with respect to $v$. Then, differentiating both sides of the last equation twice with respect to $v$, and then setting $v = 0$, we have

$$\sum c_j^2 \psi_j(x) = a + bx + cx^2 \qquad \text{for } x \in I,$$

for suitable real $a$, $b$, and $c$, thus proving the lemma.

We shall assume without proof the following result, which extends the Lévy–Cramér theorem (Theorem 1.3.13). For a proof, we refer the reader to Ramachandran (1967), pp. 134–155.

**Theorem 5.5.2.** *Let $\{f_j\}$ be a sequence of ch.f.'s and let $\{\alpha_j\}$ be a sequence of positive constants bounded away from zero. Suppose that there exists a neighborhood $I$ of the origin where none of the $f_j$ vanishes and that, for some real $\mu$* and *$\sigma > 0$,*

$$\prod f_j(t)^{\alpha_j} = \exp(i\mu t - \sigma^2 t^2/2), \qquad t \in I.$$

*Then, every $f_j$ is a normal ch.f.*

We may now formulate the main result of this section.

**Theorem 5.5.3.** *Let $\{a_j\}$, $\{b_j\}$ be two real sequences such that both of the sequences $\{a_j/b_j\colon a_j b_j \neq 0)$ and $\{b_j/a_j\colon a_j b_j \neq 0)$ are bounded. Let $\{X_j\}$ be a sequence of nondegenerate, independent r.v.'s such that $\sum a_j X_j$ and $\sum b_j X_j$ converge a.s., and further are themselves independent. Then, for every $j$ for which $a_j b_j \neq 0$, $X_j$ is normally distributed.*

***Remark.*** It is clear that the $X_j$, for $j$ such that $a_j b_j = 0$, can have arbitrary distributions. It also follows from the first paragraph of this section that $\sum a_j b_j \sigma_j^2 = 0$.

***Proof.*** Let $L_1 = \sum a_j X_j$, $L_2 = \sum b_j X_j$. The independence of $L_1$ and $L_2$ implies that

$$Ee^{i(uL_1 + vL_2)} = Ee^{iuL_1} Ee^{ivL_2} \qquad \text{for } u, v \in \mathbb{R}. \tag{5.5.3}$$

Equivalently, if $f_j$ is the ch.f. of $X_j$,

$$\prod f_j(a_j u + b_j v) = \prod f_j(a_j u) \prod f_j(b_j v) \qquad \text{for } u, v \in \mathbb{R}. \tag{5.5.4}$$

Let $I$ be a compact neighborhood of the origin such that, for $t \in I$, $Ee^{itL_1} \neq 0$, $Ee^{itL_2} \neq 0$. Then, by Lemma 5.1.3, $f_j(a_j t) \neq 0$, $f_j(b_j t) \neq 0$

for such $t$; setting $h_j = -\log |f_j|$, we have from (5.5.4) that

$$\sum h_j(a_j u + b_j v) = \sum h_j(a_j u) + \sum h_j(b_j v) \qquad \text{for } u, v \in I. \quad (5.5.5)$$

By Lemma 5.1.1, the convergence of all three series is uniform for $u, v \in I$. Letting $c_j = a_j/b_j$ for $j$ such that $a_j b_j \neq 0$, letting $\psi_j(t) = h_j(a_j t)$, and letting $\sum^*$ denote the sum restricted to such $j$ alone, we have

$$\sum{}^* \psi_j(u + c_j v) = A(u) + B(v), \qquad u, v \in I,$$

where $A$ and $B$ are nonnegative continuous functions on $I$, and so are the $\psi_j$ for all $j$. It follows from Lemma 5.5.1 that

$$\sum{}^* c_j^2 \psi_j(x) = b_0 + b_1 x + b_2 x^2, \qquad x \in I, \quad (5.5.6)$$

for some real $b_0$, $b_1$, and $b_2$. Since the $\psi_j$ are even functions vanishing at the origin, we have $b_0 = b_1 = 0$; $b_2 > 0$ since the $\psi_j$ are not identically zero and are nonnegative. Since the $c_j$ are bounded away from zero by assumption, Theorem 5.5.2 then implies that $|f_j|^2$ is a normal ch.f. for every $j$ for which $a_j b_j \neq 0$, and then we conclude as usual that so is each $f_j$ for such $j$.

***Remark.*** If the $X_j$ are also assumed to be identically distributed, then the proof becomes much easier and we need no assumptions on the coefficients. It is clear that we can ignore in our discussion all those $X_j$ corresponding to $a_j b_j = 0$ and hence, by renaming the variables, take

$$L_1 = \sum X_j, \qquad L_2 = \sum c_j X_j.$$

The convergence of $\sum c_j X_j$ a.s. implies that $\sum c_j^2 < \infty$ (Lemma 5.1.4), and Lemma 5.5.1 yields the conclusion that

$$\left(\sum c_j^2\right)\psi(x) = b_2 x^2, \qquad x \in I,$$

where $\psi = -\log |f|$, $f$ being the common ch.f. of the $X_j$, for some $b_2 > 0$. We then only need the Lévy–Cramér theorem and not the deeper decomposition result given by Theorem 5.5.2.

## NOTES AND REMARKS

The main results of this chapter, namely Theorems 5.1.5, 5.1.8, and 5.4.1, and their auxiliary results are due respectively to J. Marcinkiewicz, Yu. V. Linnik (1953a, b), and A. A. Zinger (1977). The proof in Section 5.2, of the sufficiency part of (an extended form of) Linnik's theorem, is essentially as in Riedel (1985). The proof in Section 5.3, of the necessity part of that theorem, follows Linnik—also see Kagan *et al.* (1973)—with some modifications as in Ramachandran (1977a). A generalized form of Linnik's theorem (in another direction), due to A. A. Zinger, is presented in Kagan *et al.* (1973). Lemma 5.5.1 and Theorems 5.5.2 and 5.5.3 are from Ramachandran (1967), extending or strengthening earlier results due to Yu. V. Linnik in the "finite case" and L. V. Mamai in the "infinite" case.

CHAPTER

# 6

# Independence/Identical Distribution Problems Relating to Stochastic Integrals

Let $\{X(t): t \geq 0\}$ be a stochastic process, assumed to be continuous in probability, homogeneous, and with independent increments. Let $Y_1$ and $Y_2$ be two stochastic integrals, defined in the sense of convergence in probability, with respect to the process $\{X(t)\}$ or, more generally, with respect to $\{X(\nu(t))\}$, where $\nu$ is a nonnegative, nondecreasing, right-continuous function on a given compact interval. In Section 6.1, we make the relevant definitions and state several auxiliary results needed for the later sections, proving some of them and citing references for the proofs of the others. In Section 6.2, we consider characterizations of the Wiener process (with linear mean value function) through the independence or identical distribution of $Y_1$ and $Y_2$. Sections 6.3 and 6.4 are devoted to characterizing stable and semistable processes, respectively, through the identical distribution (except for a shift) of $Y_1$ and $Y_2$; we also reexamine the results of Section 6.2 in the light of those of Sections 6.3 and 6.4. The appendix sets forth results of the Phragmén–Lindelöf type needed to establish the basic auxiliary results for Sections 6.3 and 6.4.

## 6.1. STOCHASTIC INTEGRALS

Let $T \subseteq \mathbb{R}$ be some interval, and $(\Omega, \mathcal{B}, P)$ a probability space. A map $X: T \times \Omega \to \mathbb{R}$ is called a *stochastic process*, and denoted by $\{X(t): t \in T\}$ or simply $\{X(t)\}$, if, for every $t \in T$, $X(t)$ is a r.v. It is said to be *continuous in probability* on $T$ if $X(t + \tau) \xrightarrow{P} X(t)$ as $\tau \to 0$ for every $t \in T$ (one-sided limits being taken at the endpoints where necessary). It is said to be *homogeneous* if, for every $t \in T$ and every $\tau > 0$ such that $t + \tau \in T$, the probability distribution of $X(t + \tau) - X(t)$ depends only on $\tau$. It is called a *process with independent increments* if, for every $n \in \mathbb{N}$, and for arbitrary choices of $a_j, b_j \in T$, $j = 1, \ldots, n$, such that the open intervals $(a_j, b_j)$ are pairwise disjoint, the r.v.'s $\{X(b_j) - X(a_j): j = 1, 2, \ldots, n\}$ are independent.

Throughout this chapter, the term "process" (with the quotation marks) will be used to denote a stochastic process defined on $T = [0, \infty)$, with $X(0) \equiv 0$, homogeneous, continuous in probability, and with independent increments. For $t, \tau \geq 0$, let

$$f(u, \tau) = E(e^{iu(X(t+\tau)-X(t))}) = Ee^{iuX(\tau)}, \qquad u \in \mathbb{R},$$

be the ch.f. of $X(t + \tau) - X(t)$; then, $f(u, \cdot)$ is a continuous function on $T$ for every fixed $u \in \mathbb{R}$ and $f(\cdot, 0) \equiv 1$ (since $X(0) \equiv 0$); further, the independence of increments and the homogeneity (already invoked in the preceding) imply that $f(\cdot, 1)$ is an inf. div. ch.f., with $f(\cdot, \tau) = f(\cdot, 1)^\tau$ for every $\tau > 0$. If $E(X(t))$ exists finitely for every $t \in T$, $\{E(X(t)): t \in T\}$ is called the *mean value function* (m.v.f.) of the process. A "process" $\{W(t)\}$ is called a *Wiener process* or a *Brownian motion process* if there exists a $\sigma > 0$ such that $W(t) - W(s) \sim N(0, \sigma^2|t - s|)$ for all $s, t \geq 0$. If $\{W(t)\}$ is a Wiener process and $m(\cdot)$ is a real-valued function on $T$, then $\{W(t) + m(t)\}$ is called a *Wiener process with m.v.f.* $m(\cdot)$, and a *Wiener process with linear m.v.f.* if $m(t) = ct$ for all $t \geq 0$, for some real $c$. Note that in this case the log ch.f. of $W(1)$ defined by $\phi(t) = \log f(t, 1)$ equals $-\sigma^2 t^2/2$.

Let now $\nu(\cdot)$ be a nonconstant, nonnegative, nondecreasing, right-continuous function defined on some compact interval $[a, b]$, and let $\{X(t)\}$ be a "process." Let, further, $g$ be a real-valued continuous function on $[a, b]$. Then, we define the *stochastic integral*

$$Y_g = \int_{[a,b]} g(t)\, dX(\nu(t)) \tag{6.1.1}$$

of $g$ with respect to $\{X(\nu(t))\}$ *in the sense of convergence in probability* as the limit in probability of Riemann–Stieltjes-type sums of the form

$$S(P) = \sum_{k=1}^{n} g(t_k^*)\{X(\nu(t_k)) - X(\nu(t_{k-1}))\}, \tag{6.1.2}$$

where $P$: $\{a = t_0 < t_1 < \cdots < t_n = b\}$ is an arbitrary mode of subdivision of $[a, b]$ and where $t_k^* \in [t_{k-1}, t_k]$ for every $k$, as $\|P\| = \max_k (t_k - t_{k-1}) \to 0$. For $g$ continuous as assumed, such a limit exists in probability, so that $Y_g$ is well-defined (*cf.* Lukacs, 1968, Section 5.2). We shall write $\int_a^b$ for $\int_{[a,b]}$ in what follows.

**Proposition 6.1.1.** *The ch.f. $\psi_g$ of $Y_g$ is given by*

$$\log \psi_g(u) = \int_a^b \phi(ug(t))\, d\nu(t), \qquad u \in \mathbb{R}, \tag{6.1.3}$$

*where $\phi = \log f(\cdot, 1)$ is the log ch.f. of $X(1)$.*

***Proof.*** Let $f_{S(P)}$ denote the ch.f. of $S(P)$. Then,

$$f_{S(P)}(u) = \prod_{k=1}^{n} f(ug(t_k^*))^{\nu(t_k)-\nu(t_{k-1})}, \qquad u \in \mathbb{R},$$

or, equivalently,

$$\phi_{S(P)}(u) = \sum_{k=1}^{n} \phi(ug(t_k^*))(\nu(t_k) - \nu(t_{k-1})), \qquad u \in \mathbb{R}.$$

In the limit as $\|P\| \to 0$, the right-hand side above converges to the right-hand side of (6.1.3). Since, by definition, $S(P) \to Y_g$ in probability as $\|P\| \to 0$, the left-hand side above converges to $\log \psi_g(u)$ as $\|P\| \to 0$. These two facts yield (6.1.3).

For the joint ch.f. of two stochastic integrals, we similarly have the following.

**Proposition 6.1.2.** *Let $g$, $h$ be real-valued continuous functions on $[a, b]$, and let $Y_g$, $Y_h$ be the corresponding stochastic integrals relative to $\{X(\nu(t))\}$. Then, the ch.f. $\psi_{g,h}$ of the random vector $(Y_g, Y_h)$ is given by*

$$\log \psi_{g,h}(s, u) = \int_a^b \phi(sg(t) + uh(t))\, d\nu(t), \qquad s, u \in \mathbb{R}. \tag{6.1.4}$$

Since $\nu$ is a nondecreasing, right-continuous function on $[a, b]$, we may also regard $\nu$ as a measure on the Borel subsets of $[a, b]$. It is convenient to denote the induced measure also by the same symbol $\nu$. Let us then define, for any real continuous $g$ on $[a, b]$,

$$\nu_g(x) = \nu\{t \in [a, b] : g(t) \le x\}, \qquad x \in [c, d],$$

where $c = \min g$ and $d = \max g$ on $[a, b]$. Then, $\nu_g$ is nondecreasing and right-continuous on $[c, d]$, and we have the following change of variables result.

**Theorem 6.1.3.** *The stochastic integrals $\int_a^b g(t)\, dX(\nu(t))$ and $\int_c^d t\, dX(\nu_g(t))$ have the same probability distribution. Their common log ch.f. is given by*

$$\int_a^b \phi(ug(t))\, d\nu(t) = \int_c^d \phi(ut)\, d\nu_g(t), \qquad u \in \mathbb{R}. \tag{6.1.5}$$

***Proof.*** The two integrals appearing in (6.1.5) are equal (by the rule for substitution in a Riemann–Stieltjes integral, or as easily seen directly from first principles). Proposition 6.1.1 then yields the assertion of the theorem.

We shall require the following result giving the Lévy representation for a stochastic integral.

**Theorem 6.1.4.** *Let $L(\mu, \sigma^2, M, N)$ be the Lévy representation for $\phi$, the log ch.f. of $X(1)$. Then, the log ch.f. of $\int_a^b t\, dX(v(t))$ has the Lévy representation $L(\bar{\mu}, \bar{\sigma}^2, \bar{M}, \bar{N})$, where*

$$\bar{\mu} = \int_a^b \left[ \mu t + t(1 - t^2) \int_0^\infty \frac{u^3}{(1 + u^2)(1 + t^2 u^2)} d(M(-u) + N(u)) \right] dv(t),$$

$$\bar{\sigma}^2 = \sigma^2 \int_a^b t^2\, dv(t),$$

$$\bar{M}(u) = \int_{a^-}^{b^-} -N(u/t)\, dv(t) + \int_{a^+}^{b^+} M(u/t)\, dv(t), \qquad u < 0,$$

$$\bar{N}(u) = \int_{a^-}^{b^-} -M(u/t)\, dv(t) + \int_{a^+}^{b^+} N(u/t)\, dv(t), \qquad u > 0,$$

*where* $s^+ = \max(s, 0)$, $s^- = \min(s, 0)$ *for real s.*

***Proof.*** We sketch the proof, leaving it to the reader to fill in the details (see also Riedel, 1980a). Taking $g(t) \equiv t$ in (6.1.3), we have

$$\begin{aligned} \log \psi_g(u) &= \int_a^b \phi(ut)\, dv(t) \\ &= i\mu u \int_a^b t\, dv(t) - \tfrac{1}{2}\sigma^2 u^2 \int_a^b t^2\, dv(t) \\ &\quad + \int_a^b \left( \int_{(0,\infty)} h(ut, x)\, dN(x) \right) dv(t) \\ &\quad + \int_a^b \left( \int_{(-\infty,0)} h(ut, x)\, dM(x) \right) dv(t), \end{aligned}$$

where $h(u, x) = e^{iux} - 1 - iux/(1 + x^2)$. Let us first consider the case

$0 < a < b$. Then, the third integral is easily seen to be equal to

$$\int_a^b \left( \int_{(0,\infty)} \left\{ h(u, x) + iu \left( \frac{x}{1 + x^2} - \frac{x}{1 + (x/t)^2} \right) \right\} dN(x/t) \right) dv(t)$$

$$= \int_{(0,\infty)} h(u, x)\, d\bar{N}(x)$$

$$+ iu \int_a^b t(1 - t^2) \left\{ \int_{(0,\infty)} \frac{x^3}{(1 + x^2)(1 + t^2 x^2)} dN(x) \right\} dv(t).$$

Dealing similarly with the fourth integral, and then collecting together all the linear terms, we readily obtain the stated representation in the case $0 < a < b$. Essentially the same argument applies in the case $a < b < 0$. If $a \le 0 \le b$ with at least one of these inequalities being strict, we first establish the assertion for $v$ that are constant on intervals $[-\delta, \delta]$ for arbitrary $\delta > 0$, combining the discussion of the two preceding cases. For general $v$, we consider a sequence $\{v_n\}$ converging vaguely to $v$ and such that $v_n$ is constant over $[-1/n, 1/n]$, and use the fact that

$$\int_a^b \phi(ut)\, dv_n(t) \to \int_a^b \phi(ut)\, dv(t)$$

uniformly on compact intervals of $u$ to obtain the stated conclusion in the general case.

## 6.2. CHARACTERIZATION OF WIENER PROCESSES

We begin with a discussion of the case where a stochastic integral $Y_g$, in the set-up described in Section 6.1, has the same distribution as $X(1)$.

**Theorem 6.2.1.** *Let $\{X(t): t \ge 0\}$ be a "process," and let $Y_g$ be the stochastic integral of $g$ with respect to $\{X(v(t))\}$. Suppose further that $g$ satisfies one of the two conditions below:*

(i) $|g(t)| < 1$ *for all* $t \in [a, b]$ *and* $v_g \ne \delta_0$; *or*
(ii) $|g(t)| > 1$ *for all* $t \in [a, b]$.

*If $Y_g \sim X(1)$, then $\{X(t)\}$ is a Wiener process with linear m.v.f. if and only if*

$$\int_a^b g^2(t)\, dv(t) = 1, \tag{6.2.1}$$

*and, in such a case,*

$$\textit{either} \int_a^b g(t)\, dv(t) = 1 \quad \textit{or} \quad \textit{the linear m.v.f. is identically } 0. \tag{6.2.2}$$

***Proof.*** The assumption that $Y_g \sim X(1)$ is, by Proposition 6.1.1, equivalent to

$$\int_a^b \phi(ug(t))\, d\nu(t) = \phi(u), \qquad u \in \mathbb{R}, \tag{6.2.3}$$

where (as usual) $\phi$ is the log ch.f. of $X(1)$. The necessity part of the assertion is obvious. To prove the sufficiency of (6.2.1), we may assume without loss of generality, in view of the Lévy–Cramér theorem, that $\phi$ is real and even, and also that $g \geq 0$ (otherwise, replace $g$ by $|g|$ in both (6.2.1) and (6.2.3)). Then the relations (6.2.1) and (6.2.3) imply that

$$\int_c^d (\phi(ut) - t^2\phi(u))\, d\nu_g(t) = \int_a^b (\phi(ug(t)) - g^2(t)\phi(u))\, d\nu(t) = 0,$$

where $[c, d] = g([a, b])$. Defining $\eta(\cdot)$ and the measure $\bar{\nu}_g$ according to

$$\eta(u) = \phi(u)/u^2 \qquad \text{for } u > 0,$$

$$\bar{\nu}_g[0, t] = \begin{cases} 0 & \text{for } 0 \leq t < c, \\ \displaystyle\int_c^t u^2\, d\nu_g(u) & \text{for } t \geq c, \end{cases}$$

we have $\bar{\nu}_g\{0\} = 0$ and

$$\int_{(0, d]} (\eta(ut) - \eta(u))\, d\bar{\nu}_g(t) = 0, \qquad u \in \mathbb{R}.$$

In case (i), we have $0 \leq c < d < 1$. Since $\bar{\nu}_g \neq \delta_0$, $\eta(ut) - \eta(u)$ cannot be strictly negative throughout $(0, d]$; hence, for each $u$ there exists a $t \in (0, d] \subset (0, 1)$ such that $\eta(u) \leq \eta(ut)$. A dual argument shows that there exists a $t' \in (0, 1)$ such that $\eta(ut') \leq \eta(u)$. These two facts together imply that $\eta$ is a constant (see the analog in the proof of Theorem 3.3.1), so that there exists $\sigma > 0$ such that $\phi(u) = -\frac{1}{2}\sigma^2 u^2$ for all $u \in \mathbb{R}$, and the assertion of the theorem follows immediately.

In case (ii), we have $c > 1$, and, arguing as before, we can find two sequences $\{u_n\}$, $\{u_n'\}$, both strictly increasing, with $u_0 = u_0' = u$ and $\lim_{n\to\infty} u_n = \lim_{n\to\infty} u_n' = \infty$, and such that

$$\eta(u_n) \leq \eta(u_{n+1}), \qquad \eta(u_n') \geq \eta(u_{n+1}') \qquad \forall\, n \in \mathbb{Z}_+.$$

On the other hand, the conditions of the Lévy representation $L(\mu, \sigma^2, M, N)$ for $\phi$, namely

$$\int_{(0,1)} u^2\, dN(u) < \infty, \qquad \int_{(1,\infty)} dN(u) < \infty,$$

their analogs for $M$, and the dominated convergence theorem imply that $\lim_{u\to\infty} \eta(u)$ exists and $= -\frac{1}{2}\sigma^2$. These facts then imply that $\eta(u) \equiv -\frac{1}{2}\sigma^2$, and the assertion of the theorem follows in this case as well. Equation (6.2.2) is immediate on substitution.

Our next result concerns two identically distributed stochastic integrals, and is an analog of Marcinkiewicz's theorem (Theorem 5.1.5).

**Theorem 6.2.2.** *Let $\{X(t): t \geq 0\}$ be a "process," and let $Y_g$ and $Y_h$ be the stochastic integrals of the real-valued continuous functions g and h with respect to $\{X(v(t))\}$. Suppose $X(1)$ has moments of all orders and that $\max|g| \neq \max|h|$. Then, $Y_g$ and $Y_h$ are identically distributed if and only if $\{X(t)\}$ is a Wiener process with linear m.v.f. Furthermore, in this case,*

(i) $\int_a^b g^2(t)\, dv(t) = \int_a^b h^2(t)\, dv(t)$, *and*

(ii) *either* $\int_a^b g(t)\, dv(t) = \int_a^b h(t)\, dv(t)$, *or the m.v.f. is identically* 0.

***Proof.*** The "if" part, including the statements (i) and (ii), is readily verified. The proof of the "only if" part proceeds along the lines of the corresponding part of the proof of Theorem 5.1.5. The details are as follows. Again, we may assume without loss of generality that $\phi$, the log ch.f. of $X(1)$, is real-valued and even, in view of the Lévy–Cramér theorem. The identical distribution of $Y_g$ and $Y_h$ implies that

$$\int_a^b \phi(ug(t))\, dv(t) = \int_a^b \phi(uh(t))\, dv(t), \qquad u \in \mathbb{R}.$$

$\phi$ has derivatives of all orders since $X(1)$ has moments of all orders by assumption. Taking the $2k$th derivative on both sides of the preceeding relation, and setting $u = 0$ in the resulting relation, we obtain

$$\phi^{(2k)}(0)\left\{\int_a^b g(t)^{2k}\, dv(t) - \int_a^b h(t)^{2k}\, dv(t)\right\} = 0, \qquad \forall\, k \in \mathbb{Z}_+.$$

Since $\max|g| \neq \max|h|$ on $[a, b]$, by assumption, it follows from Corollary 1.2.5 that the expression in the curly brackets in the preceding relation can vanish only for finitely many $k \in \mathbb{Z}_+$. Hence, $\phi^{(2k)}(0) = 0$ for all large $k$. Also, since $\phi$ is an even function, all of its odd order derivatives are zero. Then, by Proposition 1.3.18, $\phi$ is a polynomial, and therefore of degree two at most, by Theorem 1.3.16. Finally, we invoke the Lévy–Cramér theorem as usual to yield the result.

The final result of this section concerns the independence of two stochastic integrals characterizing the Wiener process.

**Theorem 6.2.3.** *Let $\{X(t): t \geq 0\}$ be a "process," and let $Y_g$ and $Y_h$ be the stochastic integrals of the real-valued continuous functions g and h on $[a, b]$ with respect to $\{X(v(t))\}$. Suppose g is nonvanishing on $[a, b]$ and h is nonvanishing on a set of positive v-measure on $[a, b]$. Then, $Y_g$ and $Y_h$ are independent if and only if $\{X(t)\}$ is a Wiener process with linear m.v.f. Furthermore,*

$$\int_a^b g(t)h(t)\,dv(t) = 0 \tag{6.2.4}$$

*unless $\{X(t)\}$ is degenerate.*

***Proof.*** The sufficiency part is an easy consequence of Proposition 6.1.2. To prove the necessity, we may assume without loss of generality that $\phi$ is real-valued and even, and that $|g| < 1$ (otherwise, replace $g$ by $g/(\max|g| + 1)$). Since $Y_g$ and $Y_h$ are independent by assumption, we have from Proposition 6.1.2 that (omitting the variable in $g$, $h$, and $v$ for brevity)

$$\int_a^b \phi(tg + sh)\,dv = \int_a^b \phi(tg)\,dv + \int_a^b \phi(sh)\,dv, \qquad \forall\, t, s \in \mathbb{R}. \tag{6.2.5}$$

Multiplying both sides of the equation by $(x - t)$, then integrating with respect to $t$ over $[0, x]$ (for $x > 0$), and finally interchanging the order of integration, we have

$$\begin{aligned} &\int_a^b g^{-2}\left(\int_{sh}^{xg+sh} (xg + sh - u)\phi(u)\,du\right) dv \\ &\qquad - \int_a^b g^{-2}\left(\int_0^{xg} (xg - u)\phi(u)\,du\right) dv \\ &= (x^2/2)\int_a^b \phi(sh)\,dv. \end{aligned}$$

Differentiating both sides once with respect to $s$, we obtain

$$\int_a^b g^{-2}h\left(\int_{sh}^{xg+sh} \phi(u)\,du\right) dv = x^2 A(s),$$

for a suitable $A(\cdot)$. Differentiating again with respect to $s$ and setting $s = 0$, we have

$$\int_a^b (h/g)^2(\phi(xg) - \gamma x^2 g^2)\,dv = 0$$

for some $\gamma \neq 0$, making use of the fact that $\int_a^b h^2\, d\nu > 0$ by our assumptions on $h$ and $\nu$. As in the proof of Theorem 6.2.1, we may rewrite the previous relation in the form

$$\int_c^d (\phi(xt) - \gamma x^2 t^2)\, d\bar{\nu}(t) = 0,$$

where $[c, d] = |g|([a, b]) \subset (0, 1)$ by our assumptions on $g$, and $\bar{\nu}$ is a suitable measure defined in terms of $g$ and $h$ by

$$d\nu^* = (h/g)^2\, d\nu \quad \text{on } [a, b]; \qquad \bar{\nu}(t) = \nu^*\{s\colon |g(s)| \leq t\} \quad \text{for } c \leq t \leq d.$$

Then, we may use the same argument—based on the remark following Theorem 1.3.15—as was used in the proof in the remark following Theorem 5.5.3 to conclude that $\{X(t)\}$ is a Wiener process with linear m.v.f. The relation (6.2.4) follows then on substituting the log ch.f. of $\{X(t)\}$ $(\phi(u) = i\mu u - \sigma^2 u^2/2)$ in (6.2.5).

## 6.3. IDENTICALLY DISTRIBUTED STOCHASTIC INTEGRALS AND STABLE PROCESSES

In this section, we consider characterizations of stable processes through the identical distribution, to within "shifts," of two stochastic integrals. We begin with a simple situation where one of these "integrals" reduces to an increment of the process.

**Theorem 6.3.1.** *A "process" $\{X(t)\}$ is stable if, for* some $y > 0$, *and only if for* every $y > 0$, *there exist real constants $t_y > 0$ and $s_y$ such that $\int_0^y t\, dX(t)$ has the same distribution as $X(t_y) + s_y$.*

***Proof.*** The "only if" part follows by straightforward computation using the expression for the log ch.f. of a stable law as given by Theorem 3.1.3, noting that the given condition of identical distribution is equivalent to

$$\int_0^y \phi(ut)\, dt = t_y \phi(u) + ius_y, \qquad \forall\, y > 0,\ u \in \mathbb{R},$$

which leads to the choices (in the notation of Theorem 3.1.3)

$$t_y = y^{\alpha+1}/(\alpha + 1), \qquad s_y = a((y^2/2) - t_y) \qquad \text{if } \alpha \neq 1,$$

$$t_y = y^2/2, \qquad s_y = -\tfrac{1}{2}c\lambda y^2(\log y - \tfrac{1}{2}) \qquad \text{if } \alpha = 1,$$

where $\alpha$ is the exponent of the stable process.

Turning to the "if" part, we may consider the case $y = 1$ without loss of generality. Suppose then that, for some $a > 0$ and $b \in \mathbb{R}$,

$$\int_0^1 \phi(ut)\,dt = a\phi(u) + ibu; \tag{6.3.1}$$

then, $\psi = -\operatorname{Re}\phi \geq 0$ satisfies the following "multiplicative version" of the ICFE:

$$a\psi(u) = \int_0^1 \psi(ut)\,dt \qquad \text{for } u \geq 0.$$

As in the proof of Theorem 3.4.1, we conclude that $\psi(u) = c|u|^\alpha$, where $0 < \alpha \leq 2$ necessarily; it follows that $a = 1/(1 + \alpha)$. The case $\alpha = 2$ corresponds to normal laws.

Consider then the case $0 < \alpha < 2$. If $\alpha \neq 1$, then $\frac{1}{2} - a \neq 0$; set $\gamma = b/(\frac{1}{2} - a)$ and define $\tilde{\phi}(u) = \phi(u) - i\gamma u$. Then, $\tilde{\phi}$ is a log ch.f., satisfying the ICFE

$$a\tilde{\phi}(u) = \int_0^1 \tilde{\phi}(ut)\,dt \qquad \text{for } u \geq 0. \tag{6.3.2}$$

It follows from the argument for Theorem 3.4.1(b) that

$$\tilde{\phi}(u) = -c|u|^\alpha + i\beta(\operatorname{sgn} u)|u|^\alpha + i\eta u.$$

Substituting this expression back into (6.3.2), we see that $\eta = 0$, so that

$$\phi(u) = i\gamma u - c|u|^\alpha + i\beta(\operatorname{sgn} u)|u|^\alpha.$$

We conclude that in the case $\alpha \neq 1$, $\phi$ is the log ch.f. of a stable law with exponent $\alpha$.

If $\alpha = 1$, we have to proceed somewhat differently. We now set

$$\phi(u) = -c|u| + i(\operatorname{sgn} u)\{\delta|u| \log|u| + |u|I(u)\} \qquad \text{for } u \neq 0,$$

where $I$ is a real-valued even continuous function to be determined, and $\delta$ is given by

$$b = \delta \int_0^1 t \log t\,dt \;(= -\delta/4).$$

$I$ is seen to satisfy the functional equation

$$I(u) = 2\int_0^1 I(ut)t\,dt, \qquad u > 0.$$

Proceeding as in the proof of Lemma 3.4.4 in the case $\alpha = 1$ there, we see that, for every $T > 0$ and every $0 < \varepsilon < 1$, $\sup_{0 < u \leq T} |I(\varepsilon u) - I(u)| < \infty$ and therefore, by Corollary 2.4.3, $I$ is a constant. Hence $\phi$ is the log ch.f. of a stable law with exponent $\alpha = 1$. The theorem is proven.

We pass on to a more comprehensive result due to Riedel (1980b), which characterizes stable processes by means of the identical distribution, to within shifts, of two stochastic integrals defined in terms of the process. It contains Theorem 6.3.1 as a special case (as we shall explicitly verify). The proof requires results of the type of the Phragmén–Lindelöf theorem; these are stated and proved in the appendix to this chapter.

Let $\{X(t)\}$ be a "process," and let

$$Y_j = \int_{a_j}^{b_j} g_j(t)\, dX(v_j(t)), \qquad j = 1, 2,$$

be two stochastic integrals, where $g_j$, $v_j$, defined on $[a_j, b_j]$, for $j = 1, 2$, are as per the description in Section 6.1. In view of Theorem 6.1.3, the $Y_j$ may be recast in the form

$$Y_j = \int_{c_j}^{d_j} t\, dX(\tilde{v}_j(t)), \qquad j = 1, 2,$$

for suitable $\tilde{v}_j$, $c_j$, $d_j$. Taking $a = \max\{|c_j|, |d_j| : j = 1, 2\}$, and extending the $\tilde{v}_j$ to $V_j$ defined on $[-a, a]$ in an obvious manner, we may finally take the $Y_j$ in the form

$$Y_j = \int_{-a}^{a} t\, dX(V_j(t)), \qquad j = 1, 2. \tag{6.3.3}$$

Let $W = V_1 - V_2$, and let

$$S(z) = \int_{-a}^{a} |t|^z\, dW(t). \tag{6.3.4}$$

By Theorem 6.1.4, giving the Lévy representation for a stochastic integral, the identical distribution condition $Y_1 \sim Y_2 + q$ ($q$ a real constant) is equivalent to the set of four relations that follow:

$$\int_{-a}^{a} \left[ \mu t + t(1 - t^2) \int_0^\infty \frac{u^3}{(1 + u^2)(1 + t^2u^2)}\, d(M(-u) + N(u)) \right] dW(t)$$

$$= -q,$$

$$\sigma^2 S(2) = 0, \tag{6.3.5}$$

$$L_1(u) := \int_{(-a,0)} -N(u/t)\, dW(t) + \int_{(0,a)} M(u/t)\, dW(t) = 0 \qquad \text{for } u < 0,$$

$$L_2(u) := \int_{(-a,0)} -M(u/t)\, dW(t) + \int_{(0,a)} N(u/t)\, dW(t) = 0 \qquad \text{for } u > 0.$$

We note that the function $S$ defined by (6.3.4) is analytic in $\operatorname{Re} z > 0$ and continuous in $\operatorname{Re} z \geq 0$; the same is true of the function $H$ defined by

$$H(z) := \int_{(-\infty,-1)} |u|^{-z}\, dM(u) + \int_{(1,\infty)} u^{-z}\, dN(u). \tag{6.3.6}$$

**Lemma 6.3.2.** *Suppose $Y_1 \sim Y_2 + q$, and let $H, S$ be as defined previously. Then, there exists a $K$, continuous on* $\operatorname{Re} z \leq 0$ *and analytic on* $\operatorname{Re} z < 0$, *such that*

$$H(z)S(-z) = K(z) \qquad \textit{for } \operatorname{Re} z = 0. \tag{6.3.7}$$

***Proof.*** The following sequence of equalities (except the last, which is a definition) easily follow from the relations (6.3.5): For $\operatorname{Re} z = 0$,

$$\begin{aligned}
0 &= \int_{-\infty}^{-a} |u|^{-z}\, dL_1(u) + \int_a^\infty u^{-z}\, dL_2(u) \\
&= \int_{(-a,0)} |t|^{-z}\left(\int_{-a/t}^\infty y^{-z}\, dN(y) + \int_{-\infty}^{a/t} |y|^{-z}\, dM(y)\right) dW(t) \\
&\quad + \int_{(0,a)} t^{-z}\left(\int_{-\infty}^{-a/t} |y|^{-z}\, dM(y) + \int_{a/t}^\infty y^{-z}\, dN(y)\right) dW(t) \\
&:= H(z)S(-z) - K(z), \quad \text{say;}
\end{aligned}$$

It is clear that the function $K$, so defined, has the stated properties.

If we now define another auxiliary function $\hat{H}$ by

$$\hat{H}(z) := \int_{(-1,0)} |u|^{z}\, dM(u) + \int_{(0,1)} u^{z}\, dN(u),$$

then the condition $\int_{(-1,0)} u^2\, dM(u) + \int_{(0,1)} u^2\, dN(u) < \infty$ of the Lévy representation implies that $\hat{H}$ is continuous and bounded on $\operatorname{Re} z \geq 2$ and analytic on $\operatorname{Re} z > 2$. Proceeding as in the proof of the previous lemma, we easily obtain a dual result.

**Lemma 6.3.3.** *For $K$ as in Lemma* 6.3.2,

$$\hat{H}(z)S(z) = -K(-z) \qquad \textit{for } \operatorname{Re} z \geq 2. \tag{6.3.8}$$

We proceed to a fundamental auxiliary result.

**Lemma 6.3.4.** *Let $Y_1 \sim Y_2 + q$. Suppose $S$ satisfies the two conditions:*

$$\left.\begin{array}{ll}\text{(i)} & 0 \textit{ is not a limit point of the zeros of } S \textit{ in } \operatorname{Re} z \geq 0; \\ \text{(ii)} & \overline{\lim}_{x \to 0+} x \log |S(x)| = 0.\end{array}\right\} \qquad (6.3.9)$$

*Then, $H$ is meromorphic on $\mathbb{C}$ (possibly entire), and all its poles lie in the strip $\{z: -2 < \operatorname{Re} z < 0\}$. Further, if*

$$\alpha_1 = \sup\{\operatorname{Re} z: z \text{ is a pole for } H\}, \qquad \alpha_2 = \inf\{\operatorname{Re} z: z \text{ is a pole for } H\},$$

*then $\alpha_1$ and $\alpha_2$ are themselves poles for $H$ (in particular, $-2 < \alpha_2 \leq \alpha_1 < 0$). For every $\varepsilon > 0$, $H$ is bounded on $\{z: \operatorname{Re} z \geq \alpha_1 + \varepsilon\}$, and on $\{z: \operatorname{Re} z \leq \alpha_2 - \varepsilon\}$. If $l_j$ is the order of the pole at $\alpha_j$, and if $c_j$ is the coefficient of $(z - \alpha_j)^{-l_j}$ in the Laurent expansion for $H$ around $\alpha_j$, then*

$$c_1 > 0, \qquad (-1)^{l_2+1} c_2 > 0.$$

***Remark.*** Conditions (6.3.9)(i)(ii), admittedly of a technical nature, are intended to enable us to apply certain Phragmén–Lindelöf type of arguments; the relevant auxiliary results are provided in the appendix to this chapter. The conditions (6.3.9) are trivially satisfied if $S$ is analytic at the origin—as happens for instance in the context of Theorem 6.3.1.

***Proof.*** We shall use the notation

$$D_r = \{z: |z| < r, \operatorname{Re} z < 0\}, \qquad B_r = \{z: |z| < r\}.$$

We begin by noting that, unless $H \equiv 0$ (in which case there is nothing to prove), $H(0) > 0$. By (i) of (6.3.9), $S(-z) \neq 0$ for $z \in \bar{D}_r \backslash \{0\}$, for some $r > 0$. We may take this $r$, without loss of generality, also to be such that $G(iy) = H(iy) \neq 0$ for $|y| < r$, $y$ real (in view of $H(0) \neq 0$). Consider the function

$$G(z) = K(z)/S(-z), \qquad \operatorname{Re} z \leq 0;$$

$G(iy) = H(iy) \neq 0$ for $0 < |y| < r$, $y$ real, by Lemma 6.3.2. $S(-z)$ and $K(z) = G(z)S(-z)$ are both bounded on $\bar{D}_r \backslash \{0\}$ and analytic on the interior. Taking into account (ii) of (6.3.9), Corollary 6.A.5 then implies that $G$ also is bounded on $\bar{D}_r \backslash \{0\}$. It follows that $G$ is analytic on $D_r$. Thus, $H$ extends to a function analytic on $B_r$. Then, the properties of Laplace–Stieltjes transforms of positive measures imply that $H$ is indeed analytic on $\operatorname{Re} z > -r$; and that if $\operatorname{Re} z > \alpha$ is the maximal right half-plane of analyticity for $H$, then $\alpha < 0$ and $\alpha$ is a singularity for $H$ (Theorem 1.2.8), and the integral representation for $H$ holds for $\operatorname{Re} z > \alpha$. Since $G$ is meromorphic on $\operatorname{Re} z < 0$ by its definition, it follows that $H$ is meromorphic on $\mathbb{C}$. Equation (6.3.8) then implies that $H$ is analytic on $\operatorname{Re} z < -2$ also, and that

$$H(z) = -\hat{H}(-z) \qquad \text{for } \operatorname{Re} z \leq -2.$$

Again, since $\hat{H}$ is the Laplace–Stieltjes transform of a positive measure, it follows that $H$ is also analytic on a strip of the form $\operatorname{Re} z < \beta$, that the integral representation for $\hat{H}$ holds for $\operatorname{Re} z > -\beta$, and that if $\operatorname{Re} z < \beta$ is the maximal left half-plane of analyticity for $H$, then, by Theorem 1.2.8, $\beta$ is a singularity (hence, a pole necessarily) for $H$. It follows that all the poles of $H$ lie in the strip $\beta \leq \operatorname{Re} z \leq \alpha$. $\alpha_1$ and $\alpha_2$ being as defined in the formulation of the lemma, we see that $\alpha_1 = \alpha$, $\alpha_2 = \beta$, so that $\alpha_1, \alpha_2$ are themselves poles for $H$, as stated. Further, since $H$ is bounded on $\operatorname{Re} z = 0$, and $\hat{H}$ on $\operatorname{Re} z = 2$, we have

$$-2 < \alpha_2 \leq \alpha_1 < 0.$$

Since the integral representation for $H$ holds for $\operatorname{Re} z > \alpha_1$, and for $\hat{H}$ for $\operatorname{Re} z > -\alpha_2$, it follows that, for any $\varepsilon > 0$, $H$ is bounded on $\{z: \operatorname{Re} z \geq \alpha_1 + \varepsilon\}$ as well as on $\{z: \operatorname{Re} z \leq \alpha_2 - \varepsilon\}$.

Finally, since $H(x) \geq 0$ for $x < \alpha_1$ while $\hat{H}(x) \geq 0$ for $x > -\alpha_2$, we have (the inequalities below are strict by the definition of "multiplicity" for a pole)

$$\begin{aligned} c_1 &= \lim_{x \to \alpha_1 +} (x - \alpha_1)^{l_1} H(x) > 0; \\ (-1)^{l_2+1} c_2 &= \lim_{x \to \alpha_2 -} (-1)^{l_2+1} (x - \alpha_2)^{l_2} H(x) \\ &= \lim_{x \to \alpha_2 -} (\alpha_2 - x)^{l_2} \hat{H}(-x) > 0. \end{aligned}$$

Hence, the lemma is proven.

An immediate consequence of the above lemma is the following.

**Corollary 6.3.5.** *If, in addition to the conditions imposed in Theorem 6.3.4, $S$ is assumed to have a unique zero $\alpha$ in $(0, 2)$, then $H$ has at most one pole (at $-\alpha$) in $\mathbb{C}$, and further the multiplicity of that pole is odd.*

Now, analogously to $H$, $\hat{H}$, and $S$, we introduce the functions $H_j$, $\hat{H}_j$ and $S_j$, $j = 1, 2$:

$$H_1(z) = \int_{(1,\infty)} u^{-z}\, dN(u), \qquad H_2(z) = \int_{(-\infty,-1)} |u|^{-z}\, dM(u),$$

$$\hat{H}_1(z) = \int_{(0,1]} u^{z}\, dN(u), \qquad \hat{H}_2(z) = \int_{[-1,0)} |u|^{z}\, dM(u),$$

$$S_1(z) = \int_{(0,a)} t^{z}\, dW(t), \qquad S_2(z) = \int_{(-a,0)} |t|^{z}\, dW(t),$$

all for $\operatorname{Re} z \geq 0$. Also, we define

$$\hat{S}(z) = S_1(z) - S_2(z) = \int_{-a}^{a} (\operatorname{sgn} t)|t|^z \, dW(t),$$

$$S^* = \begin{cases} S & \text{if } \hat{S} \equiv S, \text{ or if } S + \hat{S} \equiv 0, \\ S \cdot \hat{S} & \text{otherwise.} \end{cases}$$

We note that $S \equiv \hat{S}$ if and only if $S_2 \equiv 0$, and $S + \hat{S} \equiv 0$ if and only if $S_1 \equiv 0$.

Using the relations (6.3.5)—which hold if $Y_1 \sim Y_2 + q$—and proceeding as in the proof of Lemma 6.3.2, we see that, for suitable functions $\tilde{K}_1, \tilde{K}_2$, which are (like $K$) continuous on $\operatorname{Re} z \leq 0$ and analytic on $\operatorname{Re} z < 0$,

$$\left.\begin{aligned} S_1(-z)H_1(z) + S_2(-z)H_2(z) &= \tilde{K}_1(z), \\ S_2(-z)H_1(z) + S_1(-z)H_2(z) &= \tilde{K}_2(z), \end{aligned}\right\} \quad \text{for } \operatorname{Re} z = 0.$$

Solving for $H_1, H_2$ in terms of $S_1, S_2$, we see that, for suitable functions $K_1, K_2$, which are (like $K$) continuous on $\operatorname{Re} z \leq 0$ and analytic on $\operatorname{Re} z < 0$,

$$H_j(z)S^*(-z) = K_j(z) \qquad \text{for } \operatorname{Re} z = 0, j = 1, 2.$$

We also have the analog of Lemma 6.3.3, namely,

$$\tilde{H}_j(z)S^*(-z) = -K_j(-z) \qquad \text{for } \operatorname{Re} z = 2, j = 1, 2.$$

If we impose on $S^*$ the conditions (analogous to (6.3.9), and reducing to it in the cases $\hat{S} \equiv S$ or $\hat{S} + S \equiv 0$):

$$\left.\begin{aligned} &\text{(i)} \quad 0 \text{ is not a limit point of the zeros of } S^* \text{ in } \operatorname{Re} z \geq 0; \\ &\text{(ii)} \quad \overline{\lim}_{x \to 0+} x \log |S^*(x)| = 0; \end{aligned}\right\} \tag{6.3.10}$$

we see that the exact analog of Lemma 6.3.4 holds for the $H_j, j = 1, 2$, and for $H$.

We are now in a position to establish our main result.

**Theorem 6.3.6.** *Let $Y_1$ and $Y_2$ be stochastic integrals defined as in* (6.3.3), *and let conditions* (6.3.10)(i)(ii) *be satisfied. If, for some real $q$,*

$$Y_1 \sim Y_2 + q, \tag{6.3.11}$$

*then for $\{X(t)\}$ to be a stable process, it is sufficient that the following set of three conditions be satisfied:*

(i) *$S$ has a unique real zero $\alpha$ in the interval* $(0, 2]$;
(ii) *if $\alpha < 2$, the multiplicity of the zero of $S$ at $\alpha$ is at most two*; *and*
(iii) *if $\alpha < 2$, then $S^*(z) \neq 0$ for $\operatorname{Re} z = \alpha$, $z \neq \alpha$.*

*In such a case, $\alpha$ is the exponent of the stable process.*

*Conversely, if* $q = 0$ *or* $\hat{S}(1) \neq 0$, *then, in order that the only processes that satisfy* (6.3.11) *be stable processes, it is also necessary that conditions* (i)–(iii) *hold.*

***Proof of the sufficiency part.*** In the following, it will be convenient to think of $M$ as left-continuous and $N$ as right-continuous, so that $N - M(-\cdot)$ is a nonpositive, nondecreasing, right-continuous function on $(0, \infty)$, vanishing at $\infty$. Let, by (i), $\alpha$ be the unique real zero of $S$ in the interval $(0, 2]$. First, consider the case $\alpha = 2$. As pointed out previously, if (6.3.10) holds, then Lemma 6.3.4 holds for $H$. $H$ is in our present case an entire function, and the integral representation

$$H(z) = \int_{(1,\infty)} u^{-z}\, d(N(u) - M(-u))$$

holds then for all complex $z$. Since $H(z)$ is bounded for $\operatorname{Re} z \leq -2$, it then follows that $N(u) - M(-u) \equiv 0$ on $(1, \infty)$, hence $N \equiv M(-\cdot) \equiv 0$ there in view of $N \leq 0$, $M \geq 0$. Considering

$$H_r(z) := \int_{(r,\infty)} (u/r)^{-z}\, d(N(u) - M(-u))$$

in place of $H$, the same argument yields the conclusion that $N \equiv M(-\cdot) \equiv 0$ on $(r, \infty)$ for every $r > 0$, which implies that $M \equiv 0 \equiv N$, and $X(1)$ has a normal distribution.

We pass on to the case $0 < \alpha < 2$, where $\alpha$ is the (assumed) unique zero of $S$ in $(0, 2]$. If $H$ were to have no poles at all, then it would follow as above that $X(1)$ has a (nondegenerate) normal distribution. But then $\sigma^2 S(2) = 0$ from the relations (6.3.5), so that $\alpha$ is not the only zero of $S$ in $(0, 2]$, contrary to assumption. It then follows from assumption (ii) and Corollary 6.3.5 that $H$ has a unique pole, at $-\alpha$, of odd multiplicity. Assumption (ii) then implies that this must be a simple pole.

We claim that then $H(z)(z + \alpha)$ is bounded for $\operatorname{Re} z \leq 0$: for, consider the auxiliary function

$$A(z) := H(z)\frac{z + \alpha}{z - 1} = \frac{K(z)(z + \alpha)}{S(-z)(z - 1)};$$

Theorem 6.A.3 can be applied, with $f(z) = S(-z)$, $g = A$, and with the left half-plane being taken into consideration instead of the right. We conclude that $A$ is of exponential type on $\operatorname{Re} z \leq 0$. Since $H$ is bounded for $\operatorname{Re} z \leq -2$, we have $\overline{\lim}_{u\to\infty} (\log A(-u)/u) \leq 0$. By Corollary 6.A.2, therefore, $A$ is bounded for $\operatorname{Re} z \leq 0$, proving our claim. Also, $H(z)$ is bounded for $\operatorname{Re} z \geq 0$, and these two facts imply that $H(z)(z + \alpha)$ is a linear function of $z$. Hence, $H(z) = c_1/(z + \alpha) + c_2$. Since $H(z) \to 0$ as $z \to \infty$ through real values, as a consequence of its definition, it follows that $c_2 = 0$ and so $H(z) = c_1/(z + \alpha)$; and hence, by the unicity theorem for Laplace–Stieltjes transforms, we must have $N(u) - M(-u) = -c_3/u^\alpha$ for some $c_3 > 0$, for all $u \in (1, \infty)$. Again, as argued in the case $\alpha = 2$, it follows that this must hold for all $u > 0$.

In order to show that $-M(-\cdot)$ and $N$ are individually of the same form as in the preceding on $(0, \infty)$, we consider $H_1$ and $H_2$. Let $\alpha_{12} \leq \alpha_{11}$ be the analogs for $H_1$ of $\alpha_2, \alpha_1$ for $H$ in Lemma 6.3.4. Then, the right half-plane where the integral representation for $H_1$ holds obviously contains the corresponding half-plane for $H$, by their very definition; hence, we must have $\alpha_{11} \leq -\alpha$. A dual argument shows that $\alpha_{12} \geq -\alpha$. Therefore, $\alpha_{12} = \alpha_{11} = -\alpha$, and the poles of $H_1$ (if any) all lie on the line $\operatorname{Re} z = -\alpha$. Further, $H_1$ can have at most a simple pole at $-\alpha$ since $H$ has a simple pole there, in view of

$$0 \leq (x + \alpha)^2 H_1(x) \leq (x + \alpha)^2 H(x) \to 0 \qquad \text{as } x + \alpha \to 0+ \qquad (x \text{ real}).$$

By assumption (iii), $H_1$ has no other poles on $\operatorname{Re} z = -\alpha$; thus, $H_1$ is entire, or has a unique pole at $-\alpha$, which is simple. The rest of the argument proceeds as for $H$—invoking (6.3.10) instead of (6.3.9)—and we conclude finally that $N(u) = d_1/u^\alpha$ for some $d_1 \leq 0$, for all $u > 0$. A dual argument using $H_2$ (or the already established form for $N - M(-\cdot)$) shows that $M(u) = d_2/|u|^\alpha$ for some $d_2 \geq 0$, for all $u < 0$.

The sufficiency of the given conditions is hence proved.

***Proof of the necessity part.*** We establish in what follows that, if $q = 0$ or if $\hat{S}(1) \neq 0$, then there exists a nonstable process satisfying (6.3.11) if any one of the conditions (i)–(iii) does not hold.

(i) Suppose $S$ has no zero in $(0, 2]$. Then, by Lemma 6.3.4, $H$ is an entire function, $X(1)$ has a normal distribution (as argued in the first two paragraphs of the sufficiency part), and $S(2) = 0$ by (6.3.5), a contradiction to our assumption. Thus, $S$ must have a zero in $(0, 2]$.

Suppose next that $S$ has more than one zero in $(0, 2]$. Let $0 < \alpha < \beta \leq 2$ be two zeros of $S$. Define $L(\mu, \sigma^2, M, N)$ by

$$\mu = \begin{cases} 0 & \text{if} \quad q = 0, \\ -q/\hat{S}(1) & \text{if } \hat{S}(1) \neq 0, \end{cases}$$

so that $\mu\hat{S}(1) + q = 0$ in any case; if $\beta = 2$, take

$$\sigma > 0; \qquad M(-u) = -N(u) = u^{-\alpha} \qquad \text{for } u > 0;$$

if $\beta < 2$, take

$$\sigma = 0; \qquad M(-u) = -N(u) = u^{-\alpha} + u^{-\beta} \qquad \text{for } u > 0.$$

Then, (6.3.5) is satisfied, while $X(1)$ has a nonstable distribution.

(ii) Suppose $0 < \alpha < 2$ is such that $S(\alpha) = S'(\alpha) = S''(\alpha) = 0$. Define $L(\mu, \sigma^2, M, N)$ by

$$\mu \text{ as before}; \qquad \sigma = 0;$$

$$M(-u) = -N(u) = (\alpha^{-3} + \alpha^{-2} \log u + (2\alpha)^{-1} \log^2 u)u^{-\alpha} \qquad \text{for } u > 0.$$

($M$ and $N$ are easily verified to satisfy the conditions of the Lévy representation.) Straightforward computations show that (6.3.5) holds, whereas $X(1)$ has a nonstable distribution.

(iii) Suppose there exists a $w \neq \alpha$, with $\operatorname{Re} w = \alpha$, such that $S^*(w) = 0$. Then, $S^*(\bar{w}) = 0$ also ($\bar{w}$ is the conjugate of $w$).

(a) If $S(w) = 0$, define $L(\mu, \sigma, M, N)$ by

$$\mu \text{ as before}; \qquad \sigma = 0;$$

$$M(-u) = -N(u) = pu^{-\alpha} + (u^{-w} + u^{-\bar{w}}), \qquad u > 0,$$

$p$ being so chosen that the conditions of the Lévy representation are satisfied by $M$ and $N$. Then, (6.3.5) is satisfied, but $X(1)$ has a nonstable distribution.

(b) If $\hat{S}(w) = 0$, then we define $L(\mu, \sigma^2, M, N)$ by

$$\hat{S}(1)[\mu + 2\pi \operatorname{Re}(w \sec(\pi w/2))] + q = 0; \qquad \sigma = 0;$$

$$M(-u) = pu^{-\alpha} + (u^{-w} + u^{-\bar{w}}), \qquad -N(u) = pu^{-\alpha} - (u^{-w} + u^{-\bar{w}}) \qquad \text{for } u > 0,$$

$p$ being so chosen that the conditions of the Lévy representation are satisfied by $M$ and $N$. Then, (6.3.5) holds (the first relation there follows from Theorem 6.A.7), while $X(1)$ has a nonstable distribution.

This completes the proof of the necessity part of the assertion, and the theorem is proven.

Suppose we impose the restriction that the "process" be "symmetric," equivalently, that $X(1)$ has a symmetric distribution (but need not have the origin as its median), and also take $q = 0$ in (6.3.11), i.e., consider the equidistribution assumption $Y_1 \sim Y_2$. Then, a weakening of condition (iii) in Theorem 6.3.6 is possible, also assuming (6.3.9) to hold instead of (6.3.10). For, then, in discussing the case $\alpha < 2$ in the "sufficiency" part of the proof, we need only consider $H$ (rather than $H_1$ or $H_2$ as well) to arrive at the conclusion that $M(-u) = -N(u) = c/u^{\alpha}$ for $u > 0$, for some $c > 0$. We thus have the following.

**Theorem 6.3.7.** *Suppose $\{X(t)\}$ is a "process" where $X(1)$ has a symmetric distribution (with median not necessarily zero), that the stochastic integrals $Y_1, Y_2$ given by* (6.3.3) *are identically distributed, and that* (6.3.9) *is satisfied. Then, a set of necessary and sufficient conditions for $\{X(t)\}$ to be a ("symmetric") stable process is given by: the conditions* (i) *and* (ii) *of Theorem* 6.3.6 *and*

(iii′) *if $\alpha < 2$, then $S(z) \neq 0$ for $\operatorname{Re} z = \alpha$, $z \neq \alpha$.*

Specializing to Wiener processes, we have the following.

**Theorem 6.3.8.** *Let $\{X(t)\}$ be a "process," and let* (6.3.9) *hold. If $Y_1 \sim Y_2$, then a necessary and sufficient condition for $\{X(t)\}$ to be a Wiener process with linear mean value function* (*m.v.f.*) *is that $S(2) = 0$ and that $S$ does not have any other zero in* $(0, 2]$.

Imposing a moment condition is sufficient to arrive at the same conclusion:

**Corollary 6.3.9.** *If $\{X(t)\}$ is a "process" that satisfies* (6.3.9), *and if $E(X(1)^2) < \infty$, then $Y_1 \sim Y_2$ implies that $\{X(t)\}$ is a Wiener process with linear m.v.f.*

***Remark.*** In the presence of (6.3.9), this result is a strengthening of Theorem 6.2.2; we recall that, in the latter, $X(1)$ is assumed to have moments of all orders.

***Proof.*** By Theorem 1.3.12, the existence of the second moment for $X(1)$ is equivalent to

$$\int_{(-\infty, 0)} u^2 \, dM(u) + \int_{(0, \infty)} u^2 \, dN(u) < \infty.$$

Thus, $H$ is analytic here on $\operatorname{Re} z > -2$, and is therefore an entire function. The argument in the first paragraph of the proof of the "sufficiency" part of Theorem 6.3.6 then enables us to conclude that $\{X(t)\}$ is a Wiener process (with linear m.v.f.).

Finally, we examine (a) Theorem 6.3.1 in the light of Theorem 6.3.6, and (b) Theorem 6.2.1 in the light of Theorem 6.3.8.

(a) If $\{X(t)\}$ is a "process" such that $\int_0^1 t \, dX(t) \sim X(a) + b$ for some $a > 0$ and real $b$, then $S(z) = \hat{S}(z) = (1/(z + 1)) - a$, so that $S^* \equiv S$. Since $S$ is analytic at the origin, condition (6.3.9) is satisfied. Theorem 6.3.6 then shows that we must have $1/3 \leq a < 1$, and that $\{X(t)\}$ is a stable process with parameter $\alpha = (1/a) - 1$.

(b) Here,

$$S(z) = \int_a^b |g(t)|^z \, dv(t) - 1, \qquad S'(z) = \int_a^b |g(t)|^z \log |g(t)| \, dv(t),$$

for $\operatorname{Re} z > 0$, so that for $x > 0$, $S'(x) < 0$ under condition (i) of Theorem 6.2.1, and $S'(x) > 0$ under condition (ii). Equation (6.2.1) is equivalent to $S(2) = 0$, and since $S$ is strictly monotone under either of the conditions (i) and (ii), the requirements of Theorem 6.3.8 are satisfied, and we may conclude that $\{X(t)\}$ is a Wiener process with linear m.v.f. This establishes the sufficiency part of Theorem 6.2.1; the necessity part is straightforward.

## 6.4. IDENTICALLY DISTRIBUTED STOCHASTIC INTEGRALS AND SEMISTABLE PROCESSES

If $\{X(t)\}$ is a "process" such that $X(1)$ follows a semistable law with exponent $\alpha$, then we call $\{X(t)\}$ a "semistable process with exponent $\alpha$." In this section, we shall consider the problem of the preceding section, essentially of Theorem 6.3.6, in the context of the underlying process being semistable. The methods and results of Section 6.3 also prove to be applicable here—with, of course, some modifications and additional arguments.

As a prelude to the statement of the main result that follows, we recall the definitions of $\hat{S}$ and $S^*$ in Section 6.3.

**Theorem 6.4.1.** *Let* $\{X(t)\}$ *be a "process" and let* $Y_1$, $Y_2$ *be stochastic integrals defined as in* (6.3.3). *Suppose* $S^*$ *satisfies* (6.3.10). *If, for some real* $q$,

$$Y_1 \sim Y_2 + q, \tag{6.4.1}$$

*then a set of sufficient conditions for* $\{X(t)\}$ *to be a semistable process* (*with exponent* $\alpha$) *is given by*:

(i) $S$ *has a unique zero* $\alpha$ *in* $(0, 2]$;
(ii) *if* $\alpha < 2$, *the multiplicity of the zero of* $S$ *at* $\alpha$ *is at most two*; *and*
(iii) *if* $\alpha < 2$, *the zeros of* $S^*$ *on* $\operatorname{Re} z = \alpha$ *form a subset of* $\{\alpha + in\rho : n \in \mathbb{Z}\}$, *for some* $\rho > 0$.

*Conversely, if* $q = 0$ or $\hat{S}(1) \neq 0$, *then the above three conditions are also necessary for* (6.4.1) *to imply that* $\{X(t)\}$ *is a semistable process.*

***Proof.*** We adopt the notation of Section 6.3. Lemmas 6.3.2–6.3.4 and Corollary 6.3.5 continue to hold.

Sufficiency part. Let $\alpha$ be the unique zero of $S$ in $(0, 2]$, as per assumption (i). If $\alpha = 2$, we proceed as in the corresponding part of the proof of Theorem 6.3.6 and conclude that $X(1)$ has a normal distribution. If $0 < \alpha < 2$, then all the poles of $H$ (which is meomorphic on $\mathbb{C}$) lie on the line $\operatorname{Re} z = -\alpha$, and $-\alpha$ itself is a simple pole, in view of assumption (ii) and Corollary 6.3.5, as before. If $w = -\alpha + iy$ is any other pole of $H$, then it is also simple; for, the integral representation

$$H(z) = \int_1^\infty u^{-z}\, d(N(u) - M(-u))$$

holds for $\operatorname{Re} z > -\alpha$ and, for $2 \le n \in \mathbb{N}$, we have

$$|x^n H(x - \alpha + iy)| \le x^n H(x - \alpha) \to 0 \qquad \text{as } x \to 0+$$

(since $-\alpha$ is a simple pole). As in the proof of Theorem 6.3.6, all the poles of the $H_j$, $j = 1, 2$, also lie on the line $\operatorname{Re} z = -\alpha$ and, since $0 \le H_j(x) \le H(x)$ for $x > -\alpha$, it follows that any pole of $H_j$ on $\operatorname{Re} z = -\alpha$ is also necessarily simple. Assumption (iii) then enables us to apply a well-known representation theorem, namely Theorem 6.A.6, to obtain the form of $H$. For this purpose, we note the following facts. Arguing as in the discussion of $\sigma(z)$ on pp. 48–49 in Kagan *et al.* (1973)—going back in fact to Linnik (1953a)—the number of zeros of $S^*$ in every closed rectangle

$$\Gamma_n = \{\alpha/2 \le \operatorname{Re} z \le 2\alpha : n - 1 \le \operatorname{Im} z \le n\}, \qquad n \in \mathbb{Z},$$

is bounded by a number not depending on $n$; there exists a $\delta > 0$ (again, not depending on $n$) and a sequence $\{T_n\}$ with $n - 1 < T_n < n$, $n \in \mathbb{Z}$, such that the zeros of $S^*$ in $\Gamma_n$ all lie at vertical distances $>\delta$ from the line $\operatorname{Im} z = T_n$; finally, $|S^*(z)| \ge c(\delta) > 0$ for all $z$ in any of the segments

$$\{\alpha/2 \le \operatorname{Re} z \le 2\alpha : \operatorname{Im} z = T_n\}, \qquad n \in \mathbb{Z},$$

where $c(\delta)$ also does not depend on $n$. It then follows from condition (iii) and Theorem 6.A.6 ($H$ being bounded on $\operatorname{Re} z \le -2\alpha$, and on $\operatorname{Re} z \ge -\alpha/2$ as well) that

$$H(z) = H(0) + \sum_{n \in \mathbb{Z}} a_n \left( \frac{1}{z - z_n} + \frac{1}{z_n} \right),$$

where $z_n = -\alpha + in\rho$, $a_n = 0$ if $z_n$ is not a pole for $H$, and $a_n$ is the residue of $H$ at $z_n$, and so $= \lim_{x \to 0+} xH(x + z_n)$, if $z_n$ is a pole for $H$. It follows from the integral representation for $H$, valid for $\operatorname{Re} z > -\alpha$, that $|a_n| \le a_0$ for all $n \in \mathbb{Z}$. A heuristic derivation of the form of $N - M(-\cdot)$ would be as follows. Writing purely formally, we have on integration by parts in the integral representation for $H$ that, for $\operatorname{Re} z > 0$,

$$\int_1^\infty (N(u) - M(-u))u^{-z-1} \log u \, du = \frac{H(z) - H(0)}{z}$$

$$= \sum \frac{a_n}{z_n(z - z_n)} = \int_1^\infty g(u)u^{-z} \, du,$$

where

$$g(u) = \sum (a_n/z_n)u^{z_n - 1} \log u,$$

so that

$$N(u) - M(-u) = \sum (a_n/z_n)u^{z_n} = \eta(u)u^{-\alpha} \qquad \text{for } u > 1,$$

with $\eta$ satisfying $\eta(u) = \eta(ue^{2\pi/\rho})$. Then, as usual, the validity of this representation extends to all $u > 0$. To justify this conclusion, we may appeal to the complex inversion formula for Laplace transforms, as follows. Write

$$\int_1^\infty u^{-z} \, d(N(u) - M(-u)) = \int_0^\infty e^{-zu} \, dL(u),$$

where $L$ represents a finite measure (and, as customary, we shall consider it also as a function on $(0, \infty)$). Then, we have, on integration by parts, and writing $\tilde{L}(u) = L(\infty) - L(u)$,

$$\psi(z) := \frac{H(z) - H(0)}{z} = \int_0^\infty -\tilde{L}(u)e^{-zu} \, du = \sum \frac{a_n}{z_n(z - z_n)}, \qquad \operatorname{Re} z > -\alpha.$$

Appealing to the complex inversion formula (*cf.* Widder (1946), p. 77, Theorem 7.6.b; Doetsch (1974), p. 181, Theorem 27.2), we have

$$-\int_t^\infty \tilde{L}(u) \, du = \lim_{T \to \infty} \frac{1}{2\pi i} \int_{x_0 - iT}^{x_0 + iT} \frac{e^{tz}}{z} \psi(z) \, dz, \qquad \text{for } t > 0,$$

valid for any $x_0 \in (-\alpha, 0)$. Note that, for any $z$ with $\operatorname{Im} z = \tau_k := (k + \frac{1}{2})$, $k \in \mathbb{N}$,

$$|\psi(z)| \le a_0 \sum_{n=-\infty}^{\infty} \frac{1}{|\alpha - in\rho| \, |(k + \frac{1}{2} + n)\rho|} = a_0 \left( \sum_{-\infty}^{-(k+1)} + \sum_{-k}^{-1} + \sum_{0}^{\infty} \right) \le C,$$

for some constant $C$ independent of $k$; the same is true for $\operatorname{Im} z = -\tau_k$, $k \in \mathbb{N}$, as well, so that the integrals that follow converge. Now, consider the rectangular contours with vertices $-R \pm i\tau_k$, $x_0 \pm i\tau_k$, $R > 0$ arbitrary, $k \in \mathbb{N}$. As $R \to \infty$, $\psi(-R + iy) \to 0$ for any real $y$ with $|y| \le \tau_k$, so that

$$\frac{1}{2\pi i} \int_{x_0 - i\tau_k}^{x_0 + i\tau_k} \frac{e^{tz}}{z} \psi(z) \, dz = S_1(k, t) + S_2(k, t),$$

where

$$2\pi i S_1(k, t) = \int_{-\infty+i\tau_k}^{x_0+i\tau_k} - \int_{-\infty-i\tau_k}^{x_0-i\tau_k} \frac{e^{tz}}{z}\,\psi(z)\,dz,$$

and

$$\begin{aligned} S_2(k, t) &= \text{the sum of the residues of } \frac{e^{tz}}{z}\,\psi(z) \text{ at the poles } z_n \\ &\qquad\qquad\qquad\qquad\qquad \text{with } -k \le n \le k \\ &= \sum_{n=-k}^{k} (a_n/z_n)e^{(-\alpha+in\rho)t} = \eta_k(t)e^{-\alpha t}, \end{aligned}$$

where $\eta_k$ is obviously periodic with period $2\pi/\rho$. For every fixed $t > 0$, $S_1(k, t) \to 0$ as $k \to \infty$, so that

$$\int_t^\infty \tilde{L}(u)\,du = \eta(t)e^{-\alpha t},$$

where $\eta = \lim_{k\to\infty} \eta_k$ is periodic with period $2\pi/\rho$. Then, $\eta$ is differentiable, and $\tilde{L}$ is also seen to have the same kind of representation, i.e.,

$$N(u) - M(-u) = -\xi(u)/u^\alpha \qquad \text{for } u > 1,$$

where $0 \le \xi(u) = \xi(ue^{2\pi/\rho})$ for $u > 1$. Then, as before, we extend the validity of this representation to $u > 0$. Considering $H_1$ in place of $H$ and invoking (6.3.10), we see that $N(u)$ has the same form for $u > 0$, and therefore so does $-M(-u)$ (this follows also from considering $H_2$ and applying the same arguments). Thus, $X(1)$ has a semistable distribution, and the proof of the sufficiency part of the theorem is concluded.

Necessity part. If $q = 0$ or $\hat{S}(1) \ne 0$, the necessity of conditions (i) and (ii), if (6.4.1) is to be satisfied only by semistable processes, is established in exactly the same manner as in the proof of Theorem 6.3.6. We shall now show that if

$$E = \{z\colon \operatorname{Re} z = \alpha,\ S^*(z) = 0\}$$

is not a lattice (with $\alpha$ as a lattice point), then (6.4.1) is satisfied by some nonsemistable process. We enumerate $E_1$, the set of zeros of $S^*$ lying in $E$, and $E_2 = E\backslash E_1$ as follows:

$$E_1 = \{\alpha \pm i\beta_n : n \in \mathbb{Z}_+\}, \qquad \text{with } \beta_0 = 0,$$

$$E_2 = \{\alpha \pm i\gamma_n : n \in \mathbb{N}\}, \qquad \text{with } \gamma_n \ne 0 \text{ for any } n.$$

Let $\{c_n\}_{n=0}^\infty$ and $\{d_n\}_{n=1}^\infty$ be two sequences of real constants such that

$$\sum |c_n|(\alpha + |\beta_n|) + \sum |d_n|(\alpha + |\gamma_n|) < \alpha/2$$
$$\text{(in particular, } \textstyle\sum |c_n| + \sum |d_n| < \tfrac{1}{2}\text{)}.$$

Consider the distribution of $X(1)$ given by $L(\mu, \sigma^2, M, N)$ chosen as follows, depending on whether $E_2$ is empty or not (the preceding condition on the $c_n$, $d_n$ ensures that $M$ and $N$ defined in the following satisfy the conditions of the Lévy representation):

(a) If $E_2$ is empty, take $\mu$ satisfying $\mu\hat{S}(1) + q = 0$; $\sigma = 0$;

$$M(-u) = -N(u) = u^{-\alpha}\left[1 + \sum_{n=0}^{\infty} c_n(u^{i\beta_n} + u^{-i\beta_n})\right] \qquad \text{for } u > 0.$$

(b) If $E_2$ is not empty, let $\eta_n = \alpha + i\gamma_n$, take $\sigma = 0$ and $\mu$ satisfying

$$\hat{S}(1)\left\{\mu + 2\pi \sum_{n=1}^{\infty} d_n \operatorname{Re}[\eta_n \sec(\pi\eta_n/2)] + q\right\} = 0;$$

$$M(-u) = u^{-\alpha}\{1 + \sum c_n(u^{i\beta_n} + u^{-i\beta_n}) + \sum d_n(u^{i\gamma_n} + u^{-i\gamma_n})\},$$

$$-N(u) = u^{-\alpha}\{1 + \sum c_n(u^{i\beta_n} + u^{-i\beta_n}) - \sum d_n(u^{i\gamma_n} + u^{-i\gamma_n})\}, \qquad u > 0.$$

Thus, if $q = 0$ or $\hat{S}(1) \neq 0$, in either of the cases (a) or (b), there exists a nonsemistable process satisfying (6.4.1). Hence, if (6.4.1) is to be satisfied by semistable processes alone, the zeros of $S^*$ on the line $\operatorname{Re} z = \alpha$ should form a subset of some lattice with $\alpha$ as a lattice point. This completes the proof of the necessity part and of the theorem as well.

For the sake of completeness, we point out the forms that $S$, $\hat{S}$ take in the formulation of Theorem 6.3.6 and Theorem 6.4.1, with the stochastic integrals $Y_j$ left in their original forms $Y_j = \int_{a_j}^{b_j} g_j(t)\, dX(v_j(t))$, $j = 1, 2$, namely,

$$S(z) = \int_{a_1}^{b_1} |g_1(t)|^z \, dv_1(t) - \int_{a_2}^{b_2} |g_2(t)|^z \, dv_2(t),$$

$$\hat{S}(z) = \int_{a_1}^{b_1} |g_1(t)|^z (\operatorname{sgn} g_1(t)) \, dv_1(t) - \int_{a_2}^{b_2} |g_2(t)|^z (\operatorname{sgn} g_2(t)) \, dv_2(t).$$

An example that shows that Theorem 6.4.1 genuinely extends Theorem 6.3.6, and in which neither $g_1$ nor $g_2$ reduces to a constant, would be the following:

$$[a_j, b_j] = [0, 1] \quad \text{for } j = 1, 2; \qquad g_1(t) = t, \qquad g_2(t) = t - 4,$$

$$v_1 = \delta_8, \qquad v_2 = \delta_8 + 2\delta_2.$$

Then, $Y_1 \sim Y_2$ reduces to $2X(3) - 3X(2) \sim 4X(1)$. We have

$$S(z) = w(w + 1)(w - z), \qquad \hat{S}(z) = w(w^2 - w + 2),$$

where $w = 2^z$; since $S$ and $\hat{S}$ are entire functions, the conditions (6.4.10) are satisfied. $S$ has a unique real zero at 1, this zero is simple, the zeros of $S$ on the line $\operatorname{Re} z = 1$ are at the points $\{1 + 2k\pi i/\log 2 : k \in \mathbb{Z}\}$, and $\hat{S}$ has no zeros on that line (its zeros lie on the line $\operatorname{Re} z = \frac{1}{2}$). Thus, the underlying process is semistable with exponent $\alpha = 1$. It cannot be asserted to be stable, in general, in view of Theorem 6.3.6.

## APPENDIX: SOME PHRAGMÉN-LINDELÖF TYPE THEOREMS AND OTHER AUXILIARY RESULTS

Let $\mathcal{A}$ be the class of all complex-valued functions $f$ defined on the closed upper half-plane, having the following properties:

$f$ is continuous on the closed half-plane, analytic on the open half-plane, and the zeros of $f$ do not have a finite limit point.

$f \in \mathcal{A}$ is said to be of *exponential type* if, for some $A > 0$, $c > 0$,

$$M(r, f) := \max\{|f(z)| : |z| \le r,\ y \ge 0\} \le Ae^{cr} \qquad \forall\, r > 0,$$

where $z = x + iy$, equivalently, if

$$\overline{\lim_{r \to \infty}} \frac{\log M(r, f)}{r} < \infty.$$

For $f \in \mathcal{A}$, we define an associated function $\chi_f$ by: $\chi_f \equiv 1$ if $f$ has no zero; otherwise, let $\{a_n\}$ be an enumeration in order of increasing moduli of the zeros of $f$, and define

$$\chi_f(z) = \prod_{n=1}^{\infty} ((1 - z/a_n)/(1 - z/\bar{a}_n)).$$

For $a > 0$, let $\log^+ a = \max(\log a, 0)$. If $f \in \mathcal{A}$, is of exponential type, and satisfies the condition

$$\int_{-\infty}^{\infty} \frac{\log^+|f(x)|}{1 + x^2} dx < \infty, \tag{6.A.1}$$

then (*cf.* Boas, 1954, Section 6.3.6) it also satisfies the condition

$$\int_{-\infty}^{\infty} \frac{|\log|f(x)||}{1 + x^2} dx < \infty. \tag{6.A.2}$$

One can hence speak of the "Poisson integral" of $\log|f|$ defined by

$$v_f(z) = \frac{y}{\pi} \int_{-\infty}^{\infty} \frac{\log|f(t)|}{(t - x)^2 + y^2} dt.$$

It is known that (Boas, 1954, Section 6.6.2) then

$$\lim_{r\to\infty} |v_f(ir)|/r = 0, \tag{6.A.3}$$

and (Levin, 1964, Section 5.3, Lemma 5)

$$\lim_{r\to\infty}{}^{*} \frac{\log|\chi_f(ir)|}{r} = 0, \tag{6.A.4}$$

where lim* denotes (incidentally, the existence as well of) the limit as $r \to \infty$ taking all values except possibly those belonging to a set of relative measure zero.

The following result (Nevanlinna–Levin) is basic to the discussion of Phragmén–Lindelöf type theorems in this section.

**Theorem 6.A.1.** *Let $f \in \mathcal{A}$ be of exponential type and satisfy* (6.A.1) (*or* (6.A.2)). *Then,*

$$\log|f(z)| = v_f(z) + \log|\chi_f(z)| + c_f y, \tag{6.A.5}$$

*where*

$$c_f = \overline{\lim_{r\to\infty}} \log|f(ir)|/r. \tag{6.A.6}$$

***Remark.*** In Nevanlinna's formulation (Boas, 1954, Section 6.5.4), the representation (6.A.5) is obtained under (6.A.1)—with the constant $c_f$ obtained as the value of a certain integral rather than in the form (6.A.6). Levin (1964, Section 5.3, Theorem 5) obtains (6.A.5) with (6.A.6) as well, under the assumption (6.A.2). An optimal method of deriving (6.A.6) appears to be to start with Nevanlinna's proof of (6.A.5) and then appeal to the facts (6.A.3) and (6.A.4), yielding (6.A.6) immediately.

Noting that the definition of $\chi_f$ implies that $|\chi_f(z)| < 1$ if $y > 0$, we have the following, almost immediate consequences.

**Corollary 6.A.2.** *If $f \in \mathcal{A}$ is of exponential type, with $|f(t)| \le 1$ for all real $t$, and if $\overline{\lim}_{r\to\infty} \log|f(ir)|/r \le 0$, then $|f(z)| \le 1$ for all $z$ with* $\operatorname{Im} z \ge 0$.

***Proof.*** Note that (6.A.1) holds and the assertion follows immediately from (6.A.5) and (6.A.7).

**Theorem 6.A.3.** *Let $f, g \in \mathcal{A}$ be such that $f$ and $k = fg$ are bounded on the closed half-plane and $g$ is bounded on the real line. Then, $g$ is of exponential type.*

***Proof.*** Without loss of generality, we may assume 1 to be the common bound. Our assumptions imply that $k \in \mathfrak{C}$, and $f$, $k$ are of exponential type. Then, we have from (6.A.5)—since (6.A.1) holds—

$$\log |k(z)| = v_k(z) + \log |\chi_k(z)| + c_k y,$$

$$\log |f(z)| = v_f(z) + \log |\chi_f(z)| + c_f y.$$

We have, by subtraction, noting that $\chi_k = \chi_f \chi_g$,

$$\log |g(z)| = v_g(z) + \log |\chi_g(z)| + (c_k - c_f)y. \qquad (6.A.7)$$

The first two terms on the right are $\leq 0$ for $\operatorname{Im} z > 0$ (since $|g| \leq 1$ on the real line, and since $|\chi_g(z)| < 1$ for $\operatorname{Im} z > 0$) so that, for some real $c$,

$$\log |g(z)| \leq cy \qquad \text{for } y > 0.$$

Hence, the theorem is proven.

**Corollary 6.A.4.** *If, in addition to the assumptions of Theorem* 6.A.3, *we also have*

$$\overline{\lim_{r \to \infty}} \log |f(ir)|/r = 0,$$

*then g is bounded on the closed upper half-plane.*

***Proof.*** Taking 1 as the common bound, in the formulation of Theorem 6.A.3, as before, we have (6.A.7). Since $c_f = 0$ by our new assumption now, while $c_k \leq 0$, it follows that $|g(z)| \leq 1$ for $\operatorname{Im} z > 0$ (and, by assumption, for $\operatorname{Im} z = 0$).

We pass on the counterpart of Corollary 6.A.4 for semidisks. The mapping $z \to (1 - z)(1 - z^{-1})/4$ takes the open semidisk $D_+ = \{z\colon |z| < 1, y > 0\}$ onto the open upper half-plane and $\partial D_+ \backslash \{0\}$ onto the real line ($\partial D_+$, as usual, denoting the boundary of $D_+$). Then, we have, from Corollary 6.A.4, the following result.

**Corollary 6.A.5.** *Let $f$, $g$ be analytic on $D_+$, continuous and bounded on $\partial D_+ \backslash \{0\}$, and assume that the zeros of $f$ and $g$ have no limit points in $\partial D_+ \backslash \{0\}$. Suppose further that $f$ and $fg$ are bounded on $D_+$, and that $\overline{\lim}_{y \to 0+} y \log |f(iy)| = 0$. Then, $g$ is itself bounded on $\bar{D}_+ \backslash \{0\}$.*

We next note a representation theorem for meromorphic functions having only simple poles (*cf.* Titchmarsh, 1939, p. 110).

**Theorem 6.A.6.** *Let $f$ be meromorphic on $\mathbb{C}$, and let all its poles be simple. Let the poles, arranged in order of increasing moduli, be $\{a_n\}$ ($a_n \neq 0$ for any $n$), and the residue at $a_n$ be $b_n$. Suppose further that there exists a sequence $\{C_n\}$ of contours such that:*

(i) *$C_n$ encloses only $a_1, \ldots, a_n$;*
(ii) *the distance $r_n$ of $C_n$ from the origin tends to $\infty$ as $n \to \infty$, and the length of $C_n$ is $O(r_n)$; and*
(iii) $\sup\{|f(z)|: z \in C_n\} = o(r_n)$.

*Then,*

$$f(z) = f(0) + \sum_{n=1}^{\infty} b_n\left(\frac{1}{z - a_n} + \frac{1}{a_n}\right) \qquad \text{for } z \neq a_n,\ n \in \mathbb{N}.$$

(*The modification needed when the origin is also a pole is obvious.*)

Finally, we consider the evaluation of an integral, needed in the necessity parts of the proofs of Theorems 6.3.6 and 6.4.1.

**Lemma 6.A.7.** *Let* $0 \leq \operatorname{Re} w \leq 2$ *and let* $t \in \mathbb{R}$, $t \neq 0, 1$. *Then,*

$$I_w := \int_0^\infty \frac{x^{2-w}\,dx}{(1 + x^2)(1 + t^2x^2)}$$

$$= \begin{cases} \pi(|t|^{w-1} - 1)\sec(\pi w/2)/(2(1 - t^2)) & \text{if } w \neq 1, \\ -\log|t|/(1 - t^2) & \text{if } w = 1. \end{cases}$$

***Proof.*** The case $w = 1$ can be calculated directly (as a "real" integral), or obtained in the limit as $w \to 1$ from the formula for $w \neq 1$. For $w \neq 1$, first consider $\operatorname{Im} w \leq 0$. Defining

$$g(z) = \frac{z^{2-w}}{(1 + z^2)(1 + t^2z^2)}$$

and, for a fixed $\varepsilon > 0$, taking a single-valued branch of $g$ in $\operatorname{Im} z \geq 0$, $|z| \geq \varepsilon$, consider the closed contour $\Gamma$ consisting of the semicircular arcs of $|z| = R$ and $|z| = \varepsilon$ in the upper half-plane, and the segments $[-R, -\varepsilon]$ and $[\varepsilon, R]$. $G$ has poles at $i$ and at $i/|t|$, and the expression for $I_w$ follows from the residue theorem, noting that the integrals along the arcs tend to zero as $\varepsilon \to 0+$ and $R \to \infty$. For $w \neq 1$, and $\operatorname{Im} w > 0$, we need only apply the foregoing to evaluate $I_{\bar{w}}$, and then we use the fact that $I_{\bar{w}} = \bar{I}_w$.

## NOTES AND REMARKS

Theorem 6.1.4 is from Riedel (1980a). Theorems 6.2.1, 6.2.2, and 6.2.3 are extended forms of results of E. Lukacs and R. G. Laha, J. Marcinkiewicz, and V. P. Skitovich, established in Ramachandran and Rao (1970). Theorem 6.3.1, from Ramachandran (1991a), is a strong version of a result by Lukacs (1969); the latter is also proved in Riedel (1980b) using the general results established there, namely our Theorem 6.3.6. This theorem, its auxiliary results and corollaries, which comprise Section 6.3, are essentially as in Riedel (1980b). Theorem 6.4.1 is also from Ramachandran (1991a), where an example is also to be found showing that the result of Theorem 6.3.6 or Theorem 6.4.1 may hold without the technical conditions (6.3.10)(i)(ii) being satisfied; these conditions are satisfied, however, in the context of Theorem 6.3.1. Theorem 6.A.3 and Corollaries 6.A.2, 6.A.4, and 6.A.5 are due to the Leipzig school and may also be found summarized in Rossberg *et al.* (1985). Lemma 6.A.7 is from Riedel (1980b).

CHAPTER

# 7

# Distribution Problems Relating to the Arc-sine, the Normal, and the Chi-Square Laws

In this chapter, we consider three distribution problems of interest, though only one of them leads to a functional equation. We first consider an equidistribution problem: $XY \sim X + Y$, where $X$ and $Y$ are i.i.d.r.v.'s (with moments of all orders), known to be satisfied if $X$ follows a particular arc-sine law; we identify a countable class of discrete d.f.'s also having this property. The second deals with conditions for i.i.d.r.v.'s $X$, $Y$ to follow a normal law if it is assumed that, for one or more $(a, b) \in \mathbb{R}^2$, the r.v. $(aX + bY)^2/(a^2 + b^2)$ follows a chi-square law (with one degree of freedom), and is partly amenable to an approach similar to Cramér's proof of Theorem 1.3.13. The third concerns certain quadratic forms in i.i.d.r.v.'s following (possibly "noncentral") chi-square distributions, and conditions under which the basic r.v.'s are necessarily normally distributed are obtained.

## 7.1. AN EQUIDISTRIBUTION PROBLEM, AND THE ARC-SINE LAW

Consider the arc-sine law given by the p.d.f.

$$p(x) = \begin{cases} 1/(\pi\sqrt{4 - x^2}) & \text{for } |x| < 2, \\ 0 & \text{otherwise.} \end{cases} \tag{7.1.1}$$

The moment sequence of this law, as easily computed, is given by

$\{C(n): n \in \mathbb{Z}_+\}$, where $C(n)$ is defined by

$$C(0) = 1, \quad C(2j) = \binom{2j}{j}, \quad \text{and} \quad C(2j-1) = 0, \qquad j \in \mathbb{N}. \quad (7.1.2)$$

If $X$ and $Y$ are i.i.d.r.v.'s with the preceding distribution law, then all the odd order moments of $X + Y$, as well as of $XY$, are zero; the moment of order $2n$ for $XY$ is obviously $\binom{2n}{n}^2$, while that for $X + Y$ is

$$C(2n) + \binom{2n}{2n-2}C(2n-2)C(2) + \cdots + \binom{2n}{2}C(2)C(2n-2) + C(2n),$$

which simplifies to

$$\binom{2n}{n}\left[\binom{n}{0}^2 + \cdots + \binom{n}{n}^2\right] = \binom{2n}{n}^2.$$

Thus, $XY$ and $X + Y$ have the same moment sequence and, further, both have the same compact support, namely $[-4, 4]$. D.f.'s with compact supports are uniquely determined by their moment sequences; hence, we conclude that

$$XY \sim X + Y. \quad (7.1.3)$$

This fact leads naturally to two questions: (a) If $X$ and $Y$ are i.i.d.r.v.'s with moments of all orders, what are the possible distribution laws for $X$ if (7.1.3) is to hold? (b) Under what additional conditions can it be asserted that their law of distribution is an arc-sine law? In what follows, we answer these questions: The arc-sine law and a certain countable family of discrete laws qualify under (a), and the arc-sine law (7.1.1) is the only law not supported by a finite set of points, for which (7.1.3) holds (Theorem 7.1.4). We begin with some preliminary results.

For $n \in \mathbb{Z}_+$, let $m_n$ denote the $n$th moment of $X$, and $\alpha_n$ the $n$th absolute moment of $X$.

**Lemma 7.1.1.** *Suppose $X$ and $Y$ are i.i.d.r.v.'s with moments of all orders, and that* (7.1.3) *holds. Then, the d.f. of $X$ has support contained in* $[-2, 2]$.

***Proof.*** Equation (7.1.3) implies that

$$m_0 = 1, \qquad m_n^2 = \sum_{j=0}^{n} \binom{n}{j} m_j m_{n-j}, \qquad n \in \mathbb{N}. \quad (7.1.4)$$

Since $\alpha_i^{1/i} \le \alpha_j^{1/j}$ for $0 < i < j$ (Liapunov's inequality), we have

$$m_n^2 \le \sum_{j=0}^{n} \binom{n}{j} \alpha_j \alpha_{n-j} \le \alpha_n \sum_{j=0}^{n} \binom{n}{j} = 2^n \alpha_n,$$

so $m_{2k} = \alpha_{2k} \le 2^{2k}$, $k \in \mathbb{N}$. Hence, for any $\varepsilon > 0$,

$$P\{|X| > 2 + \varepsilon\} \le (2/(2 + \varepsilon))^{2k} \to 0$$

as $k \to \infty$ (Markov's inequality), whence $P\{|X| \le 2\} = 1$, and the lemma follows.

It follows that the d.f. of such a r.v. is uniquely determined by its moment sequence. We therefore turn to examine the possible forms this sequence can take. We begin by noting that (7.1.4) implies that $m_1^2 = 2m_1$, so that $m_1 = 0$ or $m_1 = 2$. We shall say that one moment sequence "branches off" from another at the $k$th moment if the two sequences agree up to the moment of order $k$, but their $(k + 1)$th moments are different. Then, we see inductively that a moment sequence satisfying (7.1.4) branches off from the moment sequence (7.1.2) at some moment, say the $k$th. If $k$ were to be odd, then (7.1.4) would imply that $m_{k+1} = C(k + 1)$ or $m_{k+1} = 2 - C(k + 1)$; the second possibility is ruled out since an even order moment cannot be negative, and the first possibility shows that there is no branching off at the $k$th moment. Thus, branching off from (7.1.2), for a moment sequence satisfying (7.1.4), is only possible at an even order moment. Let such a branching off take place at the $k$th moment, where $k$ is even; then, (7.1.4) implies that $m_{k+1} = 0$ or 2, and the second possibility in turn leads to $m_{k+2} = C(k + 2)$ or $2 - C(k + 2)$; again, the latter possibility is ruled out. Thus, if $m_{k+1} = 2$ for an even $k$, then $m_{k+2} = C(k + 2)$ necessarily.

Further analysis of possible distributions for $X$ for which (7.1.3) holds requires the following information on moment sequences (*cf.* Shohat and Tamarkin, 1943, p. 5):

**Theorem 7.1.2.** *Let $\{m_n\}$ be the moment sequence of some d.f. F. Let, for $n \in \mathbb{Z}_+$,*

$$\Delta_n := \det\begin{pmatrix} m_0 & m_1 & \cdots & m_n \\ m_1 & m_2 & \cdots & m_{n+1} \\ & & \vdots & \\ m_n & m_{n+1} & \cdots & m_{2n} \end{pmatrix}.$$

*Then, a necessary and sufficient condition for F to be supported by a set of*

*$(k+1)$ points is that*

$$\Delta_0 > 0, \ldots, \Delta_k > 0, \qquad \Delta_{k+1} = 0.$$

*Also, a necessary and sufficient condition for $F$ not to be supported by some finite set of points is that $\Delta_k > 0$ for all $k \in \mathbb{Z}_+$.*

In this context, we may remark that if $\Delta_n = 0$, then for some real $c_0, \ldots, c_n$ not all zero, we must have

$$c_0 m_j + c_1 m_{j+1} + \cdots + c_n m_{j+n} = 0, \qquad j = 0, 1, \ldots, n.$$

It follows easily that $\int (c_0 + c_1 x + \cdots + c_n x^n)^2 \, dF(x) = 0$, so that $F$ is supported by some finite set, namely, the set of real zeros of the polynomial $c_0 + c_1 x + \cdots + c_n x^n$.

Let now $\mathbf{D}_n = (d_{ij})$, $0 \le i, j \le n$, be the $(n+1) \times (n+1)$ matrix defined by $d_{ij} = C(i+j)$, and let $\hat{\mathbf{D}}_n$ be the matrix obtained from $\mathbf{D}_n$ on replacing the two entries $d_{n-1,n}$ and $d_{n,n-1}$ (both equal 0) by the number 2. Then, we have the following important auxiliary result; its proof, which is somewhat long, is deferred to the end of this section.

**Lemma 7.1.3.** *For $n \ge 1$, $\det \mathbf{D}_n = 2^n$, $\det \hat{\mathbf{D}}_n = 0$.*

As an almost immediate consequence of Lemma 7.1.3 and Theorem 7.1.2 (in fact, even the above remark), we have the following.

**Theorem 7.1.4.** *If $X$ and $Y$ are i.i.d.r.v.'s with moments of all orders and satisfy (7.1.3), and if their common d.f. $F$ is not supported by some finite set, then $X$ follows the arc-sine law.*

***Proof.*** If $X$ does not follow the arc-sine law (7.1.1)—uniquely determined by its moment sequence (7.1.2)—then, according to our discussion of the possible branchings off of its moment sequence from (7.1.2), we have that, for some $n \in \mathbb{Z}_+$,

$$m_j = C(j) \quad \text{for } 0 \le j \le 2n+2, j \ne 2n+1; \qquad m_{2n+1} = 2.$$

Then, $\Delta_{n+1} = \det \hat{\mathbf{D}}_{n+1} = 0$ by Lemma 7.1.3. The previous remark then implies that $F$ is supported by some finite set (Lemma 7.1.3 and Theorem 7.1.2 imply the stronger conclusion that $F$ is supported on a set of $(n+1)$ points), and hence the theorem.

The countable family of d.f.'s that are supported by finite sets, for which (7.1.3) holds, is given by the following.

**Theorem 7.1.5** *For every* $n \in \mathbb{N}$, *there exists a unique d.f.* $F_n$, *supported by a set of* $(n + 1)$ *points, given by*

$$F_n = (2n + 1)^{-1}\left[\delta_2 + 2 \sum_{j=1}^{n} \delta_2 \cos(2j\pi/(2n + 1))\right],$$

*with the following property: If X and Y are i.i.d.r.v.'s with* $F_n$ *as d.f., then* (7.1.3) *holds, and the moments of* $F_n$ *up to order* $(2n + 2)$ *are given by*

$$m_j = C(j) \quad \textit{for } 0 \le j \le 2n + 2,\ j \ne 2n + 1; \qquad m_{2n+1} = 2. \quad (7.1.5)$$

***Proof.*** For fixed $n \in \mathbb{N}$, let $c = (2n + 1)^{-1}$ and $x_j = 2\cos(2jc\pi)$ for $j = 0, 1, \ldots, n$. Let $A$ and $B$ be the sets of values taken (with positive probability) by $X + Y$ and $XY$, respectively, where $X$, $Y \sim F_n$ are i.i.d.r.v.'s. Note that $4 \in A$ and $P\{X + Y = 4\} = P\{XY = 4\} = c^2$. If $z \ne 4$, $z \in A$, then it is of one—possibly more than one—of the three forms:

$$2 + x_i; \qquad x_j + x_j; \qquad x_p + x_q,$$

for some $i \ge 1$, $j \ge 1$, $1 \le p < q \le n$, where $i$, $j$, $p$, $q$ are all distinct, and there is at most one such $i$ or $j$. Let $n_1$ $(=0$ or $1)$, $n_2$ $(=0$ or $1)$, and $n_3$ be the number of distinct representations of $z$ of these three forms, respectively. Then, we note the following:

$$2 + x_i = x_k^2,$$

where $k = i/2$ if $i$ is even, and $k = (2n + 1 - i)/2$ if $i$ is odd;

$$x_j + x_j = 2x_j;$$

and

$$x_p + x_q = x_r x_s,$$

where $r = (p + q)/2$, $s = |p - q|/2$ if $(p + q)$ is even, and $r = (2n + 1 - p + q)/2$, $s = (2n + 1 - p - q)/2$ if $(p + q)$ is odd, and where, again, $k$, $j$, $r$, $s$ are all different. Thus,

$$P\{X + Y = z\} = n_1 4c^2 + n_2 4c^2 + n_3 8c^2 \le P\{XY = z\}.$$

Hence, if $z \in A$, then $z \in B$ and $P\{X + Y = z\} \le P\{XY = z\}$.

We now prove, conversely, that if $w \in B$, then $w \in A$, and

$$P\{XY = w\} \le P\{X + Y = w\}.$$

If $w = 4$, this is immediate (with $=$ in place of $\le$). Let, then, $w \ne 4$. $w$ is of one—possibly more than one—of the three forms:

$$x_i^2; \qquad 2x_j; \qquad x_r x_s,$$

for some $i \geq 1$, $j \geq 1$, $1 \leq r < s \leq n$, where $i, j, r, s$ are all different and again there is at most one such $i$ or $j$. Let $m_1$ ($=0$ or 1), $m_2$ ($=0$ or 1), and $m_3$ be the number of distinct representations of $w$ according to these three forms, respectively. We then note the following:

$$x_i^2 = 2 + x_k,$$

where $k = 2i$ if $i \leq n$, $k = 2(2n + 1 - i)$ if $i > n$;

$$2x_j = x_j + x_j;$$

and

$$x_r x_s = x_p + x_q,$$

where $p = r + s$ if $r + s \leq n$, $p = 2n + 1 - r - s$ if $r + s > n$, and $q = |r - s|$ in either case, and where $i, j, p, q$ are all different. Hence, for $w \in B$, $w \neq 4$,

$$P\{XY = w\} = m_1 4c^2 + m_2 4c^2 + m_3 8c^2 \leq P\{X + Y = w\}.$$

Thus, $A = B$ and $XY \sim X + Y$, if $X$, $Y$ are i.i.d.r.v.'s with $X \sim F_n$.

$F_n$ has moments of all orders, satisfying (7.1.4) then, and its moment sequence must branch off (from (7.1.2)) at some even integer order moment, say $2k$. Lemma 7.1.3 then implies that (in the notation used in Theorem 7.1.2)

$$\Delta_j = \det \mathbf{D}_j > 0 \quad \text{for } j = 1, 2, \ldots, k, \qquad \Delta_{k+1} = \det \hat{\mathbf{D}}_{k+1} = 0.$$

Theorem 7.1.2 then implies that the corresponding d.f. is supported by a set of $(k + 1)$ points, so that we must have $k = n$. This establishes (7.1.5). The uniqueness follows from the fact that, since $\Delta_{n+1} = 0$ while $\Delta_n > 0$, there is an essentially unique column vector $\mathbf{a}$ with $\mathbf{a}' = (a_0, \ldots, a_{n+1})$ orthogonal to the row vectors of $\hat{\mathbf{D}}_{n+1}$, and, as in the remark following Theorem 7.1.2, the elements of the support of the d.f. with moments satisfying (7.1.5) are given by the roots of the equation $\sum_{j=0}^{n+1} a_j x^j = 0$; the associated probabilities are then uniquely determined by $\sum_{j=0}^{n} p_j x_j^r = C(r)$, $r = 0, 1, \ldots, n$. Since $F_n$ satisfies (7.1.5), it follows that it must be the unique such d.f. supported by a set of $(n + 1)$ points.

As an immediate consequence of Theorems 7.1.4 and 7.1.5, we have the following.

**Theorem 7.1.6.** *If $X$ and $Y$ are i.i.d.r.v.'s with moments of all orders and if $XY \sim X + Y$, then either $X$ follows the arc-sine law* (7.1.1), *or $X \sim F_n$ for some $n \in \mathbb{N}$, where $F_n$ is given by Theorem* 7.1.5.

We pass on to a proof of Lemma 7.1.3. We define the $n \times n$ matrix $\mathbf{T}_n$ and the $(n+1) \times (n+1)$ matrix $\mathbf{S}_n$ according to

$$\mathbf{T}_n = (C(2i + 2j - 2)), \qquad 1 \le i, j \le n,$$

$$\mathbf{S}_n = (C(2i + 2j)), \qquad 0 \le i, j \le n.$$

**Lemma 7.1.7.** (a) $\det \mathbf{T}_n = 2^n$; (b) $\det \mathbf{S}_n = 2^n$.

***Proof.*** (a) For $n \ge 1$, consider the lower triangular $n \times n$ matrix $\mathbf{E}_n$ given by

$$\mathbf{E}_n = \begin{pmatrix} \binom{1}{0} & 0 & \cdots & 0 \\ \binom{3}{1} & \binom{3}{0} & \cdots & 0 \\ \binom{5}{2} & \binom{5}{1} & \cdots & 0 \\ \cdots & \cdots & \cdots & \cdots \\ \binom{2n-1}{n-1} & \binom{2n-1}{n-2} & \cdots & \binom{2n-1}{0} \end{pmatrix}.$$

Using the facts that

$$\binom{5}{1} = \binom{5}{4}, \quad \binom{3}{1} = \binom{3}{2},$$

etc., and the well-known Vandermonde identity

$$\sum_{j=0}^{k} \binom{p}{j}\binom{q}{k-j} = \binom{p+q}{k},$$

we see that the inner product of the second and third rows, for instance, equals

$$\binom{3}{0}\binom{5}{1} + \binom{3}{1}\binom{5}{2}$$
$$= \tfrac{1}{2}\left[\binom{3}{0}\binom{5}{4} + \binom{3}{1}\binom{5}{3} + \binom{3}{2}\binom{5}{2} + \binom{3}{3}\binom{5}{1}\right] = \tfrac{1}{2}C(8).$$

Generalizing this argument, we see that $\mathbf{E}_n\mathbf{E}_n' = \frac{1}{2}\mathbf{T}_n$ (the prime denoting the transpose of the matrix concerned), whence $\det \mathbf{T}_n = 2^n$, since $\det \mathbf{E}_n = \det \mathbf{E}_n' = 1$.

(b) Consider the lower triangular matrix of order $(n + 1) \times (n + 1)$ given by

$$\mathbf{F}_n = \begin{pmatrix} 1/\sqrt{2} & 0 & 0 & \cdots & 0 \\ \binom{2}{1}\Big/\sqrt{2} & \binom{2}{0} & 0 & \cdots & 0 \\ \binom{4}{2}\Big/\sqrt{2} & \binom{4}{1} & \binom{4}{0} & \cdots & 0 \\ \cdots & \cdots & \cdots & \cdots & \\ \binom{2n}{n}\Big/\sqrt{2} & \binom{2n}{n-1} & \binom{2n}{n-2} & \cdots & \binom{2n}{0} \end{pmatrix}.$$

A similar argument as for (a) shows that $\mathbf{F}_n\mathbf{F}_n' = \frac{1}{2}\mathbf{S}_n$, so that

$$\det \mathbf{S}_n = 2^{n+1}(\det \mathbf{F}_n)^2 = 2^n,$$

and hence the lemma is proven.

We next prove the assertion in Lemma 7.1.3 that $\det \mathbf{D}_n = 2^n$. Consider the case $n = 2k$, $k \in \mathbb{N}$. Making an appropriate number of interchanges of the rows of $\mathbf{D}_n$ and an equal number of interchanges of its columns, we obtain the partitioned matrix

$$\begin{pmatrix} \mathbf{S}_k & 0 \\ 0 & \mathbf{T}_k \end{pmatrix},$$

whence it follows that $\det \mathbf{D}_n = \det \mathbf{S}_k \times \det \mathbf{T}_k = 2^n$.

If $n = 2k - 1$, the above procedure yields the partitioned matrix

$$\begin{pmatrix} \mathbf{S}_{k-1} & 0 \\ 0 & \mathbf{T}_k \end{pmatrix},$$

and we have $\det \mathbf{D}_n = 2^n$ in this case as well.

To prove the other assertion of Lemma 7.1.3, namely, $\det \hat{\mathbf{D}}_n = 0$, we need to develop a considerable amount of auxiliary information.

**Lemma 7.1.8.** *The cofactors of the zero entries in* $\mathbf{D}_n$ *are all zero.*

***Proof.*** We discuss the details of the case of odd $n$ and the cofactors of the entries of the first two rows of $\mathbf{D}_n$. The other cases are handled similarly. Let $n = 2k - 1$, $k \in \mathbb{N}$, and $\operatorname{adj}(\mathbf{D}_{2k-1})$ denote the adjoint of $\mathbf{D}_{2k-1}$. Let $a_1, \ldots, a_k$ be the cofactors of the zero entries in the first row of $\mathbf{D}_{2k-1}$, so that they are the even entries of the first column of $\operatorname{adj}(\mathbf{D}_{2k-1})$. Since the inner products of the first column of $\operatorname{adj}(\mathbf{D}_{2k-1})$ with the even rows of $\mathbf{D}_{2k-1}$ are all zero, we have

$$a_1 C(2) + \cdots + a_k C(2k) = 0,$$
$$a_1 C(2k) + \cdots + a_k C(4k-2) = 0.$$

This is just $\mathbf{T}_k \mathbf{a} = \mathbf{0}$, where $\mathbf{a}' = (a_1, \ldots, a_k)$. Since $\det \mathbf{T}_k \neq 0$ by Lemma 7.1.7(a), we have $\mathbf{a} = \mathbf{0}$.

Considering next the cofactors of the zero entries in the second row of $\mathbf{D}_{2k-1}$, we obtain $\mathbf{S}_k \mathbf{b} = \mathbf{0}$, where $\mathbf{b}' = (b_0, b_1, \ldots, b_k)$ is the vector with these cofactors as entries, and since $\det \mathbf{S}_k \neq 0$ by Lemma 7.1.7(b), we have $\mathbf{b} = \mathbf{0}$ as well. Hence, the lemma is proven.

We now define three $n$-dimensional column vectors $\mathbf{b}_n$, $\hat{\mathbf{b}}_n$, $\mathbf{b}_n^*$:

$$\mathbf{b}_n' = (C(n), \ldots, C(2n-2), 0); \qquad \hat{\mathbf{b}}_n' = (C(n), \ldots, C(2n-2), 2);$$
$$\mathbf{b}_n^{*\prime} = (C(n+1), \ldots, C(2n)).$$

Consider the partitioning

$$\mathbf{D}_{n+1} = \begin{pmatrix} \mathbf{D}_{n-1} & \mathbf{b}_n & \mathbf{b}_n^* \\ \mathbf{b}_n' & C(2n) & 0 \\ \mathbf{b}_n^{*\prime} & 0 & C(2n+2) \end{pmatrix}.$$

Then,

$$\det \mathbf{D}_{n+1} = \det \mathbf{D}_{n-1} \times \det\left\{\begin{pmatrix} C(2n) & 0 \\ 0 & C(2n+2) \end{pmatrix} - \begin{pmatrix} \mathbf{b}_n' \mathbf{D}_{n-1}^{-1} \mathbf{b}_n & \mathbf{b}_n' \mathbf{D}_{n-1}^{-1} \mathbf{b}_n^* \\ \mathbf{b}_n^{*\prime} \mathbf{D}_{n-1}^{-1} \mathbf{b}_n & \mathbf{b}_n^{*\prime} \mathbf{D}_{n-1}^{-1} \mathbf{b}_n^* \end{pmatrix}\right\}. \tag{7.1.6}$$

**Lemma 7.1.9.** (a) $\mathbf{b}_n^{*\prime} \mathbf{D}_{n-1}^{-1} \mathbf{b}_n = 0$ *for all* $n \in \mathbb{N}$; (b) *Let* $\gamma = C(2n) - \mathbf{b}_n' \mathbf{D}_{n-1}^{-1} \mathbf{b}_n$, $\delta = C(2n+2) - \mathbf{b}_n^{*\prime} \mathbf{D}_{n-1}^{-1} \mathbf{b}_n^*$; *then*, $\gamma\delta = 4$.

***Proof.*** (a) Write $\mathbf{b}_n^{*\prime} = (b_0^*, \ldots, b_{n-1}^*)$, $\mathbf{b}_n' = (b_0, \ldots, b_{n-1})$, and let $\mathbf{D}_{n-1}^{-1} = (d_{ij})$, $0 \le i, j \le n-1$. By Lemma 7.1.8, $d_{ij} = 0$ if $i + j$ is odd, and we also have: either

$$b_0 = b_2 = \cdots = b_1^* = b_3^* = \cdots = 0; \quad \text{or}$$
$$b_1 = b_3 = \cdots = b_0^* = b_2^* = \cdots = 0.$$

Hence, every term in the sum $\sum b_i^* b_j d_{ij}$ is zero, and we have (a).

(b) This follows from Lemma 7.1.3, relation (7.1.6) and part (a) of this lemma.

We are now in a position to prove the assertion of Lemma 7.1.2 that $\det \hat{\mathbf{D}}_n = 0$. Since

$$\hat{\mathbf{D}}_{n+1} = \begin{pmatrix} \mathbf{D}_n & \hat{\mathbf{b}}_{n+1} \\ \hat{\mathbf{b}}'_{n+1} & C(2n+2) \end{pmatrix},$$

we have

$$\det \hat{\mathbf{D}}_{n+1} = \det \mathbf{D}_n \times (C(2n+2) - \hat{\mathbf{b}}'_{n+1}\mathbf{D}_n^{-1}\hat{\mathbf{b}}_{n+1}),$$

and we need to show that the expression within the brackets is zero.

It is known (and is easily verified by straightforward multiplication) that if $\mathbf{A}$, $\mathbf{D}$ are square symmetric matrices and if all the relevant inverses exist, then

$$\begin{pmatrix} \mathbf{A} & \mathbf{B} \\ \mathbf{B}' & \mathbf{D} \end{pmatrix}^{-1} = \begin{pmatrix} \mathbf{A}^{-1} + \mathbf{F}\mathbf{E}^{-1}\mathbf{F}' & -\mathbf{F}\mathbf{E}^{-1} \\ -\mathbf{E}^{-1}\mathbf{F}' & \mathbf{E}^{-1} \end{pmatrix},$$

where

$$\mathbf{F} = \mathbf{A}^{-1}\mathbf{B}, \qquad \mathbf{E} = \mathbf{D} - \mathbf{F}'\mathbf{B} = \mathbf{D} - \mathbf{B}'\mathbf{A}^{-1}\mathbf{B}.$$

Since

$$\mathbf{D}_n = \begin{pmatrix} \mathbf{D}_{n-1} & \mathbf{b}_n \\ \mathbf{b}'_n & C(2n) \end{pmatrix},$$

we have

$$\mathbf{D}_n^{-1} = \gamma^{-1}\begin{pmatrix} \mathbf{P} & \mathbf{Q} \\ \mathbf{Q}' & 1 \end{pmatrix},$$

where $\gamma$ is as defined in Lemma 7.1.9(b),

$$\mathbf{P} = \gamma\mathbf{D}_{n-1}^{-1} + \mathbf{D}_{n-1}^{-1}\mathbf{b}_n\mathbf{b}'_n\mathbf{D}_{n-1}^{-1} \qquad \text{and} \qquad \mathbf{Q} = -\mathbf{D}_{n-1}^{-1}\mathbf{b}_n.$$

Hence

$$\begin{aligned}
&C(2n+2) - \hat{\mathbf{b}}'_{n+1}\mathbf{D}_n^{-1}\hat{\mathbf{b}}_{n+1} \\
&\quad = C(2n+2) - \gamma^{-1}(\mathbf{b}_n^{*\prime}\mathbf{P}\mathbf{b}_n^* + 4\mathbf{Q}'\mathbf{b}_n^* + 4) \\
&\quad = C(2n+2) - \gamma^{-1}(\gamma\mathbf{b}_n^{*\prime}\mathbf{D}_{n-1}^{-1}\mathbf{b}_n^* + \mathbf{b}_n^{*\prime}\mathbf{D}_{n-1}^{-1}\mathbf{b}_n\mathbf{b}'_n\mathbf{D}_{n-1}^{-1}\mathbf{b}_n^*) \\
&\qquad + 4\gamma^{-1}\mathbf{b}'_n\mathbf{D}_{n-1}^{-1}\mathbf{b}_n^* - 4\gamma^{-1} \\
&\quad = C(2n+2) - \mathbf{b}_n^{*\prime}\mathbf{D}_{n-1}^{-1}\mathbf{b}_n^* - 4\gamma^{-1} \qquad \text{(by Lemma 7.1.9(a))} \\
&\quad = 0 \qquad \text{(by Lemma 7.1.9(b))}.
\end{aligned}$$

Thus, $\det \hat{\mathbf{D}}_n = 0$ and the proof of Lemma 7.1.3, and so of Theorem 7.1.4 as well, is completed.

## 7.2. DISTRIBUTION PROBLEMS INVOLVING THE NORMAL AND THE $\chi_1^2$ LAWS

If a r.v. $X \sim \Phi$ (the standard normal d.f.), then the probability distribution of $X^2$ is referred to as a $\chi_1^2$ distribution, in words: a chi-square distribution with one degree of freedom; we write $X^2 \sim \chi_1^2$. More generally, if $X_1, \ldots, X_n$ are i.i.d.r.v.'s with distribution $\Phi$, then the probability distribution of $X_1^2 + \cdots + X_n^2$ (equivalently, the $n$-fold convolution of a $\chi_1^2$ distribution) is referred to as a $\chi_n^2$ distribution, in words: a chi-square distribution with $n$ degrees of freedom. The probability density function of a $\chi_n^2$ distribution is given by

$$p_n(x) = \begin{cases} (2^{n/2}\Gamma(n/2))^{-1}e^{-x/2}x^{n/2-1} & \text{for } x > 0, \\ 0 & \text{otherwise.} \end{cases}$$

Incidentally,

$$\sum_{j=1}^{n} (X_j - \bar{X})^2 = \left(\sum_{j=1}^{n} X_j^2\right) - n\bar{X}^2,$$

where $\bar{X} = (X_1 + \cdots + X_n)/n$, is distributed according to a $\chi_{n-1}^2$ distribution.

In this section, we shall consider a few distribution problems involving the $\chi_1^2$ and the normal laws. If $X \sim \Phi$ and if $Y = |X|$, then $Y^2 \sim \chi_1^2$ as well, showing that a r.v. need not have a standard normal d.f. if its square has a $\chi_1^2$ distribution. A trivial situation where $Y^2 \sim \chi_1^2$ implies that $Y \sim \Phi$ is when $Y$ is assumed to have a d.f. symmetric about the origin, for then $Y$ has a continuous d.f. since $Y^2$ has, and for $x > 0$,

$$\begin{aligned} P\{Y \leq -x\} &= P\{Y \geq x\} = \tfrac{1}{2}P\{|Y| \geq x\} \\ &= \tfrac{1}{2}P\{|X| \geq x\} = 1 - \Phi(x), \end{aligned}$$

where $X \sim \Phi$, and the preceding fact in turn implies that the previous relation holds for all real $x$, i.e., $Y \sim \Phi$ as well. (This fact also follows at once from Lemma 7.2.1 following since the ch.f. of $Y$ is real-valued if $Y$ has its d.f. symmetric about the origin.)

The principal result of this section asserts that if $X$ and $Y$ are i.i.d.r.v.'s with $(X + Y)^2/2 \sim \chi_1^2$, then $X \sim \Phi$. We begin with an auxiliary result. For the definition and properties of entire ch.f.'s used in what follows, we refer the reader to Lukacs (1970) or Ramachandran (1967).

**Lemma 7.2.1.** *Let $f$ be the ch.f. of the r.v. $X$. If $X^2 \sim \chi_1^2$, then* $\operatorname{Re} f(t) = \exp(-t^2/2)$, $t \in \mathbb{R}$, *and $f$ is an entire ch.f. with*

$$|f(z)| \leq 2\exp(|z|^2/2), \qquad \forall\, z \in \mathbb{C}. \tag{7.2.1}$$

***Proof.*** $X^2 \sim \chi_1^2$ implies in particular that $X$ has a continuous d.f. and, further,

$$P\{-X \leq x\} + P\{X \leq x\} = 2\Phi(x), \qquad \forall\, x \geq 0.$$

This fact in turn implies that it is also true for all real $x$. Taking the Fourier–Stieltjes transforms on both sides and noting that the ch.f. of $-X$ is $f(-\cdot) = \bar{f}$, we have

$$\int \cos tx\, dF(x) = \tfrac{1}{2}(f(-t) + f(t)) = \exp(-t^2/2), \qquad t \in \mathbb{R}. \quad (7.2.2)$$

Standard arguments then show that the leftmost expression may be differentiated under the integral sign as often as we please; evaluating the (even order) derivatives at $t = 0$, we have

$$m_{2n} := \int x^{2n}\, dF(x) = (-1)^n \frac{d^{2n}}{dt^{2n}} \exp(-t^2/2)\bigg|_{t=0}, \qquad n \in \mathbb{N}.$$

Hence,

$$\sum_{n=0}^{\infty} (-1)^n m_{2n} t^{2n}/(2n)! = e^{-t^2/2}, \qquad \forall\, t \in \mathbb{R}.$$

It follows that $\sum m_{2n} t^{2n}/(2n)!$ converges for all real $t$ (with $e^{t^2/2}$ as its sum). The inequality $m_{2n} + m_{2n-2} \geq 2|m_{2n-1}|$ shows that the series $\sum m_n t^n/n!$ converges for all real $t$. In other words, $f$ is an entire ch.f. Then, (7.2.2) implies that

$$\tfrac{1}{2}(f(ir) + f(-ir)) = \int \cosh rx\, dF(x) = \exp(r^2/2), \qquad r \in \mathbb{R}.$$

Since $M(r, f) := \max\{|f(z)|: |z| \leq r\}$ is known to be equal to $\max\{f(ir), f(-ir)\}$ for an entire ch.f. $f$ for all $r > 0$, it follows from the preceding that (7.2.1) holds.

**Corollary 7.2.2.** *If $f$ is also nonvanishing on $\mathbb{C}$, then it is a normal ch.f.*

***Proof.*** Since $f$, as a nonvanishing entire function of order two at most, in view of (7.2.1), is of the form $\exp Q$, where $Q$ is a quadratic polynomial (by Hadamard's factorization theorem, see Boas (1954), p. 22), $f$ is a normal ch.f.

**Corollary 7.2.3.** *Let $X$ and $Y$ be i.i.d.r.v.'s, and let $W = (X - Y)/\sqrt{2}$. Then, $W^2 \sim \chi_1^2$ if and only if $X - a \sim \Phi$ for some real $a$.*

***Proof.*** We need only prove the necessity. Let $f$ be the ch.f. of $W$; since $W$ has a d.f. symmetric about the origin, $f(t) = \operatorname{Re} f(t)$, and Lemma 7.2.1 implies that $W \sim \Phi$. Theorem 1.3.13 then implies that $X$ and $Y$ have normal distributions. Since $\operatorname{var} X = \operatorname{var} Y = \operatorname{var} W = 1$, our assertion follows.

We pass on to the main result of this section.

**Theorem 7.2.4.** *Let $X$ and $Y$ be i.i.d.r.v.'s, and let $W = (X + Y)/\sqrt{2}$. Then, $W^2 \sim \chi_1^2$ if and only if $X \sim \Phi$. In this case, $W \sim \Phi$ as well.*

***Proof.*** Let $f, g$ be the ch.f.'s of $X$ and $W$, respectively. By Lemma 7.2.1, $g$ is an entire ch.f. of order two at most and $\operatorname{Re} g(t) = \exp(-t^2/2)$. Since $g(t) = f(t/\sqrt{2})^2$ for $t \in \mathbb{R}$, it follows that $f$ is also an entire ch.f. of order two at most, and then the same is true of the functions

$$\int \cos zx \, dF(x) = \tfrac{1}{2}(f(z) + f(-z))$$

and

$$\int \sin zx \, dF(x) = \frac{1}{2i}(f(z) - f(-z)).$$

Let, for $t \in \mathbb{R}$,

$$u(t) = \int (\cos tx + \sin tx) \, dF(x), \qquad v(t) = \int (\cos tx - \sin tx) \, dF(x).$$

Then, the relation

$$\operatorname{Re}(f(t))^2 = \operatorname{Re} g(\sqrt{2}\, t) = e^{-t^2}, \qquad t \in \mathbb{R},$$

is equivalent to

$$u(t)v(t) = \exp(-t^2), \qquad t \in \mathbb{R}.$$

It follows that this holds for all complex $t$ as well, by analytic extension, and thus that $u$ and $v$ are both nonvanishing entire functions of order two at most, and therefore of the form $\exp Q$, where $Q$ is a quadratic polynomial. Since $u(0) = v(0) = 1$, and since $u$ and $v$ are real-valued and bounded for real values of the argument, it follows that

$$u(t) = \exp(-at^2 + bt), \qquad v(t) = \exp(-ct^2 + dt),$$

for some $a, c \geq 0$ and $b, d$ real. $u(-t) = v(t)$ implies that $a = c$, $b + d = 0$, so that

$$f(t) = (\cosh bt + i \sinh bt) \exp(-at^2).$$

Since $\operatorname{Re} f$ is a ch.f. as well, we see that $\cosh bt \exp(-at^2)$ must be a ch.f. It is integrable over $\mathbb{R}$, and therefore the corresponding d.f. is absolutely continuous, and has a continuous version of its probability density function given by (Theorem 1.3.1)

$$\begin{aligned} p(x) &= \frac{1}{2\pi} \int e^{-itx} \cosh(bt)\, e^{-at^2}\, dt \\ &= c_1(e^{(b-ix)^2/4a} + e^{(b+ix)^2/4a}) = c_2 e^{-x^2/4a} \cos(bx/2a), \end{aligned}$$

for suitable $c_1$, $c_2 > 0$. The requirement that $p(x) \geq 0$ for all $x \in \mathbb{R}$ then implies that $b = 0$ necessarily, so that $f$ corresponds to a normal d.f. with zero mean. Since $\operatorname{var} X = \operatorname{var} Y = \operatorname{var} W = 1$, it then follows that $X, Y, W \sim \Phi$.

***Remark.*** The following "quick proof" that $b = 0$ is of some interest. As above, $\cosh(bt)\exp(-at^2)$ is a ch.f. Note that

$$\cosh(bt) = \prod_{j=1}^{\infty} \left\{1 + \frac{4b^2t^2}{(2j-1)^2\pi^2}\right\},$$

and that $(1 + \alpha^2t^2)^{-1}$ is a ch.f. for any real $\alpha$, and so is any convergent infinite product of such ch.f.'s whose limit function is continuous at the origin (the Lévy–Cramér continuity theorem). This implies that $(\cosh(bt))^{-1}$ is a ch.f. as well. Thus, if $b \neq 0$, the normal ch.f. $\exp(-at^2)$ would be the product of two nonnormal ch.f.'s, contrary to the Lévy–Cramér theorem on the normal law (Theorem 1.3.13). Hence, we must have $b = 0$.

An open question in this context is: If, for some real $a$, $b$, with $ab \neq 0$, $|a| \neq |b|$,

$$(aX + bY)^2/(a^2 + b^2) \sim \chi_1^2,$$

is $X \sim \Phi$ necessarily? All that we can assert, in view of Lemma 7.2.1, is that $aX + bY$, and therefore $X$, has an entire ch.f. of order two at most.

We consider now some related results.

**Proposition 7.2.5.** *Let $X$ and $Y$ be independent (not necessarily identically distributed), and $W = (aX + bY)/c$, where $ab \neq 0$ and $c = \sqrt{a^2 + b^2}$. If $X^2$, $Y^2$, $W^2$ are all $\chi_1^2$ r.v.'s, then at least one of $X$ and $Y$ is distributed as $\Phi$.*

***Proof.*** Our assumptions and Lemma 7.2.1 imply that

$$\operatorname{Re} f(t) = \operatorname{Re} g(t) = \operatorname{Re}\{f(at/c)g(bt/c)\} = \exp(-t^2/2), \tag{7.2.3}$$

where $f$ and $g$ are respectively the ch.f.'s of $X$ and $Y$. Hence,

$$\operatorname{Re} f(at/c) \times \operatorname{Re} g(bt/c) = \exp(-t^2/2) = \operatorname{Re}\{f(at/c)g(bt/c)\}.$$

Since, for any ch.f. $h$, $\operatorname{Re} h = (h + h(-\cdot))/2$, we have from the previous equality that

$$(f(at/c) - f(-at/c))(g(bt/c) - g(-bt/c)) = 0.$$

Thus, at least one of these factors must vanish at any given $t \in \mathbb{R}$, and therefore at least one of them vanishes at some sequence of points tending to zero. Let $f(at/c) = f(-at/c)$ for some such sequence $\{t_n\}$ of values of $t$. Then, (7.2.3) implies that $f(t) = \exp(-t^2/2)$ at all points of such a sequence, namely $\{at_n/c\}$, and Theorem 1.3.15 enables us to conclude then that $X \sim \Phi$. In the complementary case, we conclude that $Y \sim \Phi$, and the proposition follows.

**Proposition 7.2.6.** *Let $X$ and $Y$ be independent r.v.'s with the same (finite) variance, and let $W_1 = (aX + bY)/c$, $W_2 = (aX - bY)/c$, where $a, b$ are real numbers with $ab \neq 0$, and $c = \sqrt{a^2 + b^2}$. If both $W_1^2$ and $W_2^2$ are $\chi_1^2$ r.v.'s, then at least one of $X$ and $Y$ is distributed as $\Phi$.*

***Proof.*** By our hypothesis and Lemma 7.2.1, $f, g, h_1, h_2$ being the ch.f.'s of $X, Y, W_1, W_2$, respectively, we have

$$\operatorname{Re}\{f(at/c)g(bt/c)\} = \operatorname{Re} h_1(t) = \exp(-t^2/2),$$

$$\operatorname{Re}\{f(at/c)g(-bt/c)\} = \operatorname{Re} h_2(t) = \exp(-t^2/2).$$

On adding these two relations, we have (since $g(-u) = \overline{g(u)}$ for $u \in \mathbb{R}$)

$$\operatorname{Re} f(at/c) \operatorname{Re} g(bt/c) = \exp(-t^2/2).$$

Since the two factors on the left are ch.f.'s (in fact, of symmetric d.f.'s, being real-valued), the Lévy–Cramér theorem (Theorem 1.3.13) implies that they both correspond to normal laws with zero means. Then, the assumption that $X$ and $Y$ have the same variance leads to

$$\operatorname{Re} f(t) = \operatorname{Re} g(t) = \exp(-t^2/2),$$

so that $X^2 \sim \chi_1^2$, $Y^2 \sim \chi_1^2$. It then follows from the previous proposition that at least one of $X$ and $Y$ is $\sim \Phi$.

The following is an immediate consequence.

**Corollary 7.2.7.** *Let $X$ and $Y$ be i.i.d.r.v.'s. Then, $X \sim \Phi$ if and only if, for some real $a, b$ with $ab \neq 0$, and with $c = \sqrt{a^2 + b^2}$, both $(aX + bY)^2/c^2$ and $(aX - bY)^2/c^2$ have a $\chi_1^2$ distribution.*

***Remark.*** If $|a| = |b|$, we need only require that either one of these two r.v.'s has a $\chi_1^2$ distribution.

## 7.3. QUADRATIC FORMS, NONCENTRAL $\chi^2$ LAWS, AND NORMALITY

In the preceding section, we have seen how the $\chi_n^2$ ($n \in \mathbb{N}$) distributions arise, and also identified their probability density functions $p_n$. If, instead of $X_1, \ldots, X_n$ being i.i.d. and $\sim \Phi$, they are independent and $X_j \sim N(\mu_j, \sigma)$, then the distribution of $(\sum_{i=1}^n X_i^2)/\sigma^2$ is called a noncentral chi-square distribution with $n$ degrees of freedom and noncentrality parameter $\lambda = (\sum_{i=1}^n \mu_i^2)/\sigma^2$. We denote such a d.f. as $\chi^2(n, \lambda)$; its p.d.f. is given by

$$e^{-\lambda/2} \sum_{r=0}^{\infty} (1/r!)(\lambda/2)^r p_{n+2r},$$

where $p_n$ denotes the p.d.f. of the $\chi_n^2$ density function.

In this section, we shall be concerned with the problem of characterizing the normal law (as the common d.f. of the "basic" r.v.'s concerned) through hypotheses of the form that one or more quadratic forms in i.i.d.r.v.'s have a (possibly noncentral) $\chi^2$ distribution.

An important preliminary result relevant in this context is the following, due to H. Sakamoto and O. Carpenter (*cf.* Rao, 1973, p. 186):

**Theorem 7.3.1.** *Let* $\mathbf{X}' = (X_1, \ldots, X_N)$ *be a random vector with i.i.d. components and with* $X_1 \sim \Phi$. *Let* $\mathbf{A}$ *be a real, symmetric, nonnegative definite* $N \times N$ *matrix of rank* $r > 0$, *and* $\mathbf{b}$ *a constant real N-dimensional* (*column*) *vector. Then,* $Q = (\mathbf{X} + \mathbf{b})'\mathbf{A}(\mathbf{X} + \mathbf{b}) \sim \chi^2(r, \lambda)$ (*for some* $\lambda \geq 0$) *if and only if* $\mathbf{A}$ *is idempotent, i.e.,* $\mathbf{A}^2 = \mathbf{A}$. *In this case,* $\lambda = \mathbf{b}'\mathbf{A}\mathbf{b}$.

We shall consider (i) a single quadratic form in symmetrically distributed i.i.d.r.v.'s, and (ii) two quadratic forms in i.i.d.r.v.'s without the restriction of symmetric distribution, from the point of view of deducing the normality of the basic r.v.'s from the hypothesis that the quadratic forms have (possibly noncentral) $\chi^2$ distributions.

Our principal result of type (i) is the following.

**Theorem 7.3.2.** *Let* $\mathbf{X}' = (X_1, \ldots, X_N)$ *be a random vector with i.i.d. components, with* $X_1$ *having a symmetric d.f., and let* var $X_1 = \sigma^2$ $(<\infty)$. *Let* $\mathbf{A}$ *be a real symmetric, nonnegative definite, idempotent* $N \times N$ *matrix, of rank* $r > 0$, *and* $\mathbf{b}$ *a real N-vector. Then,* $X_1 \sim N(0, \sigma^2)$ *if and only if* $Q/\sigma^2 = (\mathbf{X} + \mathbf{b})'\mathbf{A}(\mathbf{X} + \mathbf{b})/\sigma^2 \sim \chi^2(r, \lambda)$ *for a suitable* $\lambda \geq 0$. *The parameter* $\lambda$ *then equals* $\mathbf{b}'\mathbf{A}\mathbf{b}/\sigma^2$.

***Proof.*** The "only if" part is the same as the "if" part of Theorem 7.3.1 (except that $\sigma = 1$ in the latter).

To prove the "if" part, let us assume that $Q/\sigma^2 \sim \chi^2(r, \lambda)$. Then, $E(Q/\sigma^2) = r + \lambda$ as well as $r + \mathbf{b}'\mathbf{A}\mathbf{b}/\sigma^2$; hence, $\lambda$ is as stated. Let $\mathbf{C}$ be an $N \times N$ orthogonal matrix such that $\mathbf{CAC}' = \mathbf{D}$, a diagonal matrix with its first $r$ ($=\operatorname{rank} \mathbf{A}$) diagonal elements positive and the rest zero. Then, $Q = \mathbf{Y}'\mathbf{D}\mathbf{Y}$, where $\mathbf{Y} = \mathbf{C}(\mathbf{X} + \mathbf{b})$, so that $Q = \sum_{j=1}^{r} d_j Y_j^2$. Since $Q$ has moments of all orders, so does $Y_1$; in turn, every $X_j$ that figures in $Y_1$, and hence $X_1$, has moments of all orders (recalling that if $U$ and $V$ are independent, then $E|U + V|^p < \infty$ for some $p > 0$ if and only if $E|U|^p$, $E|V|^p < \infty$). Let $M_n$ and $m_n$ denote the $n$th moments of $Q$ and $X_1$, respectively; $m_n = 0$ for odd $n$ since $X_1$ has a symmetric d.f., and we have

$$M_n = c_n m_{2n} + P_n(\mathbf{b}'\mathbf{A}\mathbf{b}; m_1, \ldots, m_{2n-2}), \qquad n \in \mathbb{N}, \qquad (7.3.1)$$

where $c_n = \sum_{j=1}^{N} a_{jj}^n$ and $P_n$ is a polynomial. Since $\mathbf{A}$ is a nonnegative definite matrix of positive rank, it follows that $a_{jj} \geq 0$ for all $j$ and $>0$ for at least one $j$, so that $c_n > 0$. Thus, one can successively solve for $m_2, m_4, \ldots$ uniquely, from the known values of $M_n$, for the moments of $\chi^2(r, \lambda)$ with $\lambda = \mathbf{b}'\mathbf{A}\mathbf{b}/\sigma^2$. The uniqueness of the solution then implies that the moments $m_2, m_4, \ldots$ must agree with the corresponding moments of the $N(0, \sigma^2)$ distribution. Since a normal d.f. is uniquely determined by its moments, we must have $X_1 \sim N(0, \sigma^2)$. Hence, the theorem is proven.

We consider three special cases. The first is one where $Q$ is the sample variance based on a single sample; the second deals with the squared norm of the projection of a random vector onto a linear subspace, in the "general linear model" set-up with i.i.d. error components; the third deals with the residual sum of squares in the same set-up as in the preceding.

**Corollary 7.3.3.** *Let $\mathbf{X}' = (X_1, \ldots, X_N)$ be a random vector with i.i.d. components, and let $EX_1 = \mu$, $\operatorname{var} X_1 = \sigma^2$. Suppose further that $X_1$ has a d.f. symmetric about $\mu$. Then, $Q = \sum_{j=1}^{N} (X_j - \bar{X})^2 \sim c\chi_{N-1}^2$ for some constant $c$ if and only if $X_1 \sim N(\mu, \sigma^2)$. In such a case, $c = \sigma^2$.*

***Proof.*** Define $\mathbf{e}' = (1, 1, \ldots, 1)$, $\mathbf{b} = \mu\mathbf{e}$, and $\mathbf{Y} = \mathbf{X} - \mathbf{b}$, and let $\mathbf{A}$ be the $N \times N$ matrix with all diagonal elements equal to $1 - (1/N)$ and all off-diagonal elements equal to $-1/N$. Then, $\mathbf{A}^2 = \mathbf{A}$ and $\mathbf{Ab} = 0$, $\sum (X_j - \bar{X})^2 = (\mathbf{Y} + \mathbf{b})'\mathbf{A}(\mathbf{Y} + \mathbf{b})$, $r = N - 1$, and Theorem 7.3.2 yields the desired conclusion. The "noncentrality parameter," being equal to $\mathbf{b}'\mathbf{A}\mathbf{b}/\sigma^2$, is zero since $\mathbf{Ab} = \mathbf{0}$.

**Corollary 7.3.4.** *Let* $\mathbf{X}' = (X_1, \ldots, X_N)$ *be a random vector with i.i.d. symmetrically distributed components and let* $\operatorname{var} X_1 = \sigma^2$. *For a fixed N-vector* $\mathbf{b}$, *let* $\mathbf{Y} = \mathbf{X} + \mathbf{b}$, *and let* $\mathbf{Y}_L = \mathbf{PY}$ *be the orthogonal projection of* $\mathbf{Y}$ *onto a linear subspace* $L$ *of dimension* $r > 0$ *in* $\mathbb{R}^N$. *Then,* $X_1 \sim N(0, \sigma^2)$ *if and only if* $\|\mathbf{Y}_L\|^2/\sigma^2 \sim \chi^2(r, \lambda)$ *for some* $\lambda \geq 0$. *In such a case,* $\lambda = \mathbf{b}'\mathbf{P}\mathbf{b}/\sigma^2$.

***Proof.*** Theorem 7.3.2 applies, with $\mathbf{A} = \mathbf{P}'\mathbf{P}$. We note that $\mathbf{P}' = \mathbf{P} = \mathbf{P}^2$, and

$$(\mathbf{X} + \mathbf{b})'\mathbf{A}(\mathbf{X} + \mathbf{b}) = (\mathbf{PY})'(\mathbf{PY}) = \mathbf{Y}_L'\mathbf{Y}_L = \|\mathbf{Y}_L\|^2, \qquad \mathbf{b}'\mathbf{A}\mathbf{b} = \mathbf{b}'\mathbf{P}\mathbf{b}.$$

**Corollary 7.3.5.** *Under the hypotheses of Corollary* 7.3.4, *let further* $\mathbf{b} \in L$. *Then,* $\|\mathbf{Y} - \mathbf{Y}_L\|^2 \sim \sigma^2\chi^2_{N-r}$ *if and only if* $X_1 \sim N(0, \sigma^2)$.

***Proof.*** Corollary 7.3.4 applies, with $L^\perp$, the orthogonal complement of $L$, in place of $L$ there. We have $\dim L^\perp = N - r$, $\mathbf{Y} - \mathbf{Y}_L = \mathbf{Y}_{L^\perp}$, and since $\mathbf{b} \in L$, $\mathbf{P}_{L^\perp}\mathbf{b} = \mathbf{0}$, so that the $\lambda$ in Corollary 7.3.4 is zero here.

In the preceding results, the basic r.v.'s were assumed to have d.f.'s symmetric about some point. Such an assumption can be dropped if we assume that *two* homogeneous quadratic forms in the basic r.v.'s have chi-square distributions with appropriate degrees of freedom. We consider in what follows sufficient conditions for such an assumption to imply the normality of the basic r.v.'s.

We first establish some notation and certain auxiliary results. In the rest of this section, we shall assume that the random vector $\mathbf{X}' = (X_1, \ldots, X_N)$ has nondegenerate i.i.d. components and that $EX_1 = 0$, $\operatorname{var} X_1 = \sigma^2$. $\mathbf{A}$ is an $N \times N$ real symmetric nonnegative definite matrix, $\mathbf{b}$ a real $N$-vector, and $Q = (\mathbf{X} + \mathbf{b})'\mathbf{A}(\mathbf{X} + \mathbf{b})$. $M_n$ and $m_n$ will denote the $n$th moments of $Q$ and of $X_1$, respectively.

**Lemma 7.3.6.** *Suppose* $X_1$ *has moments of all orders. Then:*

(a) *$Q$ has moments of all orders;*

(b) *the coefficient of* $m_{2n}$ *in the expression for* $M_n$ *as a polynomial in the* $m_j$, $0 \leq j \leq 2n$, *is*

$$\lambda(n; \mathbf{A}) := \sum_{j=1}^{N} a_{jj}^n; \tag{7.3.2}$$

(c) *if* $\mathbf{b} = \mathbf{0}$, *the coefficient of* $m_{2p+1}^2$ *in the polynomial expression for* $M_{2p+1}$ *in terms of the* $m_j$, $0 \leq j \leq 4p + 2$ *is*

$$\rho(2p + 1; \mathbf{A}) := \sum_{q=0}^{p} \frac{(2p + 1)!}{(2q + 1)!\{(p - q)!\}^2} \sum_{j>i} \{(2a_{ij})^{2q+1}(a_{ii}a_{jj})^{p-q}\}. \tag{7.3.3}$$

***Proof.*** (a) This follows from the fact that $Q^n$ is a polynomial of degree $2n$ in the $X_j$'s.

(b) Since $M_n = E\{\sum_{i,j} a_{ij}(X_i + b_i)(X_j + b_j)\}^n$, it follows that the coefficient of $m_{2n}$ is equal to $\sum a_{jj}^n$.

(c) If $\mathbf{b} = \mathbf{0}$, $Q = \sum_{i,j} a_{ij} X_i X_j$ and the contributions to $m_{2p+1}^2$ come (only) from the terms in the expansion for $EQ^{2p+1}$ that correspond to

$$E\{(a_{ii}X_i^2)^{p-q}(a_{jj}X_j^2)^{p-q}(2a_{ij}X_iX_j)^{2q+1}\}, \qquad j > i,\ 0 \le q \le p.$$

Equation (7.3.3) is merely a restatement of this fact.

If now $f$ is the ch.f. of a r.v. with moments up to some order $l$, and $\phi = \log f$ in a neighborhood of the origin, then we have

$$f(t) = \sum_{n=0}^{l} m_n \frac{(it)^n}{n!} + |t|^l \varepsilon(t),$$

$$\phi(t) = \sum_{n=1}^{l} k_n \frac{(it)^n}{n!} + |t|^l \delta(t),$$

where $\varepsilon(t)$, $\delta(t) \to 0$ as $t \to 0$. The $k_r$, $1 \le r \le l$, are called the (first $r$) *cumulants* of the r.v., or of its d.f. On identifying the coefficients of $(it)^n$ on both sides of the relation $f = e^{\phi}$, we obtain (for *any* d.f. with moments up to order $l$)

$$m_r = \sum \frac{r!}{n_1! \cdots n_r!} \left(\frac{k_1}{1!}\right)^{n_1} \cdots \left(\frac{k_r}{r!}\right)^{n_r}, \qquad r = 1, \ldots, l, \tag{7.3.4}$$

where the summation extends over all nonnegative integer $r$-tuples $(n_1, \ldots, n_r)$ such that $\sum_{j=1}^{r} jn_j = r$. We also see from $\phi = \log f$ that $k_r$ can be expressed as a polynomial in the $m_j$, $1 \le j \le r$, with the same property (see, for instance, Lukacs, 1970, p. 27).

Suppose we denote by $\mathcal{P}_r(x_1, \ldots, x_r)$ the set of all polynomials in the real variables $x_1, \ldots, x_r$ with real coefficients, of the form

$$\sum c(n_1, \ldots, n_r) x_1^{n_1} \cdots x_r^{n_r}; \qquad \sum_{j=1}^{r} jn_j = r,\ n_j \ge 0.$$

Then, we may write

$$k_r \in \mathcal{P}_r(m_1, \ldots, m_r); \qquad m_r \in \mathcal{P}_r(k_1, \ldots, k_r). \tag{7.3.5}$$

The families $\mathcal{P}_r(x_1, \ldots, x_r)$ have an important closure property, which is easily established through straightforward computation:

$$\left.\begin{array}{l} \text{If } p \in \mathcal{P}_r(q_1, \ldots, q_r) \text{ and if, for } 1 \le i \le r, \\ q_i \in \mathcal{P}_{li}(t_1, \ldots, t_{li}), \text{ then } p \in \mathcal{P}_{lr}(t_1, \ldots, t_{lr}). \end{array}\right\} \tag{7.3.6}$$

We are now in a position to establish the following.

**Lemma 7.3.7.** *Let $Q = \mathbf{X}'\mathbf{A}\mathbf{X}$, and let $K_n$ and $k_n$ be the nth cumulants of $Q$ and $X_1$, respectively. Then, recalling the definitions of $\lambda(n; \mathbf{A})$ and $\rho(2p+1; \mathbf{A})$ in (7.3.2) and (7.3.3), we have:*

(a) $K_n \in \mathcal{P}_{2n}(k_1, \ldots, k_{2n})$, $n \in \mathbb{N}$;

(b) *the coefficient of $k_{2n}$ in the expression for $K_n$ as a polynomial in the $k_j$, $1 \le j \le 2n$, is $\lambda(n; \mathbf{A})$;*

(c) *the coefficient of $k_{2p+1}^2$ in the expression for $K_{2p+1}$ as a polynomial in the $k_j$ is given by*

$$\theta(2p+1; \mathbf{A}) := \tfrac{1}{2}\binom{4p+2}{2p+1}\lambda(2p+1; \mathbf{A}) + \rho(2p+1; \mathbf{A}), \qquad p \in \mathbb{Z}_+. \tag{7.3.7}$$

***Proof.*** (a) This follows at once from (7.3.5), (7.3.6), and the fact that $M_r \in \mathcal{P}_{2r}(m_1, \ldots, m_{2r})$.

(b) It is obvious that the coefficient of $k_{2n}$ in the expression for $m_{2n}$ as a polynomial in the $k_j$ is unity; also, by Lemma 7.3.6(b), the coefficient of $m_{2n}$ in the expression for $M_n$ as a polynomial in the $m_j$ is $\lambda(n; \mathbf{A})$; hence, (b) is proven.

(c) We have from (7.3.4) that the coefficient of $k_{2p+1}^2$ in the expression for $m_{4p+2}$ as a polynomial in the $k_j$ is $\frac{1}{2}\binom{4p+2}{2p+1}$, and that the coefficient of $k_{2p+1}$ in the expression for $m_{2p+1}$ as a polynomial in the $k_j$ is unity. The coefficients of $m_{4p+2}$ and $m_{2p+1}^2$ in the expression for $M_{2p+1}$ as a polynomial in the $m_j$ are $\lambda(2p+1; \mathbf{A})$ and $\rho(2p+1; \mathbf{A})$, respectively. Hence, the coefficient of $k_{2p+1}^2$ in the expression for $M_{2p+1}$, and hence in that for $K_{2p+1}$ as a polynomial in the $k_j$, is $\theta(2p+1; \mathbf{A})$, as given by (7.3.7). Hence, the lemma is proven.

Suppose now that $X_1$ has moments of all orders and that $EX_1 = 0$, $\operatorname{var} X_1 = \sigma^2$. Then, $k_1 = 0$, $k_2 = \sigma^2$. For $Q = \mathbf{X}'\mathbf{A}\mathbf{X}$, we have, by Lemma 7.3.7(a),

$$K_l = \sum c_{2l}(n_2, \ldots, n_{2l}; \mathbf{A}) k_2^{n_2} \cdots k_{2l}^{n_{2l}},$$

where the summation $\sum$ is taken over all $(n_2, \ldots, n_{2l})$, $\sum_{j=2}^{2l} jn_j = 2l$. Denoting by $k_j^*$ and $K_j^*$ the cumulants of $X_1$ and of $\mathbf{X}'\mathbf{A}\mathbf{X}$, respectively, where $X_1 \sim N(0, \sigma^2)$, we have

$$k_2^* = k_2 = \sigma^2, \qquad \text{and} \qquad k_n^* = 0 \quad \text{for } n > 2;$$

$$K_l^* = c_{2l}(l, 0, \ldots, 0; \mathbf{A}) k_2^l,$$

so that we have

$$K_l = K_l^* + \sum^* c_{2l}(n_2, \ldots, n_{2l}; \mathbf{A}) k_2^{n_2} \cdots k_{2l}^{n_{2l}}, \tag{7.3.8}$$

where the summation $\sum^*$ is taken over all $(n_2, \ldots, n_{2l})$ with

$$n_2 \neq l, \qquad \sum_{j=2}^{2l} j n_j = 2l.$$

We proceed to our second principal result. It provides a sufficient condition for the normality of $X_1$ to be equivalent to two homogeneous quadratic forms in the $X_j$ having chi-square distributions simultaneously.

**Theorem 7.3.8.** *Let* $\mathbf{A}, \mathbf{B}$ *be two constant symmetric, idempotent, non-negative definite matrices, of positive ranks* $r$ *and* $s$, *respectively, and suppose that*

$$\frac{\rho(2p+1; \mathbf{A})}{\sum a_{jj}^{2p+1}} \neq \frac{\rho(2p+1; \mathbf{B})}{\sum b_{jj}^{2p+1}}, \qquad \forall p \in \mathbb{Z}_+. \tag{7.3.9}$$

*Then,* $X_1 \sim N(0, \sigma^2)$ *if and only if, for some constants* $\sigma_{\mathbf{A}}, \sigma_{\mathbf{B}} > 0$, *we have simultaneously*

$$\mathbf{X}'\mathbf{A}\mathbf{X} \sim \sigma_{\mathbf{A}} \chi_r^2 \qquad \text{and} \qquad \mathbf{X}'\mathbf{B}\mathbf{X} \sim \sigma_{\mathbf{B}} \chi_s^2. \tag{7.3.10}$$

*In such a case,* $\sigma_{\mathbf{A}} = \sigma_{\mathbf{B}} = \sigma^2$.

***Proof.*** The "only if" part is immediate. Let (7.3.10) hold. By equating the expected values, we have $\sigma_{\mathbf{A}} = \sigma_{\mathbf{B}} = \sigma^2$, so that $\mathbf{X}'\mathbf{A}\mathbf{X}/\sigma^2 \sim \chi_r^2$ and $\mathbf{X}'\mathbf{B}\mathbf{X}/\sigma^2 \sim \chi_s^2$. Then, as in the proof of Theorem 7.3.2, $X_1$ has moments of all orders. Applying (7.3.8) to $Q_{\mathbf{A}} = \mathbf{X}'\mathbf{A}\mathbf{X}$, we see that $K_l = K_l^* =$ the $l$th cumulant of the $\sigma^2\chi_r^2$ distribution, so that

$$\sum^* c_{2l}(n_2, \ldots, n_{2l}; \mathbf{A}) k_2^{n_2} \cdots k_{2l}^{n_{2l}} = 0, \tag{7.3.11a}$$

with $\sum^*$ taken over all $(n_2, \ldots, n_{2l})$ with $n_2 \neq l$, $\sum j n_j = 2l$. Similarly, we have from $\mathbf{X}'\mathbf{B}\mathbf{X} \sim \sigma^2\chi_s^2$ that

$$\sum^* c_{2l}(n_2, \ldots, n_{2l}; \mathbf{B}) k_2^{n_2} \cdots k_{2l}^{n_{2l}} = 0. \tag{7.3.11b}$$

Let $H$ denote the set of all infinite-dimensional real vectors $\mathbf{k} = (k_3, k_4, \ldots)$ satisfying the two relations (7.3.11a, b). Since $\mathbf{0} \in H$, $H$ is nonempty. We proceed to show that, if (7.3.9) holds, then $\mathbf{0}$ is the only element of $H$. Let $H_p$ be the subset of $H$ with its elements $\mathbf{k}$ such that

$$k_3 = k_5 = \cdots = k_{2p+1} = 0, \qquad k_4 = k_6 = \cdots = k_{4p+2} = 0,$$

i.e., $\mathbf{k}$ is of the form

$$\mathbf{k} = (0, 0, \ldots, 0, k_{2p+3}, 0, k_{2p+5}, 0, \ldots, k_{4p+1}, 0, k_{4p+3}, k_{4p+4}, \ldots).$$

We shall show that: (i) $H \subseteq H_1$; (ii) $H_p \subseteq H_{p+1}$ for $p \geq 1$. It will follow that $H = \{\mathbf{0}\}$ so that $k_2 = \sigma^2$, $k_n = 0$ for $n > 2$, i.e., $X_1 \sim N(0, \sigma^2)$ as in the proof of Theorem 5.1.5.

(i) Let $\mathbf{k} \in H$. Setting $l = 2$ in (7.3.11a), we have ($\sum^*$ consists of only one term)

$$c_4(0, 0, 1; \mathbf{A})k_4 = 0.$$

That $c_4(0, 0, 1; \mathbf{A}) = \lambda(2; \mathbf{A}) > 0$ implies $k_4 = 0$. Setting $l = 3$ in (7.3.11a), and taking the fact that $k_4 = 0$ into account, we have

$$\sum{}^* c_6(n_2, n_3, 0, n_5, n_6; \mathbf{A})k_2^{n_2}k_3^{n_3}k_5^{n_5}k_6^{n_6} = 0,$$

where

$$n_2 \neq 3, \qquad 2n_2 + 3n_3 + 5n_5 + 6n_6 = 6.$$

The only vectors $(n_2, n_3, n_5, n_6)$ satisfying these conditions are $(0, 0, 0, 1)$ and $(0, 2, 0, 0)$, and we have

$$c_6(0, 0, 0, 1; \mathbf{A})k_6 + c_6(0, 2, 0, 0; \mathbf{A})k_3^2 = 0,$$

and similarly, from (7.3.11b),

$$c_6(0, 0, 0, 1; \mathbf{B})k_6 + C_6(0, 2, 0, 0; \mathbf{B})k_3^2 = 0.$$

Our assumption (7.3.9) for $p = 1$ then implies that $k_3 = k_6 = 0$. Hence, $H \subseteq H_1$.

(ii) Let $\mathbf{k} \in H_p$ for some $p \geq 1$. We first show that then $k_{4p+4} = 0$. Setting $l = 2p + 2$ in (7.3.11a), we have

$$\sum{}^* c_{4p+4}(n_2, 0, 0, \ldots, n_{2p+3}, 0, n_{2p+5}, 0, \ldots, n_{4p+3}, n_{4p+4}; \mathbf{A}) \times k_2^{n_2} \cdot k_{2p+3}^{n_{2p+3}} \cdot k_{2p+5}^{n_{2p+5}} \cdot \cdots \cdot k_{4p+3}^{n_{4p+3}} \cdot k_{4p+4}^{n_{4p+4}} = 0, \tag{7.3.12a}$$

and an analog for $\mathbf{B}$ in place of $\mathbf{A}$ from (7.3.11b), where the summation runs over all $(n_2, n_{2p+3}, n_{2p+5}, \ldots, n_{4p+3}, n_{4p+4})$ satisfying

$$n_2 \neq 2p + 2, \qquad \sum jn_j = 4p + 4. \tag{7.3.12b}$$

The only vector satisfying (7.3.12b) has $n_{4p+4} = 1$ and all other $n_j = 0$. The corresponding coefficient $c_{4p+4}(0, \ldots, 1)$ is the coefficient of $k_{4p+4}$ in the polynomial expression for $K_{2p+2}$ (for $\mathbf{X}'\mathbf{AX}$) in terms of the $k_j$; this, by Lemma 7.3.7(b), is $\lambda(2p + 2; \mathbf{A})$. Hence, (7.3.12a) reduces to $\lambda(2p + 2; \mathbf{A})k_{4p+4} = 0$, whence our assertion.

We next show that if $\mathbf{k} \in H_p$, then $k_{2p+3} = k_{4p+6} = 0$ as well, using the already established fact that $k_{4p+4} = 0$. Setting $l = 2p + 3$ in (7.3.11a), we have

$$\sum{}^{*} c_{4p+6}(n_2, 0, 0, \ldots, n_{2p+3}, 0, n_{2p+5}, 0, \ldots, n_{4p+5}, n_{4p+6}; \mathbf{A}) \times k_2^{n_2} \cdot k_{2p+3}^{n_{2p+3}} \cdot k_{2p+5}^{n_{2p+5}} \cdot \cdots \cdot k_{4p+5}^{n_{4p+5}} \cdot k_{4p+6}^{n_{4p+6}} = 0, \tag{7.3.13a}$$

where the summation runs over all vectors $(n_2, n_{2p+3}, \ldots, n_{4p+5}, n_{4p+6})$ satisfying

$$n_2 \neq 2p + 3, \qquad \sum jn_j = 4p + 6. \tag{7.3.13b}$$

There are only two vectors satisfying (7.3.13b), namely, the one where $n_{4p+6} = 1$ and all the others are zero, and the one where $n_{2p+3} = 2$ and all the others are zero. The respective $c_{4p+6}(\cdot; \mathbf{A})$ values are the coefficients of $k_{4p+6}$ and $k_{2p+3}^2$, respectively, in the expression for $K_{2p+3}$ of $Q_\mathbf{A} = \mathbf{X}'\mathbf{A}\mathbf{X}$ as a polynomial in the $k_j$ and, thus, by Lemma 7.3.7(b), (c), these are $\lambda(2p + 3; \mathbf{A})$ and $\theta(2p + 3; \mathbf{A})$, respectively. Thus, (7.3.13a) and its analog for $\mathbf{B}$ in place of $\mathbf{A}$ give

$$\lambda(2p + 3; \mathbf{A})k_{4p+6} + \theta(2p + 3; \mathbf{A})k_{2p+3}^2 = 0,$$

$$\lambda(2p + 3; \mathbf{B})k_{4p+6} + \theta(2p + 3; \mathbf{B})k_{2p+3}^2 = 0.$$

Our assumption (7.3.9) then implies that $k_{4p+6} = k_{2p+3} = 0$, thus concluding our proof of the fact that $H_p \subset H_{p+1}$, and hence of the theorem itself.

We remark that, in view of our definition of $\rho(2p + 1; \mathbf{A})$, condition (7.3.9) is implied by the following condition, more convenient to handle: For all $p \geq q \geq 0$, the expressions

$$d(p, q; \mathbf{A}, \mathbf{B}) := \sum_{j>i} \left\{ \frac{(2a_{ij})^{2q+1}(a_{ii}a_{jj})^{p-q}}{\sum_l a_{ll}^{2p+1}} - \frac{(2b_{ij})^{2q+1}(b_{ii}b_{jj})^{p-q}}{\sum_l b_{ll}^{2p+1}} \right\} \tag{7.3.14}$$

have the same sign, and at least one of them is nonzero.

We conclude this section with the examination of some special cases of Theorem 7.3.8. These concern the following situations: (i) $Q_\mathbf{A}$, $Q_\mathbf{B}$ are two sample variances; (ii) $Q_\mathbf{A}$ is the squared sample mean, and $Q_\mathbf{B}$ is the sample variance; (iii) $Q_\mathbf{A}$, $Q_\mathbf{B}$ are the squares of two linear forms in two i.i.d.r.v.'s.

**Corollary 7.3.9.** *Let $\{X_j\}$ be a sequence of i.i.d.r.v.'s with* $\operatorname{var} X_1 = \sigma^2$. Let $Q_m = \sum_{j=1}^m (X_j - \bar{X}_m)^2$ *and* $Q_n = \sum_{j=1}^n (X_j - \bar{X}_n)^2$, *where* $n > m \geq 2$. *Then,* $Q_m \sim \sigma_m \chi_{m-1}^2$ *and* $Q_n \sim \sigma_n \chi_{n-1}^2$ *for some* $\sigma_m, \sigma_n > 0$ *if and only if* $X_1$ *is normally distributed. In such a case,* $\sigma_m = \sigma_n = \sigma^2$.

***Proof.*** If $EX_1 = \mu$, let $\mathbf{b}'$ be the $n$-vector $(\mu, \ldots, \mu)$ and let $\mathbf{Y} = \mathbf{X} - \mathbf{b}$. As in the proof of Corollary 7.3.3, let $\mathbf{C}_k$ denote the $k \times k$ matrix with all elements on its principal diagonal equal to $1 - 1/k$ and all off-diagonal elements equal to $-1/k$. Let $\mathbf{B} = \mathbf{C}_n$ and

$$\mathbf{A} = \begin{pmatrix} \mathbf{C}_m & \mathbf{0} \\ \mathbf{0} & \mathbf{0} \end{pmatrix};$$

then, $Q_m = \mathbf{Y}'\mathbf{A}\mathbf{Y}$, $Q_n = \mathbf{Y}'\mathbf{B}\mathbf{Y}$, $\mathbf{Ab} = \mathbf{Bb} = \mathbf{0}$, $r = m - 1$, and $s = n - 1$. Finally, we note that

$$\begin{aligned} d(p, q; \mathbf{B}, \mathbf{A}) &= \binom{n}{2} \frac{2^{2q+1}}{n(n-1)^{2q+1}} - \binom{m}{2} \frac{2^{2q+1}}{m(m-1)^{2q+1}} \\ &= 2^{2q}((n-1)^{-2q} - (m-1)^{-2q}), \end{aligned}$$

so that (7.3.14) is satisfied and Theorem 7.3.8 applies.

**Corollary 7.3.10.** *Let $X_1, \ldots, X_N$ be i.i.d.r.v.'s with $EX_1 = \mu$, $\operatorname{var} X_1 = \sigma^2$. Then, $N(\bar{X}_N - \mu)^2 \sim \sigma^2\chi_1^2$ and $\sum_{j=1}^N (X_j - \bar{X}_N)^2 \sim \sigma^2\chi_{N-1}^2$ if and only if $X_1 \sim N(\mu, \sigma^2)$.*

***Proof.*** Let $\mathbf{A}$ be the $N \times N$ matrix with every entry $= 1/N$, and $\mathbf{B} = \mathbf{C}_N$ as defined in the preceding. Then, $\mathbf{A}, \mathbf{B}$ have ranks $r = 1$, $s = N - 1$, respectively, and

$$d(p, q; \mathbf{B}, \mathbf{A}) = 2^{2q+1}\{(N-1)^{-(2q+1)} - 1\}/N.$$

Thus, (7.3.14) is satisfied and Theorem 7.3.8 applies.

**Corollary 7.3.11.** *Let $X_1$ and $X_2$ be i.i.d.r.v.'s with $EX_1 = 0$, $\operatorname{var} X_1 = 1$. Let $a_1, a_2, b_1, b_2$ be real numbers such that*

$$a_1^2 + a_2^2 = b_1^2 + b_2^2 = 1; \qquad a_1 b_2 - a_2 b_1 \neq 0, \qquad a_1 b_1 - a_2 b_2 \neq 0.$$

*Then, $(a_1X_1 + a_2X_2)^2 \sim \chi_1^2$ and $(b_1X_1 + b_2X_2)^2 \sim \chi_1^2$ if and only if $X_1 \sim N(0, 1)$.*

***Proof.*** Take

$$\mathbf{A} = \begin{pmatrix} a_1^2 & a_1 a_2 \\ a_1 a_2 & a_2^2 \end{pmatrix}, \qquad \mathbf{B} = \begin{pmatrix} b_1^2 & b_1 b_2 \\ b_1 b_2 & b_2^2 \end{pmatrix}.$$

Then, the ranks of $\mathbf{A}$ and $\mathbf{B}$ are each 1, so that $r = s = 1$, and

$$d(p, q; \mathbf{A}, \mathbf{B}) = -\frac{2^{2q+1}\{(a_1 b_2)^{2p+1} - (a_2 b_1)^{2p+1}\}\{(a_1 b_1)^{2p+1} - (a_2 b_2)^{2p+1}\}}{(a_1^{4p+2} + a_2^{4p+2})(b_1^{4p+2} + b_2^{4p+2})},$$

so that (7.3.14) is satisfied and Theorem 7.3.8 applies.

A proof of Corollary 7.3.9, examining the moments rather than the cumulants of the basic r.v.'s, is also of independent interest. Suppose it is given then that

$$Q_l = \sum_{j=1}^{l} (X_j - \bar{X}_l)^2 \sim \sigma_l \chi_{l-1}^2 \qquad \text{and} \qquad Q_n = \sum_{j=1}^{n} (X_j - \bar{X}_n)^2 \sim \sigma_n \chi_{n-1}^2.$$

Then, as usual, we conclude that $X_1$ has moments of all orders, and since $Q_l, Q_n$ are translation-invariant statistics, we may assume without loss of generality that $m_1 := EX_1 = 0$. Since the $N(0, m_2)$ distribution is uniquely determined by its moment sequence, it suffices to show that

$$m_k = (k-1)!!m_2^{k/2} \quad \text{for even } k; \qquad m_k = 0 \quad \text{for odd } k$$

($n!! = n(n-2)(n-4)\cdots 3\cdot 1$ for odd positive integers $n$). Since

$$Q_n = \left(1 - \frac{1}{n}\right)\sum_{j=1}^{n} X_j^2 - \frac{2}{n}\sum_{i<j} X_i X_j,$$

we have

$$EQ_n^N = a_{N,n} m_{2N} + b_{N,n} m_N^2 + \sum_{k=0}^{N-1} P_{k,N,n} m_{N+k} + R_{N,n}, \tag{7.3.15}$$

where

$$a_{N,n} = n(1 - 1/n)^N,$$

$$b_{N,n} = \binom{n}{2} \sum_{\substack{0 \le j \le N \\ N-j \text{ even}}} \binom{N}{j, (N-j)/2, (N-j)/2} \times \left(-\frac{2}{n}\right)^j \left(1 - \frac{1}{n}\right)^{(N-j)/2} \left(1 - \frac{1}{n}\right)^{(N-j)/2} \tag{7.3.16}$$

$\left(\binom{N}{r, s, t}\right.$ with $0 \le r, s, t; r+s+t = N$ standing for the trinomial coefficient $\left.N!/(r!s!t!)\right)$, and $P_{k,N,n}, R_{N,n}$ are polynomials in $m_2, \ldots, m_{N-1}$. In particular, the first of these has the form

$$P_{k,N,n} = \sum c_{j_2,\ldots,j_{N-1}} m_2^{j_2} \cdots m_{N-1}^{j_{N-1}}, \tag{7.3.17}$$

the summation being over all $(N-2)$-vectors $(j_2, \ldots, j_{N-1})$ with non-negative integer entries such that $\sum_{r=2}^{N-1} rj_r = N - k$, the coefficients of course depending on $k$, $N$, $n$. Since $\sigma_m = \sigma_n = \sigma^2 = m_2$ necessarily, we obtain from our principal assumption that

$$a_{N,n} m_{2N} + b_{N,n} m_N^2 = m_2^N \rho_{N,n} - \sum_{k=0}^{N-1} P_{k,N,n} m_{N+k} - R_{N,n}, \tag{7.3.18a}$$

$$a_{N,l} m_{2N} + b_{N,l} m_N^2 = m_2^N \rho_{N,l} - \sum_{k=0}^{N-1} P_{k,N,l} m_{N+k} - R_{N,l}, \tag{7.3.18b}$$

where $\rho_{N,r}$ is the $N$th moment of a $\chi^2_{r-1}$ r.v. We introduce the following inductive hypothesis:

$$H_N: \begin{cases} m_k = 0 & \text{for odd} \quad k \leq N - 1, \\ m_k = (k-1)!!m_2^{k/2} & \text{for even} \quad k < 2N. \end{cases}$$

Let $N \geq 2$ and suppose $H_N$ holds. It follows from (7.3.17) that $P_{k,N,n}$ is zero for odd $N + k$: For, if $N + k$ is odd, so is $N - k$, and so every term in (7.3.17) must contain at least one factor of the form $m_j$ for an odd $j$, $j \leq N - 1$, which is zero by $H_N$. The right members of the relations (7.3.18a, b) are thus uniquely determined (by $H_N$) in terms of $m_2$.

For odd $N$, we have

$$b_{N,n}/a_{N,n} = -\sum_{\substack{1 \leq j \leq N, \\ j \text{ odd}}} \binom{N}{j, (N-j)/2, (N-j)/2} \{2/(n-1)\}^{j-1},$$

so that $b_{N,n}/a_{N,n}$ is a strictly monotone function of $n$ when $N$ is odd. For such $N$, the matrix

$$\begin{pmatrix} a_{N,n} & b_{N,n} \\ a_{N,l} & b_{N,l} \end{pmatrix}$$

has an inverse, implying that $m_{2N}$ and $m_N^2$ are uniquely determined by $m_2$. Since $m_{2N}$, $m_N$ of a normal distribution (with second moment $m_2$) satisfy (7.3.18), we get $m_{2N} = (2N-1)!!m_2^N$, and $m_N = 0$. For even $N$, $H_N$ asserts that $m_N = (N-1)!!m_2^{N/2}$, and so either of the relations (7.3.18a, b) implies that $m_{2N} = (2N-1)!!m_2^N$ is the only possibility. Thus, $H_{N+1}$ holds whether $N$ is odd or even, and Corollary 7.3.9 thus is proven.

## NOTES AND REMARKS

The results of Section 7.1 are due to R. M. Norton (1975, 1978) and (in final form) to R. Shantaram (1978, 1980). The simplification given here of the proof of Lemma 7.1.7 is due to K. Balasubramanian and R. B. Bapat. Theorem 7.2.4 was conjectured by S. Geisser (1973) and was established by Ramachandran (1975) among others. Propositions 7.2.5 and 7.2.6 and Corollary 7.2.7 are from Geisser (1973)—also see the papers cited there. Theorems 7.3.1 and 7.3.8 and the auxiliary results in Section 7.3 are from H. Ruben (1978), and constitute the general or final forms of results established by him in a series of earlier papers, now stated as corollaries. The proof of Corollary 7.3.9, identifying the moments instead of the cumulants to establish normality, is due to L. Bondesson (1977).

CHAPTER

# 8

# Integrated Cauchy Functional Equations on $\mathbb{R}$

In Chapter 2, we considered the ICFE on $\mathbb{Z}_+$ and on $\mathbb{R}_+$ and various related questions. The present chapter is concerned with the analogs of those results for $\mathbb{Z}$ and $\mathbb{R}$. In Section 8.1, we take up the ICFE on $\mathbb{R}$ and on $\mathbb{Z}$ (in that order), including a discussion of simultaneous ICFE's on $\mathbb{R}$, the main result of the section (and the chapter) being Theorem 8.1.6. Section 8.2 is concerned with a proof of the same theorem using the Krein–Milman theorem of functional analysis. Section 8.3 relates to a variant of the ICFE on $\mathbb{R}$, which requires the Wiener–Hopf decomposition, and the solution also leads to another proof of Theorem 8.1.6.

## 8.1. THE ICFE ON $\mathbb{R}$ AND ON $\mathbb{Z}$

We begin with two auxiliary results on subsemigroups and subgroups of $\mathbb{Z}$ and of $\mathbb{R}$. By the greatest common divisor (g.c.d.) of a subset $E$ of $\mathbb{R}$, we shall mean the g.c.d. of the set $\{|x|: x \in E\}$.

**Proposition 8.1.1.** *Let $A$ be a subset of $\mathbb{Z}$ that contains at least one positive and one negative element. Let $S$ be the semigroup generated by $A$, i.e.,*

$$S = \{x_1 + \cdots + x_n: x_i \in A,\ i = 1, \ldots, n,\ n \in \mathbb{N}\}. \qquad (8.1.1)$$

*Then, $S$ is in fact a subgroup of $\mathbb{Z}$: $S = d\mathbb{Z}$ for some $d \in \mathbb{N}$.*

***Proof.*** Let $A^+ = A \cap \mathbb{N}$ and $A^- = A \cap (-\mathbb{N})$. Let $d_1$, $d_2 > 0$ be the g.c.d.'s of $A^+$ and $A^-$, respectively. We shall assume without loss of generality that 1 is the g.c.d. of $d_1$ and $d_2$, and show that $S = \mathbb{Z}$. Then, there exist $p_1, p_2 \in \mathbb{Z}$ such that

$$p_1 d_1 + p_2(-d_2) = 1. \qquad (8.1.2)$$

Noting that $d_1$ and $d_2$ are (already) the g.c.d.'s of some *finite* subsets of $A^+$ and $A^-$, respectively, it follows from Theorem 1.1.2 that there exists a $k \in \mathbb{N}$ such that, for all $n \geq k$, $nd_1$ and $n(-d_2) \in S$. Now, the integers $p_1$ and $p_2$ in (8.1.2) are either both nonnegative or both nonpositive. In the first case,

$$n = (np_1)d_1 + np_2(-d_2) \in S, \qquad \text{for } n \geq k. \qquad (8.1.3)$$

Since $S$ obviously contains arbitrarily large positive and negative integers, (8.1.3) implies that $S = \mathbb{Z}$. In the second case,

$$-n = n(-p_1)d_1 + n(-p_2)(-d_2) \in S, \qquad \text{for } n \geq k,$$

and the same argument applies. Hence, the proposition is proven.

**Proposition 8.1.2.** *Let $A$ be a subset of $\mathbb{R}$ containing at least one positive and one negative element. Let $S$ be the semigroup generated by $A$, as defined by* (8.1.1). *Then, the closure $\bar{S}$ of $S$ is a closed subgroup of $\mathbb{R}$, i.e., either $S = d\mathbb{Z}$ for some $d \in \mathbb{R}_+$ or $\bar{S} = \mathbb{R}$.*

***Proof.*** If $A$ has a g.c.d., call it $d$; then, it follows from Proposition 8.1.1 that $S = d\mathbb{Z}$. If the elements of $A$ have no common divisor, then at least one of the following conditions is satisfied:

(i) There exists a sequence $\{x_n\}$ of members of $A$, with $x_1 > 0$, $x_2 < 0$, and such that, for all $n$, $\{x_1, \ldots, x_n\}$ has a g.c.d., call it $d_n$, and $\lim_{n\to\infty} d_n = 0$. This implies that $\bigcup_n (d_n\mathbb{Z}) \subset S$, and hence $\bar{S} = \mathbb{R}$.

(ii) There exist $x_1 > 0$ and $x_2 < 0$ in $A$ such that $x_1/x_2$ is irrational. Assume without loss of generality that $|x_2| < x_1$. Then, there exists $k \in \mathbb{Z}_+$ such that

$$0 < |x_1 + kx_2| < \tfrac{1}{2}|x_2|.$$

Let $x_3 = x_1 + kx_2$; then, $x_3 \in S$ and the set $\{x_1, x_2, x_3\}$ has no g.c.d. Inductively, we can find a sequence $\{x_n\}$ of distinct elements of $S$ such that $\{x_1, \ldots, x_n\}$ has no g.c.d., $n \in \mathbb{N}$, and

$$0 < |x_{n+1}| < \tfrac{1}{2}|x_n|, \qquad n \geq 2.$$

The facts that $\lim_{n\to\infty} x_n = 0$ and that $S$ contains arbitrarily large positive and negative elements imply that $S$ is dense in $\mathbb{R}$.

We now take up the consideration of the main result of this section and chapter (Theorem 8.1.6), dealing with solutions of the equation

$$f(x) = \int_{\mathbb{R}} f(x+y)\, d\sigma(y), \qquad \forall\, x \in \mathbb{R}, \tag{8.1.4}$$

with $f \geq 0$ and $\sigma$ a positive Borel measure on $\mathbb{R}$. We shall find it convenient to refer to (8.1.4) as "the equation $f = f \bullet \sigma$ on $\mathbb{R}$." We begin with two auxiliary results.

**Lemma 8.1.3.** *Let $f$ be a (strictly) positive solution of* (8.1.4) *and let* $\operatorname{supp} \sigma = \mathbb{R}$. *Also suppose that, for some $\eta \geq 0$, the inequalities*

$$e^{-\eta y} \leq f(x+y)/f(x) \leq 1, \qquad \forall\, x \in \mathbb{R},\ y \geq 0, \tag{8.1.5}$$

*hold. For fixed $\lambda \in [0, 1)$, let $\psi_\lambda(x) = f(x) - \lambda f(x+1)$. Then, for every $y > 0$, $\psi_\lambda(x+y)/\psi_\lambda(x)$ is uniformly continuous (in $x$) and there exist $0 \leq \alpha_\lambda \leq \beta_\lambda$ such that*

$$\begin{aligned} g_\lambda(y) &:= \sup_x \psi_\lambda(x+y)/\psi_\lambda(x) = e^{-\alpha_\lambda y}, \\ h_\lambda(y) &:= \inf_x \psi_\lambda(x+y)/\psi_\lambda(x) = e^{-\beta_\lambda y}. \end{aligned} \tag{8.1.6}$$

*In particular, these assertions hold for $\lambda = 0$, i.e., for $\psi_0 = f$ itself.*

***Remark.*** See also Corollary 8.1.5 in this context.

***Proof.*** $f = f \bullet \sigma$ on $\mathbb{R}$ obviously implies that $\psi_\lambda = \psi_\lambda \bullet \sigma$ on $\mathbb{R}$ as well. The inequalities (8.1.5) imply that $f$ is decreasing and also continuous; since $f > 0$ and $\lambda < 1$, it follows that $\psi_\lambda$ is (strictly) positive. From (8.1.5), we have the obvious estimates

$$c_\lambda e^{-\eta y} \leq \psi_\lambda(x+y)/\psi_\lambda(x) \leq c_\lambda^{-1}, \tag{8.1.7}$$

where $c_\lambda = (1-\lambda)/(1-\lambda e^{-\eta})$. For $0 < \delta < \min(1, y)$, we have

$$\frac{\psi_\lambda(x+y+\delta)}{\psi_\lambda(x+\delta)} - \frac{\psi_\lambda(x+y)}{\psi_\lambda(x)} = H(x, y, \lambda, \delta)/K(x, \lambda, \delta),$$

where

$$\begin{aligned} &H(x, y, \lambda, \delta) \\ &\quad = \left\{\frac{f(x+y+\delta)}{f(x+y)} - \frac{f(x+\delta)}{f(x)}\right\} f(x) f(x+y) \\ &\qquad - \lambda \left\{\frac{f(x+y+1+\delta)}{f(x+y+1)} - \frac{f(x+\delta)}{f(x)}\right\} f(x) f(x+y+1) \\ &\qquad + \lambda^2 \left\{\frac{f(x+y+1+\delta)}{f(x+y+1)} - \frac{f(x+1+\delta)}{f(x+1)}\right\} f(x+1) f(x+y+1), \end{aligned}$$

so that

$$|H(x, y, \lambda, \delta)| \leq (1 - e^{-\eta\delta})\{f(x)f(x + y) + \lambda f(x)f(x + y + 1) + \lambda^2 f(x + 1)f(x + y + 1)\}.$$

Also,

$$K(x, \lambda, \delta) = f(x)f(x + \delta)\{1 - \lambda f(x + 1)/f(x)\}\{1 - \lambda f(x + \delta + 1)/f(x + \delta)\} \geq f(x)f(x + \delta)(1 - \lambda)^2.$$

It follows that, for such $\delta$,

$$\left| \frac{\psi_\lambda(x + y + \delta)}{\psi_\lambda(x + \delta)} - \frac{\psi_\lambda(x + y)}{\psi_\lambda(x)} \right| \leq (1 - e^{-\eta\delta})(1 + \lambda + \lambda^2)/(1 - \lambda)^2 =: \rho_\lambda(\delta), \tag{8.1.8}$$

which establishes the uniform continuity assertion. Let now $c$ denote $g_\lambda(y)$ for a fixed $y > 0$. Then—proceeding as in the proof of Theorem 2.2.4 but using the uniform continuity of $\psi_\lambda(\cdot + y)/\psi_\lambda(\cdot)$ here in place of the convexity assumption on the function there—we have

$$c - \psi_\lambda(x + y)/\psi_\lambda(x) = \int_{\mathbb{R}} \frac{c\psi_\lambda(x + t) - \psi_\lambda(x + t + y)}{\psi_\lambda(x)} d\sigma(t),$$

and since the integrand on the right is $\geq 0$, this is

$$\geq \int_{(y-\delta, y+\delta)} \cdots = \frac{c\psi_\lambda(x + y + \tau) - \psi_\lambda(x + 2y + \tau)}{\psi_\lambda(x)} \sigma(y - \delta, y + \delta),$$

for some $\tau \in (-\delta, \delta)$, by the mean value theorem; this again is

$$= \frac{\psi_\lambda(x + y + \tau)}{\psi_\lambda(x)} \{c - \psi_\lambda(x + 2y + \tau)/\psi_\lambda(x + y + \tau)\}\sigma(y - \delta, y + \delta)$$

$$\geq e^{-\eta(y+\delta)} c_\lambda \{c - \psi_\lambda(x + 2y)/\psi_\lambda(x + y) - \rho_\lambda(\delta)\}\sigma(y - \delta, y + \delta),$$

where $\rho_\lambda(\delta)$ is as defined in (8.1.8). It follows from the arbitrariness of $\delta > 0$ that, given $\varepsilon > 0$, there exists $\varepsilon' > 0$ such that

$$c - \psi_\lambda(x + 2y)/\psi_\lambda(x + y) < \varepsilon \qquad \text{if } c - \psi_\lambda(x + y)/\psi_\lambda(x) < \varepsilon'.$$

Thus, $g_\lambda(2y) \geq g_\lambda(y)^2$, and the reverse inequality is obvious. Then, we proceed as in the proof of Theorem 2.2.4 to show that $g_\lambda(y) = \exp(-\alpha_\lambda y)$ for all $y > 0$, for some $\alpha_\lambda \geq 0$. A dual argument shows that $h_\lambda(y) = \exp(-\beta_\lambda y)$ for all $y > 0$, for some $\beta_\lambda(\geq \alpha_\lambda) \geq 0$.

**Lemma 8.1.4.** *Let $f$, $\sigma$, $\alpha_\lambda$, $\beta_\lambda$ be as in Lemma* 8.1.3. *Then,*

$$\int_{\mathbb{R}} e^{-\alpha_\lambda y}\, d\sigma(y) = \int_{\mathbb{R}} e^{-\beta_\lambda y}\, d\sigma(y) = 1, \qquad \forall\, \lambda \in [0, 1).$$

***Proof.*** We begin by noting that $\sigma(I) < \infty$ for every compact interval $I$: In fact, since $f > 0$ and decreases, and $f = f \bullet \sigma$, we have $\sigma(-\infty, x] \leq f(0)/f(x)$ for every $x \in \mathbb{R}$.

We shall consider the case of $\lambda = 0$ as typical; the same argument applies for all $\lambda \in (0, 1)$ as well. $\psi_0 = f$ itself, and, writing $\alpha$, $\beta$, $g$ instead of $a_0, \beta_0, g_0$, we have

$$\sup_x \frac{f(x+1)}{f(x)} = e^{-\alpha}, \qquad \inf_x \frac{f(x+1)}{f(x)} = e^{-\beta}.$$

The argument in Lemma 8.1.3 shows that, given $\varepsilon > 0$, there exists $\varepsilon' > 0$ such that, if $e^{-\alpha} - \varepsilon' < f(x+1)/f(x)$, then $e^{-2\alpha} - \varepsilon < f(x+2)/f(x)$. By induction, it follows that, for every $\varepsilon > 0$ and $k \in \mathbb{N}$, there exists an $x = x(\varepsilon, k)$ such that

$$f(x+n)/f(x) > e^{-n\alpha} - \varepsilon \qquad \text{for } n = 1, \ldots, 2k.$$

If $n - 1 < y \leq n$, $n = 1, \ldots, 2k$, then, since $f$ decreases, we have, for that $x$,

$$\begin{aligned} f(x+y)/f(x) &= \{f(x+n)/f(x)\} \cdot \{f(x+y)/f(x+n)\} \\ &> (e^{-n\alpha} - \varepsilon)e^{\alpha(n-y)} \geq e^{-\alpha y} - \varepsilon e^{\alpha}. \end{aligned}$$

Hence, again for the same $x$,

$$\begin{aligned} 1 &\geq \int_{(-k,k]} \frac{f(x+k+t)}{f(x+k)}\, d\sigma(t) = \frac{f(x)}{f(x+k)} \int_{(-k,k]} \frac{f(x+k+t)}{f(x)}\, d\sigma(t) \\ &> \left\{\int_{(-k,k]} (e^{-\alpha(k+t)} - \varepsilon e^{\alpha})\, d\sigma(t)\right\} e^{\alpha k} \\ &= \int_{(-k,k]} e^{-\alpha t}\, d\sigma(t) - \varepsilon e^{\alpha(k+1)} \sigma(-k, k]. \end{aligned}$$

$\sigma(-k, k] < \infty$; since $\varepsilon > 0$ and $k \in \mathbb{N}$ are arbitrary, it follows that $\int_{\mathbb{R}} e^{-\alpha t}\, d\sigma(t) \leq 1$. Similarly we show that $\int_{\mathbb{R}} e^{-\beta t}\, d\sigma(t) \leq 1$ also—a fact we need to prove that $\int_{\mathbb{R}} e^{-\alpha t}\, d\sigma(t) \geq 1$. To establish the latter fact, arguing as before we see that, for every $\varepsilon > 0$ and $k \in \mathbb{N}$, there exists an $x = x(\varepsilon, k)$ such that

$$f(x+y)/f(x) < e^{-\alpha y} + \varepsilon e^{\alpha} \qquad \text{for } -k \leq y \leq 0.$$

For this $x$, we have

$$1 = \int_{\mathbb{R}} \{f(x+y)/f(x)\}\, d\sigma(y) = \int_{(-\infty,-k)} + \int_{[-k,0)} + \int_{(0,\infty)} \cdots$$

$$\leq \int_{(-\infty,-k)} e^{-\beta y}\, d\sigma(y) + \int_{[-k,0]} (e^{-\alpha y} + \varepsilon e^{\alpha})\, d\sigma(y) + \int_{(0,\infty)} e^{-\alpha y}\, d\sigma(y).$$

Since $\varepsilon > 0$ is arbitrary and $\sigma[-k, 0]$ is finite, the preceding inequality holds with $\varepsilon = 0$ as well. Since $\int e^{-\beta y}\, d\sigma(y) < \infty$ (being in fact $\leq 1$), the first integral on the right tends to zero as $k \to \infty$, and we have $\int e^{-\alpha y}\, d\sigma(y) \geq 1$. Hence, $\int e^{-\alpha y}\, d\sigma(y) = 1$; similarly, $\int e^{-\beta y}\, d\sigma(y) = 1$, and the lemma is proven.

**Corollary 8.1.5.** *If $\alpha = \beta$, then $\alpha_\lambda = \beta_\lambda = \alpha$ for all $\lambda \in [0, 1)$. If $\alpha < \beta$, then $\alpha_\lambda = \alpha$ and $\beta_\lambda = \beta$ for all such $\lambda$.*

***Proof.*** The first assertion is immediate, from the fact that if $\alpha = \beta$, then $f(x + y) = f(x)e^{-\alpha y}$ for all $x \in \mathbb{R}$ and $y \geq 0$. If $\alpha < \beta$, then, since $\alpha$ and $\beta$ are the only real numbers such that $\int e^{-\alpha y}\, d\sigma(y) = \int e^{-\beta y}\, d\sigma(y) = 1$, Lemma 8.1.4 implies that $\alpha_\lambda = \alpha$ or $\beta$, and $\beta_\lambda = \alpha$ or $\beta$. If, for some $\lambda \in [0, 1)$, $\alpha_\lambda = \beta_\lambda$ (while $\alpha < \beta$), then denoting their common value by $\xi$, we have

$$f(x+y) - e^{-\xi y}f(x) = \lambda\{f(x+y+1) - e^{-\xi y}f(x+1)\} = \cdots$$
$$= \lambda^n\{f(x+y+n) - e^{-\xi y}f(x+n)\} \to 0 \qquad \text{as } n \to \infty,$$

since the last expression is in absolute value $\leq 2f(x)\lambda^n$. Hence, we must have $\alpha = \beta(=\xi)$—a contradiction. Hence, the second assertion of the corollary is proven.

We proceed to establish the main result, namely, Theorem 8.1.6 (taken together with Theorem 8.1.7).

**Theorem 8.1.6.** *Let $f$ be a nonnegative real-valued function on $\mathbb{R}$, locally integrable with respect to Lebesgue measure $\omega$ on $\mathbb{R}$, and $\sigma$ a positive $\sigma$-finite Borel measure such that* (8.1.4)—*$f = f \bullet \sigma$—holds a.e. $[\omega]$ on $\mathbb{R}$. In the nontrivial cases with $\sigma(0) < 1$, we have*

$$f(x) = p_1(x)e^{\alpha_1 x} + p_2(x)e^{\alpha_2 x} \qquad \textit{for a.e. } [\omega]\ x \in \mathbb{R},$$

*where $p_1 \geq 0$, $p_2 \geq 0$, and both have every element of* supp $\sigma$ *as a period, and*

$$\int e^{\alpha_1 y}\, d\sigma(y) = \int e^{\alpha_2 y}\, d\sigma(y) = 1.$$

***Remark.*** There exist at most two such $\alpha$'s. If no such $\alpha_j$ exist, then the only solution is the trivial solution: $f = 0$ a.e. $[\omega]$ on $\mathbb{R}$.

***Proof.*** If $\operatorname{supp} \sigma$ is contained in $[0, \infty)$, then Theorem 2.2.4 applied to $f(\cdot + x_0)$ provides the given solution (with $\alpha_1 = \alpha_2$) on every interval $[x_0, \infty)$, $x_0 \in \mathbb{R}$, and hence on $\mathbb{R}$ itself. A dual argument applies if $\operatorname{supp} \sigma$ is contained in $(-\infty, 0]$. Hence, we need only consider here cases where

$$\sigma(-\infty, 0) > 0 \qquad \text{and} \qquad \sigma(0, \infty) > 0.$$

Since $\sigma(0) < 1$ by assumption, we may further assume without loss of generality that $\sigma(0) = 0$ by the usual reduction, suitably modifying $f$ and $\sigma$. Then, the $\sigma$-finiteness of $\sigma$ and the local integrability of $f(\geq 0)$ imply, as in the proof of Theorem 2.2.4, that there exists a $\beta$ such that

$$\int_0^\infty e^{-\beta t} f(t)\, dt < \infty, \qquad \text{so} \qquad \int_x^\infty e^{-\beta t} f(t)\, dt < \infty \quad \forall\, x \in \mathbb{R},$$

and it follows via Fubini's theorem that if

$$\tilde{f}(x) = \int_x^\infty \int_y^\infty e^{-2\beta t} f(t)\, dt\, dy, \qquad d\tilde{\sigma}(y) = e^{2\beta y}\, d\sigma(y),$$

then $\tilde{f}$ is (strictly) positive, decreasing, and convex; $\tilde{\sigma}$ is $\sigma$-finite; and $\tilde{f} = \tilde{f} \bullet \tilde{\sigma}$; to prove the positivity of $\tilde{f}$, we may proceed as in the proof of Lemma 2.2.3(c), with the interval $(0, \delta]$ of the argument there being replaced here by the interval $(-\infty, \delta]$. Once a representation of the stated form is established for $\tilde{f}$, that a similar representation holds for $f$ can be seen directly by differentiation. The periodicity of $p_1$ and $p_2$ will imply, taking 1 to be a period, that, since $\tilde{f}$ and $\tilde{f}(\cdot + 1)$ are both differentiable, so are $p_1$ and $p_2$. We may therefore assume without loss of generality that, in Eq. (8.1.4), $f$ is strictly positive, decreasing, and convex, and that $\sigma$ is a $\sigma$-finite positive Borel measure. It follows that $f = f \bullet \sigma^n$ for every $n \in \mathbb{N}$, and hence that $f = f \bullet \nu$, where $\nu = \sum_{n=0}^\infty 2^{-n} \sigma^n$ (with $\sigma^0 = \delta_0$ as per our usual convention). By Lemma 8.1.2, $\operatorname{supp} \nu = d\mathbb{Z}$ for some $d > 0$ or $= \mathbb{R}$ itself. (As already remarked in the context of Lemma 8.1.4, since $f$ is positive and decreasing, $f = f \bullet \sigma$ implies that, for every real $x$, $\sigma(-\infty, x] \leq f(0)/f(x)$, and in particular $\sigma(I) < \infty$ for every compact interval.) We shall assume in what follows, without loss of generality, that $\operatorname{supp} \sigma$ itself is as just discussed.

*We shall consider now the case* $\operatorname{supp} \sigma = \mathbb{R}$. Theorem 8.1.7 and the remark following it indicate the modifications needed for the discussion of the case $\operatorname{supp} \sigma = d\mathbb{Z}$ for some $d > 0$.

It follows as in the proof of Theorem 2.2.4 that there exist $0 \leq \alpha \leq \beta$ such that

$$g(y) = \sup_x \frac{f(x+y)}{f(x)} = \begin{cases} e^{-\alpha y} & \text{for } y \geq 0, \\ e^{-\beta y} & \text{for } y \leq 0; \end{cases}$$

or, equivalently,

$$h(y) = \inf_x \frac{f(x+y)}{f(x)} = \begin{cases} e^{-\beta y} & \text{for } y \geq 0, \\ e^{-\alpha y} & \text{for } y \leq 0. \end{cases}$$

If $\alpha = \beta$, the assertion of the theorem is immediate. Let $\alpha < \beta$. By Lemma 8.1.4 and Corollary 8.1.5, we have then that

$$\text{for every } \lambda \in [0, 1), \qquad g_\lambda(y) = e^{-\alpha y}, \quad h_\lambda(y) = e^{-\beta y}, \qquad \forall\, y \geq 0,$$

where $g_\lambda$ and $h_\lambda$ are defined by (8.1.6). For convenience in writing and without loss of generality—consider $f(x)e^{\alpha x}$ instead of $f(x)$—we may assume that $\alpha = 0$, and denote the corresponding $\beta$ by $\theta$. We then have, for every $\lambda \in [0, 1)$,

$$f(x+y) - \lambda f(x+y+1) \geq \{f(x) - \lambda f(x+1)\}e^{-\theta y}, \qquad \forall\, x \in \mathbb{R},\ y > 0.$$

Now a review of the foregoing proof shows that the same arguments as for $t = 1$ apply if, instead of $\psi_\lambda$, we consider the function $\psi_{\lambda,t}(x) := f(x) - \lambda f(x+t)$ for arbitrary fixed $t > 0$. Hence, we have

$$f(x+y) - \lambda f(x+y+t) \geq \{f(x) - \lambda f(x+t)\}e^{-\theta y}, \qquad \forall\, \lambda \in [0, 1),\ x \in \mathbb{R},\ y > 0,\ t \geq 0.$$

By continuity, this inequality holds also for $\lambda = 1$, i.e.,

$$f(x+y) - f(x+y+t) \geq \{f(x) - f(x+t)\}e^{-\theta y}, \qquad \forall\, x \in \mathbb{R},\ y \geq 0,\ t \geq 0. \tag{8.1.9a}$$

Replacing $x$ by $x - y$, we have

$$f(x) - f(x+t) \geq \{f(x-y) - f(x+t-y)\}e^{-\theta y}, \qquad \forall \text{ such } x, y, t.$$

This may be rewritten as

$$f(x+t+y) - f(x+y) \geq \{f(x+t) - f(x)\}e^{-\theta y}, \qquad \forall\, x \in \mathbb{R},\ t \geq 0,\ y < 0. \tag{8.1.9b}$$

It follows from (8.1.9a) that, for $y > 0$, and arbitrary $n \in \mathbb{N}$,

$$\begin{aligned} f(x) - f(x+y) &= \sum_{k=1}^{n} \{f(x + (k-1)y/n) - f(x + ky/n)\} \\ &\geq [f(x) - f(x + y/n)] \sum_{k=0}^{n-1} e^{-\theta k y/n} \\ &= \{f(x) - f(x + y/n)\}(1 - e^{-\theta y})/(1 - e^{-\theta y/n}), \end{aligned}$$

so that, for any $x_0$ such that $f'(x_0)$ exists ($f$ being convex, $f'$ exists a.e. $[\omega]$), we have, on letting $n \to \infty$ in the preceding,

$$f(x_0) - f(x_0 + y) \geq -f'(x_0)(1 - e^{-\theta y})/\theta \qquad \forall\, y \geq 0.$$

Similarly, (8.1.9b) leads to

$$f(x_0) - f(x_0 + y) \geq -f'(x_0)(1 - e^{-\theta y})/\theta \qquad \forall\, y \leq 0,$$

as well. Therefore, since $f = f \bullet \sigma$, $\int d\sigma(y) = 1 = \int e^{-\theta y}\, d\sigma(y)$, we have

$$\int \{f(x_0) - f(x_0 + y) + f'(x_0)(1 - e^{-\theta y})/\theta\}\, d\sigma(y) = 0,$$

where the integrand is $\geq 0$. Since $\operatorname{supp} \sigma = \mathbb{R}$ by assumption, it follows that the integrand $= 0$ for all $y \in \mathbb{R}$. Taking $x_0 = 0$ without loss of generality, we have $f(y) = A + Be^{-\theta y}$ for all real $y$, $A$ and $B$ constants ($\geq 0$), proving the theorem in the case $\operatorname{supp} \sigma = \mathbb{R}$.

The discrete case $\operatorname{supp} \sigma = d\mathbb{Z}$ for some $d > 0$ has to be dealt with essentially separately. The assertion of Theorem 8.1.6 there follows almost at once from the following proposition—also see the remark at the end of its proof. No ideas essentially different from the proof for the case $\operatorname{supp} \sigma = \mathbb{R}$ are involved: Historically, the discrete case precedes and motivates the discussion of the other case.

**Theorem 8.1.7.** *Let $\{w_n\}_{n=-\infty}^{\infty}$, with $w_0 < 1$, and $\{v_n\}_{n=-\infty}^{\infty}$ be two sequences of nonnegative real constants such that*

$$v_m = \sum_{n=-\infty}^{\infty} v_{m+n} w_n \qquad \forall\, m \in \mathbb{Z}. \tag{8.1.10}$$

*Let $A = \{n \in \mathbb{Z} : w_n > 0\}$. If the group generated by $A$ is $d\mathbb{Z}$, then $v_m = B(m)b^m + C(m)c^m$ for $m \in \mathbb{Z}$, where $b > 0$ and $c > 0$ are such that*

$$\sum w_n b^n = \sum w_n c^n = 1,$$

*and $B$ and $C$ have period $d$. Also, $v_m = 0$ for all $m$ if no such $b$, $c$ exist.*

***Remark.*** As easily seen, the case $w_0 \geq 1$ is of no interest and is therefore omitted from our discussion.

***Proof.*** We shall assume that $A$ contains at least one positive and one negative element; in the complementary cases, Theorem 2.2.4 applies and $v_m = B(m)b^m$ for the unique $b > 0$ such that $\sum w_n b^n = 1$ if such $b$ exists, or $v_m = 0$ for all $m$ if no such $b$ exists. By iteration of (8.1.10), we see that

$$v_m = \sum v_{m+n} w_n^{(k)}, \qquad \forall\, m \in \mathbb{Z},\ k \in \mathbb{N},$$

for a suitable sequence $\{w_n^{(k)}\}$, where $w_n^{(k)} > 0$ for every $n$ in $A_k = \{x_1 + \cdots + x_k : x_i \in A, i = 1, \ldots, k\}$. Consequently,

$$v_m = \sum v_{m+n} w_n^*, \qquad \forall\, m \in \mathbb{Z},$$

for the sequence $\{w_n^*\}$ given by $w_n^* = \sum_{k=1}^{\infty} 2^{-k} w_n^{(k)}$, so that $w_n^* > 0$ for every $n \in \bigcup_k A_k = d\mathbb{Z}$, by Proposition 8.1.1. Again without loss of generality then, we shall take $d = 1$, so that $w_n > 0$ for every $n \in \mathbb{Z}$, in (8.1.10). The assumption that $w_0 < 1$ will still be made, since the solutions in the complementary cases are trivial. If then $v_m = 0$ for some $m$, it is clear that $v_n = 0$ for all $n$. Thus, the nontrivial solutions correspond to $w_0 < 1$ *and* $v_m > 0$ for all $m$.

The sequence $\{v_{m+1}/v_m\}$ is easily seen to be bounded; let

$$b := \sup_m \frac{v_{m+1}}{v_m} \qquad \text{and} \qquad c := \inf_m \frac{v_{m+1}}{v_m}.$$

If $b = c$, the theorem is immediate. We need therefore only consider the case $c < b$. Then, proceeding as in the proof of Lemma 8.1.4, we see that

$$\sup_m \frac{v_{m+n}}{v_m} = b^n, \qquad \inf_m \frac{v_{m+n}}{v_m} = c^n, \qquad \text{for } n \in \mathbb{Z}_+;$$

and that

$$\sum w_n b^n = \sum w_n c^n = 1.$$

Writing $\lambda = c/b$, $p_n = w_n b^n$, $t_m = v_m b^{-m}$, we have

$$t_m = \sum t_{m+n} p_n; \qquad \sum p_n = \sum p_n \lambda^n = 1;$$

$$\sup_m \frac{t_{m+n}}{t_m} = 1; \qquad \inf_m \frac{t_{m+n}}{t_m} = \lambda^n.$$

Let $u_m = t_m - t_{m+1}$. Then, $u_m \geq 0$ and $u_m = \sum u_{m+n} p_n$. $u_m = 0$ for some $m$ would imply that $u_n = 0$ for all $n$, so that $v_m = v_0 b^m$ for all $m$, contrary to our assumption that $b$ and $c$ are distinct. Hence, $u_m > 0$ for all $m$, and, again by the same reasoning as for the $v_m$, we must have $\inf_m(u_{m+1}/u_m) = 1$ or $\lambda$, 1 and $\lambda$ being the only positive constants $x$ satisfying the equation $\sum p_n x^n = 1$. But, $\inf_m(u_{m+1}/u_m) = 1$ would mean that $\sup_m(u_{m+1}/u_m) = 1$ as well (since the latter must also be $= 1$ or $\lambda$). But, then $u_m = \text{constant} = \lim_{n\to\infty} u_n = 0$ (since $\{t_n\}$ decreases to a finite limit as $n \to \infty$). Hence, we have $\inf_m(u_{m+1}/u_m) = \lambda$, and so $\inf_m(u_{m+n}/u_m) = \lambda^n$

for $n \in \mathbb{N}$. Therefore, we easily have:

$$\text{for } n \in \mathbb{N}, \quad t_m - t_{m+n} = u_m + \cdots + u_{m+n-1} \geq u_m(1-\lambda^n)/(1-\lambda);$$

$$\text{for } -n \in \mathbb{N}, \quad t_{m+n} - t_m = u_{m-1} + \cdots + u_{m+n+1} \leq \lambda u_{m-1}(\lambda^n - 1)/(1-\lambda).$$

Since $\sum_{n\geq 1} p_n(t_m - t_{m+n}) = \sum_{n \leq -1} p_n(t_{m+n} - t_m)$, we have from the preceding that $u_m \sum_{n\geq 1} p_n(1 - \lambda^n) \leq \lambda u_{m-1} \sum_{n\leq -1} p_n(\lambda^n - 1)$. But the two series in the previous relation are equal in view of $\sum p_n = \sum p_n \lambda^n \,(=1)$. Hence, $u_m \leq \lambda u_{m-1}$, so that $\sup_m(u_{m+1}/u_m) = \lambda$ as well. Therefore, $u_m = u_0 \lambda^m$ for all $m$, whence $t_m = B + C\lambda^m$ for all $m$, where $B = \lim_{n\to\infty} t_n$, and the theorem is proven.

***Remark.*** Reverting to the case $\operatorname{supp} \sigma = d\mathbb{Z}$ of Theorem 8.1.6, and taking $d = 1$ without loss of generality, fix $x \in [0, 1)$ and define $w_n = \sigma\{n\}$, $v_n = f(x + n)$ to obtain the conclusion of Theorem 8.1.6 in this case; $p_1$ and $p_2$ can be taken arbitrarily on $[0, 1)$ and periodic with period 1. The condition of local integrability of $f$ can obviously be dropped in this case, while that of the $\sigma$-finiteness of $\sigma$ is trivially satisfied.

The following is an immediate consequence of Theorem 8.1.6—applied to the function $f - c$ with $c < \inf f$.

**Corollary 8.1.8.** *Let $f$ be a bounded real-valued continuous function on $\mathbb{R}$, and $\sigma$ a probability measure on $\mathbb{R}$ such that*

$$f(x) = \int_{\mathbb{R}} f(x + y)\, d\sigma(y), \qquad \forall\, x \in \mathbb{R}.$$

*Then, $f$ has every element of* $\operatorname{supp} \sigma$ *as a period and, in particular, $f$ is a constant if* $\operatorname{supp} \sigma$ *has two incommensurable periods.*

We conclude this section with the analogs for $\mathbb{R}$ of Theorem 2.4.2 and Corollary 2.4.4. It is convenient to introduce the following notation: For $\rho > 0$,

$$A(\rho) = \{n\rho : n \in \mathbb{Z}\}, \qquad B(\rho) = \{2n\rho : n \in \mathbb{Z}\}, \qquad C(\rho) = \{(2n - 1)\rho : n \in \mathbb{Z}\}.$$

Given two positive Borel measures $\mu$ and $\nu$ on $\mathbb{R}$, there are the following three possibilities concerning their supports:

***Case 1.*** $\operatorname{supp}(\mu + \nu) \not\subset A(\rho)$ for any $\rho > 0$.

In the complementary situation, $\operatorname{supp}(\mu + \nu) \subset A(\rho)$ for some $\rho > 0$, which we shall take to be the largest of such $\rho$. Then, we have:

***Case 2.*** At least one of the sets $\operatorname{supp} \mu \cap C(\rho)$, $\operatorname{supp} \nu \cap B(\rho)$ is nonempty.

*Case 3.* $\operatorname{supp}\mu \subset B(\rho)$ *and* $\operatorname{supp}\nu \subset C(\rho)$.

Then, we have the following result.

**Theorem 8.1.9.** *Let $\mu$ and $\nu$ be positive subprobability measures on $\mathbb{R}$ such that $\mu + \nu$ is a probability measure, and let $f$ be a bounded continuous real-valued function on $\mathbb{R}$ such that $f = f \bullet (\mu - \nu)$ on $\mathbb{R}$. Then:*

(a) $f(x) \equiv 0$ *in Cases* 1 *and* 2; *and*
(b) $f(x + \rho) = -f(x)$ *for all* $x \in \mathbb{R}$ *in Case* 3.

***Proof.*** Proceeding along the lines of the proofs of Theorems 2.4.1 and 2.4.2, respectively, we establish first that $f(x + y) = f(x)$ for all $y \in \operatorname{supp}\mu$ and $= -f(x)$ for all $y \in \operatorname{supp}\nu$, and then the more explicit form of that statement given by the theorem.

**Corollary 8.1.10.** *Let $\mu$ and $\nu$ be positive $\sigma$-finite Borel measures on $\mathbb{R}$. Let $g$ and $h$, neither equal to 0 a.e. on $\mathbb{R}$, be nonnegative functions on $\mathbb{R}$, locally integrable with respect to $\omega$, and satisfying the simultaneous equations*

$$g = g \bullet \mu + h \bullet \nu, \qquad h = h \bullet \mu + g \bullet \nu, \qquad a.e.\ [\omega]\ on\ \mathbb{R}.$$

*Then,*

$$g(x) = p_1(x)e^{\alpha_1 x} + p_2(x)e^{\alpha_2 x},$$

$$h(x) = q_1(x)e^{\alpha_1 x} + q_2(x)e^{\alpha_2 x},$$

*for a.a.* $[\omega]$ $x \in \mathbb{R}$, *where the $\alpha_j$ satisfy*

$$\int e^{\alpha_j y}\, d(\mu + \nu)(y) = 1, \qquad j = 1, 2, \tag{8.1.11}$$

*and the $p_j, q_j$ are nonnegative functions with the following properties:*

(i) $p_j = q_j =$ *constant, for* $j = 1, 2$, *in Case* 1;
(ii) $p_j = q_j$, $j = 1, 2$, *and each has period* $\rho$, *in Case* 2;
(iii) $p_j(\cdot + \rho) = q_j$, $q_j(\cdot + \rho) = p_j$, *in Case* 3; *in particular*, $p_j, q_j$ *have period* $2\rho$.

***Proof.*** As usual, we may assume without loss of generality that $g$ and $h$ are continuous. If $k = g + h$, then $k = k \bullet (\mu + \nu)$ on $\mathbb{R}$, so that, by Theorem 8.1.6,

$$k(x) = l_1(x)e^{\alpha_1 x} + l_2(x)e^{\alpha_2 x},$$

where the $\alpha_j$ satisfy (8.1.11) and the $l_j$ have every element of $\operatorname{supp}(\mu + \nu)$ as a period. Further, proceeding as in the proof of Theorem 2.4.1, we easily check, using the fact that $0 \le g, h \le k$ and the preceding representation for $k$, that,

$$\text{with } \sigma = \mu + \nu^2 * \left(\sum_{n=0}^{\infty} \mu^n\right), \qquad g = g \bullet \sigma, \qquad h = h \bullet \sigma.$$

Therefore, again by Theorem 8.1.6, and noting that $\int e^{\alpha_j y}\, d\sigma(y) = 1$ for $j = 1, 2$, we have

$$g(x) = p_1(x)e^{\alpha_1 x} + p_2(x)e^{\alpha_2 x},$$

$$h(x) = q_1(x)e^{\alpha_1 x} + q_2(x)e^{\alpha_2 x},$$

where the $p_j, q_j$ ($j = 1, 2$) have every element of $\operatorname{supp} \sigma$ as a period. Fix $\tau \in \operatorname{supp} \mu$ (or $\operatorname{supp} \sigma$), so that $\tau$ is a period for the $p_j, q_j$. Assuming without loss of generality that $\alpha_2 = 0$, we have

$$\bar{g}(x) := (g(x + \tau) - g(x))e^{-\alpha_1 x} = p_1(x)(e^{\alpha_1 \tau} - 1),$$

$$\bar{h}(x) := (h(x + \tau) - h(x))e^{-\alpha_1 x} = q_1(x)(e^{\alpha_1 \tau} - 1).$$

Defining

$$d\bar{\mu}(y) = e^{\alpha_1 y}\, d\mu(y), \qquad d\bar{\nu}(y) = e^{\alpha_1 y}\, d\nu(y),$$

$$r = (\bar{g} + \bar{h})/2, \qquad \text{and} \qquad s = (\bar{g} - \bar{h})/2,$$

we see that $\bar{\mu}$ and $\bar{\nu}$ are subprobability measures whose sum is a probability measure, and $r$ and $s$ are bounded continuous functions satisfying

$$r = r \bullet (\bar{\mu} + \bar{\nu}), \qquad s = s \bullet (\bar{\mu} - \bar{\nu}).$$

By Corollary 8.1.8, $r$ has every element of $\operatorname{supp}(\mu + \nu)$ as a period, and, by Theorem 8.1.9, $s(x) = 0$ in Cases 1 and 2, and has the property $s(x + \rho) = -s(x)$ in Case 3. The relations

$$p_1(e^{\tau\alpha_1} - 1)\,(=\bar{g}) = r + s, \qquad q_1(e^{\tau\alpha_1} - 1)\,(=\bar{h}) = r - s$$

then imply statements (i)–(iii) of the corollary for $p_1, q_1$. Similarly for $p_2, q_2$. The corollary is proven.

## 8.2. A PROOF USING THE KREIN–MILMAN THEOREM

Let $C$ be a convex set in a linear space. A point $x \in C$ is an *extreme point* of $C$ if the representation

$$x = \lambda x_1 + (1 - \lambda)x_2, \qquad \text{for some } x_1, x_2 \in C,\ \lambda \in (0, 1),$$

is possible only if $x_1 = x_2 = x$. The Krein–Milman theorem says: If $C$ is a compact convex subset of a locally convex linear topological space, then

$C$ is the closed convex hull of its extreme points. For a discussion of the concepts involved, see, for instance, Dunford and Schwartz (1953), and also the appendix to Chapter 9.

We present in this section a proof of Theorem 8.1.6 using the Krein–Milman theorem in the case $\operatorname{supp}\sigma = \mathbb{R}$; the discrete case $\operatorname{supp}\sigma \subset d\mathbb{Z}$, for some $d > 0$, can be handled similarly.

We have seen in Section 8.1 that if a nonnegative solution $\neq 0$ a.e. $[\omega]$ exists for the equation $f = f \bullet \sigma$ on $\mathbb{R}$, under the assumptions on $f$ and $\sigma$ made there, then there also exists a strictly positive, decreasing, and convex solution $\tilde{f}$ for a suitably modified form $\tilde{f} = \tilde{f} \bullet \bar{\sigma}$ of that equation. Renaming the modified form itself as $f = f \bullet \sigma$, we have seen that, for a solution $f$ with these properties there exist $(0 \leq)\ \alpha \leq \beta$ such that

$$\sup_x \frac{f(x+y)}{f(x)} = e^{-\alpha y}, \qquad \inf_x \frac{f(x+y)}{f(x)} = e^{-\beta y}, \qquad \forall\, y \geq 0. \qquad (8.2.1)$$

Further, we also see, from the appropriate part of the proof of Lemma 8.1.3, that

$$\int_{\mathbb{R}} e^{-\alpha y}\, d\sigma(y) \leq 1, \qquad \int_{\mathbb{R}} e^{-\beta y}\, d\sigma(y) \leq 1. \qquad (8.2.2)$$

We shall not need to prove the corresponding equalities; these will be part of the final consequences of the argument that follows.

Let $B = \{t : \int_{\mathbb{R}} e^{-ty}\, d\sigma(y) \leq 1\}$; $B$ is nonempty, by virtue of (8.2.2). Let $\gamma = \inf B$, and $\delta = \sup B$. Then,

$$e^{-\delta y} \leq f(x+y)/f(x) \leq e^{-\gamma y}, \qquad \forall\, x \in \mathbb{R},\ y \geq 0, \qquad (8.2.3)$$

and

$$\int e^{-\gamma y}\, d\sigma(y) \leq 1, \qquad \int e^{-\delta y}\, d\sigma(y) \leq 1. \qquad (8.2.4)$$

Now, let $\mathcal{H}$ be the class of all functions $h$, strictly positive on $\mathbb{R}$, satisfying $h = h \bullet \sigma$ on $\mathbb{R}$ and the inequalities (8.2.3), and with $h(0) = 1$; this last condition is meaningful to impose since an $h$ satisfying (8.2.3) is necessarily continuous. Our above argument shows that, if any nontrivial solution exists for $f = f \bullet \sigma$ on $\mathbb{R}$ at all, then $\mathcal{H}$ is nonempty. Let $h_0$ be defined on $\mathbb{R}$ according to

$$h_0(x) = \begin{cases} e^{-\gamma x} & \text{for } x \geq 0, \\ e^{-\delta x} & \text{for } x \leq 0, \end{cases}$$

and let $\mathcal{A}$ be the set of all real-valued functions $g$ on $\mathbb{R}$ such that $0 \leq g \leq h_0$. Then, under the topology of pointwise convergence, $\mathcal{A}$, being equal to $\times_{x \in \mathbb{R}} [0, h_0(x)]$, is compact, by Tychonoff's theorem. As a closed subset of $\mathcal{A}$, $\mathcal{H}$ is then compact and is also obviously convex. We shall see below that the set of extreme elements of $\mathcal{H}$ comprises at most two exponential functions, so that the Krein–Milman theorem immediately yields Theorem 8.1.6 (in the case supp $\sigma = \mathbb{R}$).

Fix $h \in \mathcal{H}$ and $\lambda$ such that $0 < \lambda < e^{\gamma}$, and consider

$$h_1(x) := \frac{h(x) + \lambda h(x+1)}{1 + \lambda h(1)}, \qquad h_2(x) := \frac{h(x) - \lambda h(x+1)}{1 - \lambda h(1)}. \tag{8.2.5}$$

$h_1$ and $h_2$ are positive and satisfy $h_j = h_j \bullet \sigma$ for $j = 1, 2$. Since $\mathcal{H}$ is convex, it is obvious that $h_1 \in \mathcal{H}$. To see that $h_2 \in \mathcal{H}$ as well, we first note that the relevant parts of the proofs of Lemmas 8.1.3 and 8.1.4 show that there exist $\alpha_\lambda \le \beta_\lambda$ such that

$$\left.\begin{aligned} &e^{-\beta_\lambda y} \le h_2(x+y)/h_2(x) \le e^{-\alpha_\lambda y}, \qquad \forall\, x \in \mathbb{R},\ y \ge 0, \\ &\int e^{-\alpha_\lambda y}\, d\sigma(y) \le 1, \qquad \int e^{-\beta_\lambda y}\, d\sigma(y) \le 1. \end{aligned}\right\} \tag{8.2.6}$$

It follows that $[\alpha_\lambda, \beta_\lambda] \subset [\gamma, \delta]$, and then that $h_2 \in \mathcal{H}$. We must remark here that, though (8.2.3) is apparently not identical with (8.1.5), we need only obvious modifications of the argument in Lemma 8.1.3 to derive (8.2.6) from it.

With $h_1$ and $h_2$ as in (8.2.5), we have $h = ph_1 + (1-p)h_2$, where $0 < p = \frac{1}{2}(1 + \lambda h(1)) < 1$ and $h_1, h_2 \in \mathcal{H}$. Therefore, if $h$ is an extreme element of $\mathcal{H}$, we must have $h_1 = h_2 = h$, leading to $h(x+1) = h(x)h(1)$ for all $x$. But, the unit increment in the definitions (8.2.5) may be replaced by an arbitrary $t > 0$ without affecting the conclusions (8.2.6) for the corresponding $h_2$. Hence, we have $h(x+t) = h(x)h(t)$ for all $t > 0$, $x \in \mathbb{R}$, so that $h(x) = h(1)^x$ for all $x \in \mathbb{R}$; i.e., $h$ is an exponential function. Writing $h(x) = e^{-\theta x}$, we have from $h = h \bullet \sigma$ that $\int e^{-\theta y}\, d\sigma(y) = 1$. There can be at most two real $\theta$ satisfying this condition, and our assertion follows. (If $\theta_1 < \theta_2$ are such that $\int e^{-\theta_j y}\, d\sigma(y) = 1$, it is clear that we must have $\alpha = \alpha_\lambda = \gamma = \theta_1$ and $\beta = \beta_\lambda = \delta = \theta_2$.)

## 8.3. A VARIANT OF THE ICFE ON $\mathbb{R}$ AND THE WIENER–HOPF TECHNIQUE

In this section, we first consider solutions of ICFE's of the form

$$f(x) = \int_{\mathbb{R}} f(x+y)\, d\sigma(y) \qquad \text{for } x \ge 0,$$

with $f(\ge 0)$ and $\sigma$ subject to the familiar restrictions. The result can be used to obtain another proof of Theorem 8.1.6.

For $n \in \mathbb{N}$ and $x_1, \ldots, x_n \in \mathbb{R}$, we define $s_0 = 0$, $s_n = x_1 + \cdots + x_n$. If $\sigma$ is any positive $\sigma$-finite Borel measure on $\mathbb{R}$, we adopt the notation:

$$\left.\begin{aligned}
\sigma^{(n)} &= \text{the product measure (on } \mathbb{R}^n\text{) of } n \text{ copies of } \sigma;\\
\sigma^{n} &= \text{the } n\text{-fold convolution of } \sigma \text{ with itself if defined};\\
\sigma_1^+ &= \sigma_1^- = \sigma.\\
&\text{For } 1 < n \in \mathbb{N} \text{ and for any Borel subset } B \text{ of } \mathbb{R},\\
\sigma_n^+(B) &= \sigma^{(n)}\{(x_1, \ldots, x_n) \in \mathbb{R}^n: s_m \geq 0 \text{ for } 1 \leq m < n, s_n \in B\};\\
\sigma_n^-(B) &= \sigma^{(n)}\{(x_1, \ldots, x_n) \in \mathbb{R}^n: s_m < 0 \text{ for } 1 \leq m < n, s_n \in B\};\\
\rho(B) &= \sum_{n=1}^{\infty} \sigma_n^+((-\infty, 0) \cap B);\\
\tau(B) &= \sum_{n=1}^{\infty} \sigma_n^-([0, \infty) \cap B);
\end{aligned}\right\} \tag{8.3.1}$$

and finally, for an arbitrary positive Borel measure $\mu$ on $\mathbb{R}$, $S(\mu)$ will denote the *closed* subgroup generated by the support of $\mu$.

In analogy with the terminology of the theory of random walks, where $\sigma$ is a probability measure (*cf.* Feller, 1971, p. 412), we may call $\rho$ and $\tau$, respectively, the descending (strict) ladder measure and ascending (weak) ladder measure induced on $\mathbb{R}$ by $\sigma$.

An immediate, interesting, and important reformulation of $\sigma_n^-$ is given by the following.

**Lemma 8.3.1.** *For any Borel set $B \subset \mathbb{R}$,*

$$\sigma_n^-(B) = \sigma^{(n)}\{(x_1, \ldots, x_n) \in \mathbb{R}^n : s_m > s_n \text{ for } 1 \leq m < n, s_n \in B\}. \tag{8.3.2}$$

***Proof.*** Rename the variables $x_1, \ldots, x_n$ according to $\tilde{x}_k = x_{n+1-k}$ for $k = 1, \ldots, n$ (i.e., list them in reverse order); noting then that $\tilde{s}_k := \tilde{x}_1 + \cdots + \tilde{x}_k = s_n - s_{n-k}$ for all such $k$, that

$$\begin{aligned}
&\{(x_1, \ldots, x_n) \in \mathbb{R}^n : \tilde{s}_1 < 0, \ldots, \tilde{s}_{n-1} < 0, \tilde{s}_n \in B\}\\
&\qquad = \{(x_1, \ldots, x_n) \in \mathbb{R}^n : s_n < s_k \text{ for } 1 \leq k < n, s_n \in B\},
\end{aligned}$$

and that the integrals over this set with respect to $d\sigma(\tilde{x}_1) \cdots d\sigma(\tilde{x}_n)$ and with respect to $d\sigma(x_1) \cdots d\sigma(x_n)$ coincide, we obtain the relation (8.3.2).

***Remark.*** A dual statement (which, however, we shall not use in the sequel) holds of course for $\sigma_n^+$.

**Proposition 8.3.2.** (i) $S(\tau) \supseteq S(\sigma)$ *if* $\sigma(0, \infty) > 0$.
(ii) $S(\rho) \supseteq S(\sigma)$ *if* $\sigma(-\infty, 0) > 0$.

***Proof.*** (i) Let $A^+ = \operatorname{supp} \sigma \cap (0, \infty)$, and $A^- = \operatorname{supp} \sigma \cap (-\infty, 0)$. Obviously, $A^+ \subseteq \operatorname{supp} \tau$. If now $A^-$ is empty, then $\operatorname{supp} \sigma$ $(= A^+ \cup A^-)$ is itself contained in $\operatorname{supp} \tau$. Let $x \in A^-$. Since $A^+$ is nonempty by assumption, let $y \in A^+$, and let $n$ be the least positive integer such that $x + ny \geq 0$. If $x + ny = 0$, then $x = -ny \in S(\tau)$ since $y \in \operatorname{supp} \tau$. If $x + ny > 0$, then it belongs to $\operatorname{supp} \sigma_{n+1}^- \cap (0, \infty)$, and so to $\operatorname{supp} \tau$; since $y \in \operatorname{supp} \tau$, it then follows that $x \in S(\tau)$. Hence, (i) is proven.

(ii) This assertion is proved in a dual manner, with in fact a further simplification: In the latter half of the argument, we need only consider now an $x \in A^+$, a $y \in A^-$, and the least positive integer $n$ such that $x + ny < 0$.

**Lemma 8.3.3.** *Let $\rho$ and $\tau$ be as defined in* (8.3.1). *Then, $\tau$ is finite for compact sets if* $\operatorname{supp} \sigma \cap (-\infty, 0)$ *is nonempty and if $\rho$ is finite for compact sets; a dual statement holds for $\rho$ in relation to $\tau$.*

***Proof.*** We first note that, for any $a > 0$ and $b < c < 0$,

$$\tau[0, a)\sigma(b, c) \leq \tau[0, a + c) + \rho(b, 0). \tag{8.3.3}$$

Indeed, for any $n \in \mathbb{N}$, we have (invoking Lemma 8.3.1 twice)

$$\begin{aligned} \sigma_n^-[0, a)\sigma(b, c) &= \sigma^{(n+1)}\{(x_1, \ldots, x_{n+1}) : s_m > s_n \text{ for } 1 \leq m < n, \\ &\qquad s_n \in [0, a),\ x_{n+1} \in (b, c)\} \\ &= \sigma^{(n+1)}\{(x_1, \ldots, x_{n+1}) : s_m > s_{n+1} \text{ for } 1 \leq m < n, \\ &\qquad s_n \in [0, a),\ x_{n+1} \in (b, c)\} \\ &= \sigma^{(n+1)}\{(x_1, \ldots, x_{n+1}) : s_m > 0 \text{ for } 1 \leq m < n,\ s_{n+1} \in (b, 0)\} \\ &= \sigma_{n+1}^-[0, a + c) + \sigma_{n+1}^+(b, 0), \end{aligned}$$

and (8.3.3) follows upon summing over $n \in \mathbb{N}$.

Let now $s \in (-\infty, 0) \cap \operatorname{supp} \sigma$, and $y \in \mathbb{R}_+$. Then, noting that $\tau(-\infty, 0) = 0$, we have from (8.3.3) that

$$\tau(-\infty, y + ns/2)\sigma(3s/2, s/2) \leq \tau(-\infty, y + (n + 1)s/2) + \rho(3s/2, 0).$$

It follows by (reverse) induction that $\tau(-\infty, y)$ is finite for every $y \in \mathbb{R}_+$. Hence, the first statement of the lemma is proven; the second follows by a dual argument.

**Lemma 8.3.4 (The Wiener–Hopf Decomposition).** *Let $\rho$ and $\tau$ be as defined in* (8.3.1). *If either of them is finite for compact sets* (*so is the other, by Lemma* 8.3.3), *then,*

$$\text{for } B \subseteq (-\infty, 0), \qquad \rho(B) = \sigma(B) + (\rho * \tau)(B), \tag{8.3.4a}$$

$$\text{for } B \subseteq [0, \infty), \qquad \tau(B) = \sigma(B) + (\rho * \tau)(B), \tag{8.3.4b}$$

*so that*

$$\rho + \tau = \sigma + \rho * \tau \qquad \text{on Borel subsets of } \mathbb{R}. \tag{8.3.4c}$$

***Proof.*** For any finite sequence $s_1, \ldots, s_n$ of real numbers, there exists a least integer $m$ for which $s_m = \min(s_1, \ldots, s_n)$. This fact easily leads to the decomposition of $\mathbb{R}^n$ given by

$$\mathbb{R}^n = \bigcup_{m=1}^{n} \{(x_1, \ldots, x_n) \in \mathbb{R}^n : s_1 > s_m, \ldots, s_{m-1} > s_m,$$
$$s_{m+1} \geq s_m, \ldots, s_n \geq s_m\}$$

—a relation that we shall use shortly. By the same reasoning, we also have, for $B \subseteq (-\infty, 0)$.

$$\sigma_{n+1}^{+}(B) = \sigma^{(n+1)}\Bigg(\bigcup_{m=1}^{n} \{(x_1, \ldots, x_n) \in \mathbb{R}^n : s_1 > s_m, \ldots, s_{m-1} > s_m,$$
$$s_m \geq 0, s_{m+1} \geq s_m, \ldots, s_n \geq s_m, s_{n+1} \in B\}\Bigg)$$
$$= \sum_{m+1}^{n} \sigma^{(n+1)}\{(x_1, \ldots, x_n) \in \mathbb{R}^n : s_1 > s_m, \ldots, s_{m-1} > s_m,$$
$$s_m \geq 0, \tilde{s}_1 \geq 0, \ldots, \tilde{s}_{n-m} \geq 0, \tilde{s}_{n+1-m} \in B - s_m\},$$

where $\tilde{s}_k = s_{m+k} - s_m$ for $1 \leq k \leq n + 1 - m$,

$$= \sum_{m=1}^{n} \int_{\mathbb{R}_+} \sigma_{n+1-m}^{+}(B - y)\, d\sigma_m^{-}(y).$$

Since $\rho, \tau$ are finite on compact sets, so are $\sigma_j^{+}((-\infty, 0) \cap \cdot)$ and $\sigma_j^{-}((0, \infty) \cap \cdot)$; they are $\sigma$-finite, and their convolution is defined. Thus, if $B \subseteq (-\infty, 0)$, we have

$$\rho(B) = \sigma_1^{+}(B) + \sum_{n=1}^{\infty} \sigma_{n+1}^{+}(B) = \sigma(B) + \sum_{j=1}^{\infty} \sum_{k=1}^{\infty} \int_{\mathbb{R}_+} \sigma_j^{+}(B - y)\, d\bar{\sigma}_k(y),$$

and since the sum of a double series of nonnegative terms is the same whether summed along rows, along columns, or along diagonals, the last expression is equal to

$$\sigma(B) + \int_{\mathbb{R}_+} \rho(B - y)\, d\tau(y),$$

whence (8.3.4a). A dual argument yields (8.3.4b). Equation (8.3.4c) follows at once from (8.3.4a) and (8.3.4b).

**Corollary 8.3.5.** *If $\rho(K) < \infty$ (respectively, if $\tau(K) < \infty$) for all compact $K$, then the same is true of $\tau$ (respectively, $\rho$), and of $\rho$ and $\rho * \tau$ as well, and we have*

$$\sigma = \rho + \tau - \rho * \tau \tag{8.3.5}$$

*on (relatively) compact subsets of* $\mathbb{R}$.

***Proof.*** The assertion follows at once from Lemma 8.3.3 and (8.3.4c).

We now come to the main result of this section.

**Theorem 8.3.6.** *Let $\sigma \geq 0$ be a $\sigma$-finite Borel measure on $\mathbb{R}$, and let $\rho$ and $\tau$ be as defined in* (8.3.1). *Let $f$ be a nonnegative, Borel measurable function, locally integrable with respect to Lebesgue measure on $\mathbb{R}$, such that*

$$f(x) = \int_{\mathbb{R}} f(x + y)\, d\sigma(y) \qquad \text{for a.a. } x \geq 0. \tag{8.3.6}$$

*If $f(x) \not\equiv 0$ a.e. for $x \geq \inf(\operatorname{supp} \sigma)$, then $f$ admits the representation*

$$f(x) = \int_{(-\infty, 0)} f(x + y)\, d\rho(y) + p(x)e^{\alpha x} \qquad \text{for a.a. } x \geq 0, \tag{8.3.7}$$

*where $p \equiv 0$ if no real $\alpha$ exists such that $\int_{\mathbb{R}_+} e^{\alpha y}\, d\tau(y) = 1$, and $p$ is periodic with every element of* $\operatorname{supp} \sigma$ *as period if such an $\alpha$ (necessarily unique) exists.*

***Proof.*** If $\sigma(-\infty, 0) = 0$, the theorem reduces to Theorem 2.2.4. If $\sigma(0, \infty) = 0$, then the representation holds trivially, with $p \equiv 0$ and $\rho = \sigma$. Hence, we need only consider here cases where $\sigma(-\infty, 0) > 0$ and $\sigma(0, \infty) > 0$. Also, arguing as in Lemma 2.2.2, we see that there exists a real $\beta$ such that $0 < \tilde{f}(x) = \int_x^\infty e^{-\beta y} f(y)\, dy < \infty$ for all $x \in \mathbb{R}$; then, $\tilde{f}$ satisfies an equation of the form (8.3.6) with $d\sigma(y)$ replaced by $e^{\beta y}\, d\sigma(y)$. We may therefore assume without loss of generality that, in (8.3.6), $f$ is a (strictly) positive, continuous, and decreasing function. Now, we have from (8.3.6)—used repeatedly and invoking Fubini's theorem—that, for $x \geq 0$,

$$
\begin{aligned}
f(x) &= \int_{(-\infty,0)} f(x+y)\,d\sigma_1^+(y) + \int_{\mathbb{R}_+} f(x+y)\,d\sigma_1^+(y) \\
&= \int_{(-\infty,0)} f(x+y)\,d\sigma_1^+(y) + \int_{(-\infty,0)} f(x+y)\,d\sigma_2^+(y) \\
&\quad + \int_{\mathbb{R}_+} f(x+y)\,d\sigma_2^+(y) \\
&= \cdots \\
&= \sum_{k=1}^{n} \int_{(-\infty,0)} f(x+y)\,d\sigma_k^+(y) + \int_{\mathbb{R}_+} f(x+y)\,d\sigma_n^+(y). \qquad (8.3.8)
\end{aligned}
$$

Equation (8.3.8) implies that the sequence $\{\int_{\mathbb{R}_+} f(x+y)\,d\sigma_n^+(y)\}$ is therefore decreasing in $n$ for any fixed $x \geq 0$, and so converges to a limit, say $\tilde{f}(x)$, so that we have

$$
f(x) = \int_{(-\infty,0)} f(x+y)\,d\rho(y) + \tilde{f}(x), \qquad x \geq 0. \qquad (8.3.9)
$$

(A digression: Now, we have from (8.2.1) and (8.2.4c) that

$$
\int f(x+y)\,d\rho(y) + \int f(x+y)\,d\tau(y) = f(x) + \int f(x+y)\,d(\rho * \tau)(y).
$$

For $x \geq 0$, $\int f(x+y)\,d\rho(y) < \infty$, by (8.3.9); *if* we have $\int f(x+y)\,d\tau(y) < \infty$ also for $x \geq 0$, then (8.3.9) and the preceding relation imply that $\tilde{f}(x) = \int \tilde{f}(x+y)\,d\tau(y)$ for $x \geq 0$, and so $\tilde{f}(x)$ is of the form $e^{\alpha x}p(x)$, as claimed, in view of Theorem 2.2.4. However, Example 8.3.7 following shows that it is possible that, while (8.3.1) holds, $\int f(x+y)\,d\tau(y) = \infty$ for every $x \geq 0$. Hence, to cover the general case, we have to argue as in the following.)

By Proposition 1.2.1, (8.3.9) implies that $\rho(K) < \infty$ for compact $K$. Then, by Lemma 8.3.3, $\tau$ has the same property.

For any Borel set $A \subseteq \mathbb{R}$, let $\sigma_{n,A}^+$ be the measure obtained upon replacing "$s_1 \geq 0$" by "$s_1 \in \mathbb{R}_+ \cap A$" in the definition (8.3.1) of $\sigma_n^+$. For fixed $x \in \mathbb{R}$ and $A$, $\{\int_{\mathbb{R}_+} f(x+y)\,d\sigma_{n,A}^+(y)\}$ is a decreasing sequence of nonnegative real numbers, and so converges. Consequently, for each $c > 0$, $x \to \mathbb{R}_+$,

$$
f(x) = \lim_{n\to\infty} \int_{\mathbb{R}_+} f(x+y)\,d\sigma_{n,[0,c]}^+(y) + \lim_{n\to\infty} \int_{\mathbb{R}_+} f(x+y)\,d\sigma_{n,(c,\infty)}^+(y).
$$

Since $\sigma_n^+(B) \le \sigma(B)$ obviously, we have

$$\int_{\mathbb{R}_+} f(x+y)\, d\sigma_{n,(c,\infty)}^+(y) \le \int_{(c,\infty)} f(x+y)\, d\sigma(y) \to 0 \qquad \text{as } c \to \infty.$$

Hence,

$$f(x) = \lim_{c\to\infty} \lim_{n\to\infty} \int_{\mathbb{R}_+} f(x+y)\, d\sigma_{n,[0,c]}^+(y), \qquad x \ge 0. \tag{8.3.10}$$

Again, let $\sigma_{n,c}^-$ be the measure obtained by replacing "$s_1 > s_n$" by "$c \ge s_1 > s_n$" in the formula for $\sigma_n^-$ given by Lemma 8.3.1. Using the decomposition of $\mathbb{R}^n$ derived in the proof of Lemma 8.3.4, and Lemma 8.3.1, we have, as in the proof of Lemma 8.3.4, for $x \ge 0$,

$$\int_{\mathbb{R}_+} f(x+y)\, d\sigma_{n,[0,c]}^+(y)$$
$$= \sum_{m=1}^{n} \int_{[0,c]} \left\{ \int_{\mathbb{R}_+} f(x+y+t)\, d\sigma_{n-m}^+(t) \right\} d\sigma_{m,c}^-(y) \tag{8.3.11}$$

(with $\sigma_0^+ = \delta_0$). Since $\tau$ is finite for compact sets, and so regular, $\tau^{(c)}(\cdot) := \sum_{n=1}^{\infty} \sigma_{n,c}^-(\mathbb{R}_+ \cap \cdot)$ increases to the limit $\tau$ as $c \to \infty$. Also, by (8.3.8),

$$0 \le \int_{\mathbb{R}_+} f(x+y+t)\, d\sigma_n^+(t) \le f(x), \qquad \text{for } x, y \in \mathbb{R}_+,\ n \in \mathbb{N}.$$

Then, we have from (8.3.10), (8.3.11), and the dominated convergence theorem that

$$\hat{f}(x) = \lim_{c\to\infty} \int_{[0,c]} \hat{f}(x+y)\, d\tau^{(c)}(y) = \int_{\mathbb{R}_+} \hat{f}(x+y)\, d\tau(y), \qquad x \ge 0.$$

It follows then from Theorem 2.2.4 that $\hat{f}(x) = p(x)e^{\alpha x}$, where $p$ has every element of supp $\tau$ as a period, or $f \equiv 0$, according to whether there exists an $\alpha$ such that $\int e^{\alpha y}\, d\tau(y) = 1$ or not; in the former case, Proposition 8.3.2 implies that in fact $p$ has every element of supp $\sigma$ as a period. Hence, the representation (8.3.7) holds, and the theorem is proven.

We now consider an example to illustrate the possibility referred to in the course of the preceding proof.

***Example 8.3.7.*** Let $\sigma$ be the probability measure with $\{-1, 0, 1, \ldots\}$ as support, and with its Laplace–Stieltjes transform (LST) given by

$$L(t) = \int e^{-tx}\, d\sigma(x) = 1 + \alpha e^t (1 - e^{-t})^\beta, \qquad t \ge 0,$$

where $\alpha > 0$, $1 < \beta < 2$, and $0 < \alpha\beta \leq 1$. Since $\sigma$ is a probability measure, it follows from their definitions (as is well-known) that $\rho$ and $\tau$ are necessarily subprobability measures. It is also then clear without explicit computation that $\rho$ has to be of the form $\rho = p\delta_{-1}$, with $0 < p \leq 1$. Then, in view of (8.3.5), $\tau$ has its LST given by

$$L^*(t) = \frac{M(t) - pe^t}{1 - pe^t} = 1 - \alpha\frac{(1 - e^{-t})^\beta}{p - e^{-t}}.$$

$L^*$, as the LST of a subprobability measure, should be defined (and bounded) for all $t \geq 0$; this fact implies that we must have $p = 1$, so that $\rho = \delta_{-1}$ and

$$L^*(t) = 1 - \alpha(1 - e^{-t})^{\beta-1}.$$

If

$$f(x) = \begin{cases} x + 1 & \text{for } x \geq 0, \\ 0 & \text{otherwise,} \end{cases}$$

then, noting that $\sigma$ has zero expectation, a simple computation shows that $f$ satisfies (8.3.6). However, $\tau$ does not have finite expectation, and as a consequence of this fact $\int f(x + y)\, d\tau(y) = \infty$ for all $x \geq 0$. This concludes the discussion of the example.

Theorem 8.3.6 immediately leads to the following consequences, as well as to a proof of Theorem 8.1.6 given at the end of this section.

**Corollary 8.3.8.** *Let $k \in \mathbb{R}_+$, $\sigma$ a positive $\sigma$-finite Borel measure on $\mathbb{R}$, and $f \geq 0$ a locally integrable function on $\mathbb{R}_+$ such that*

$$f(x + k) = \int_{\mathbb{R}_+} f(x + y)\, d\sigma(y) \qquad \text{for a.a. } x \geq 0.$$

*Let $\hat{\rho}$, $\hat{\tau}$ be the measures obtained upon replacing* 0 *by $k$ in the definitions* (8.3.1). *Then, $f$ admits the representation*

$$f(x) = \int_{[0,k)} f(x + y)\, d\tilde{\rho}(y) + p(x)e^{\alpha x} \qquad \text{for a.a. } x \geq k,$$

*where $p \equiv 0$ if there exists no $\alpha$ such that $\int e^{\alpha y}\, d\hat{\tau}(y) = 1$, and $p$ has every element of* (supp $\sigma$) $-$ *$k$ as period if there exists such an $\alpha$ (necessarily unique).*

***Proof.*** If $\tilde{f}$ and $\tilde{\sigma}$ are defined on $[-k, \infty)$ according to $\tilde{f}(x) = f(x + k)$, $d\tilde{\sigma}(y) = d\sigma(y + k)$, then the given equation reduces to the form (8.3.6). Now, let $\tilde{\rho}$, $\tilde{\tau}$ correspond to $\tilde{\sigma}$ as per the definitions (8.3.1). Then, we have the representation

$$\tilde{f}(x) = \int_{-k}^{0} \tilde{f}(x+y)\,d\tilde{\rho}(y) + \tilde{p}(x)e^{\alpha x} \qquad \text{for a.a. } x \geq 0,$$

where $\tilde{p} \equiv 0$ or has every element of supp $\tilde{\sigma}$ as period according as there exists an $\alpha$ with $\int e^{\alpha y}\,d\tilde{\tau}(y) = 1$ or not. Our assertion follows immediately upon translating the function and measures concerned by $k$.

**Corollary 8.3.9.** *Let $\{v_n\}_{n=-\infty}^{\infty}$ and $\{p_n\}_{n=-\infty}^{\infty}$ be nonnegative real sequences satisfying*

$$v_m = \sum_{n=-\infty}^{\infty} v_{m+n} p_n, \qquad m \in \mathbb{Z}_+.$$

*Defining $A = \{n : p_n > 0\}$, $n_0 = \inf A$ (possibly $-\infty$), and $d$ as the g.c.d. of $A$, if $v_m \neq 0$ for $m \geq n_0$, then $v_m$ admits the representation*

$$v_m = \sum_{n=n_0}^{-1} v_{m+n}\rho_n + \xi_m r^m, \qquad m \in \mathbb{Z}_+,$$

*where $\xi_m \equiv 0$ if there exists no $r$ such that $\sum_{n=0}^{\infty} \tau_n r^n = 1$, and $\xi_m$ has period $d$ if there exists such an $r$ (necessarily unique), where $\{\rho_n\}$, $\{\tau_n\}$ are defined on $-\mathbb{N}$ and $\mathbb{Z}_+$, respectively, analogously to* (8.3.1).

We next consider $\alpha$ such that $\int e^{\alpha y}\,d\mu(y) = 1$ for $\mu = \sigma$, $\rho$, or $\tau$, regardless of whether the ICFE (8.3.6) is satisfied or not.

**Proposition 8.3.10.** *Let $\sigma$ be a positive $\sigma$-finite Borel measure on $\mathbb{R}$ and let $\rho$ and $\tau$ be defined as in* (8.3.1). *Suppose further that either $\rho$ or $\tau$ is finite for compact sets. If $\int e^{\alpha y}\,d\tau(y) = 1$ for some $\alpha \in \mathbb{R}$, then $\int e^{\alpha y}\,d\sigma(y) = 1$ and $\int e^{\alpha y}\,d\rho(y) \leq 1$. A dual assertion holds as well, with the roles of $\rho$ and $\tau$ interchanged.*

***Proof.*** For $t \in \mathbb{R}$, denote $\int_{\mathbb{R}} e^{ty}\,d\sigma(y)$ by $t_\sigma$, and let $t_\rho$, $t_\tau$ be defined similarly. We note first that

$$t_\rho > 1 \Rightarrow t_\tau < \infty; \qquad t_\tau > 1 \Rightarrow t_\rho < \infty. \tag{8.3.12}$$

For, let $x > 0$ be such that $\int_{-x}^{0} e^{ty}\,d\rho(y) = \gamma > 1$. By our hypothesis and Lemma 8.3.3, both $\rho$ and $\tau$ are finite for compact sets, (8.3.5) holds, and consequently

$$t_\sigma + t_\rho t_\tau = t_\rho + t_\tau. \tag{8.3.13}$$

Hence, for $(0\le)\ x < z < \infty$,

$$\gamma \int_x^z e^{ty}\, d\tau(y) \le \left(\int_{-x}^0 e^{ty}\, d\rho(y)\right)\left(\int_x^z e^{ty}\, d\tau(y)\right)$$

$$\le \int_0^z e^{ty}\, d(\rho * \tau)(y) \le \int_0^z e^{ty}\, d\tau(y),$$

in view of (8.3.4b), so that

$$(\gamma - 1)\int_0^x e^{ty}\, d\tau(y) \le \gamma \int_0^x e^{ty}\, d\tau(y) \qquad \forall\, z > x.$$

Hence, $t_\tau < \infty$. The proof of the second assertion in (8.3.12) is similar.

We note next that, for every $t \in \mathbb{R}$, either $t_\rho \le 1$ or $t_\tau \le 1$. For, if both were to be $>1$, then (8.3.12) would imply that $1 < t_\rho, t_\tau < \infty$; but, then we have from (8.3.13) that

$$1 - t_\sigma = (1 - t_\rho)(1 - t_\tau), \tag{8.3.14}$$

and we must have $t_\sigma < 1$. Define $d\tilde{\sigma}(y) = e^{ty}\, d\sigma(y)$ and define $\tilde{\rho}, \tilde{\tau}$ correspondingly. Then (as ladder-height measures corresponding to a subprobability measure), $\tilde{\rho}$ and $\tilde{\tau}$ are themselves subprobability measures, so that we must have both $t_\rho$ and $t_\tau < 1$. This contradiction shows that we must have either $t_\rho \le 1$ or $t_\tau \le 1$. If, for some $\alpha \in \mathbb{R}$, we have $\alpha_\tau = 1$, then (8.3.14) implies that $\alpha_\sigma = 1$ as well. Since $t_\tau > 1$ for $t > \alpha$, we have $t_\rho \le 1$ for $t > \alpha$ and so, by the monotone convergence theorem, $\alpha_\rho \le 1$. Hence, the proposition is proven.

From (8.3.14) and the preceding, we have the following result.

**Corollary 8.3.11.** *Let $\sigma$ be a positive $\sigma$-finite Borel measure on $\mathbb{R}$, let $\rho$ and $\tau$ be defined as in* (8.3.1), *and let one of them* (*consequently, both*) *be finite for compact sets. Then, $\int e^{\alpha y}\, d\sigma(y) = 1$ for some real $\alpha$ if and only if at least one of $\int e^{\alpha y}\, d\rho(y)$ and $\int e^{\alpha y}\, d\tau(y)$ is also* $=1$.

Finally, we consider a proof of Theorem 8.1.6 based on Theorem 8.3.6 and Theorem 2.2.4. We need only consider the case where $\operatorname{supp}\sigma \cap (0, \infty)$ and $\operatorname{supp}\sigma \cap (-\infty, 0)$ are both nonempty, since otherwise the desired conclusion follows from Theorem 2.2.4 itself. By assumption, $f = f \bullet \sigma$ on $\mathbb{R}$ now, where $f$ and $\sigma$ are as in Theorem 8.1.6. For arbitrary $(0>)\ k \in \mathbb{R}$, $f(\cdot + k)$ satisfies (8.3.6), so (8.3.7) holds for a.a. $x \ge k$; asssuming without loss of generality hereafter that $f$ is also continuous, we therefore have

$$f(x) = (f \bullet \rho)(x) + r(x), \qquad \forall\, x \in \mathbb{R}, \tag{8.3.15}$$

where $r(x) = p_0(x)e^{\alpha x}$, with $\alpha$ and $p_0$ having the properties stated for $\alpha$ and $p$ in Theorem 8.3.6. Iterating (8.3.15), we have that, for every $n \in \mathbb{N}$,

$$f = f \bullet \rho^n + r \bullet (\delta_0 + \rho + \cdots + \rho^{n-1}) \qquad \text{pointwise on } \mathbb{R}.$$

Now, $(r \bullet \rho)(x) = \int p_0(x+y)e^{\alpha y}\,d\rho(y)\,e^{\alpha x} = p_1(x)e^{\alpha x}$, where $p_1$ has every element of supp $\sigma$ as a period since $p_0$ has; similarly, by induction, $(r \bullet \rho^j)(x) = p_j(x)e^{\alpha x}$, where each $p_j$ has the same property. Writing $p(x) = \sum_{j=0}^{\infty} p_j(x)$, $x \in \mathbb{R}$, we see that $p$ has the same property, and that $f(x) = g(x) + p(x)e^{\alpha x}$, where $g(x) = \lim_{n\to\infty}(f \bullet \rho^n)(x)$ exists on $\mathbb{R}$ and satisfies $g = g \bullet \rho$ on $\mathbb{R}$. It follows from Theorem 2.2.4 that $g(x) = q(x)e^{\beta x}$, where $\int e^{\beta y}\,d\rho(y) = 1$ and $q$ has every element of supp $\rho$, and so of supp $\sigma$, as a period (by Proposition 8.3.2). Finally, by Proposition 8.3.10, we must have $\int e^{\alpha y}\,d\sigma(y) = \int e^{\beta y}\,d\sigma(y) = 1$. This completes the proof of Theorem 8.1.6.

## NOTES AND REMARKS

The proofs of Theorems 8.1.6 and 8.1.7 and of the auxiliary results given in Section 8.1 are from Ramachandran (1984) and Ramachandran and Prakasa Rao (1984). Theorem 8.1.9 and Corollary 8.1.10 are from Ramachandran *et al.* (1988). The proof of Theorem 8.1.6 using the Krein–Milman theorem, given in Section 8.2, is from Lau and Rao (1984). A proof of Theorem 8.1.6 appealing to Theorem 2.2.4 and the Wiener–Hopf decomposition of a probability measure on $\mathbb{R}$ was established in Ramachandran (1987), and is close in spirit to the proof of that result in Section 8.3. The results of Section 8.3 are essentially as in Alzaid *et al.* (1988); this subsumes an unpublished (1984) paper by Lau and Rao, in which Theorem 8.3.6 was first formulated.

CHAPTER

# 9

# Integrated Cauchy Functional Equations on Semigroups of $\mathbb{R}^d$

In this chapter, we continue our investigation of the ICFE, with subsemigroups of $\mathbb{R}^d$, $d \in \mathbb{N}$, now as the underlying sets (in what follows, the prefix "sub" will be omitted for convenience in writing). The real analysis method used in the previous chapters on $\mathbb{R}_+$ or $\mathbb{R}$ does not apply to this general case. Our approach here is functional-analytic. Following closely the method of Dény (1960), we identify the extremal elements of the solution set and then apply a representation theorem of Choquet's to obtain the general solution. The special case of $\mathbb{R}$, amenable to the application of the (simpler) Krein–Milman theorem, has already been dealt with in Section 8.3 by that method.

The results here can be extended to locally compact metrizable abelian semigroups. However, we will restrict our attention to semigroups of $\mathbb{R}^d$ for simplicity. Dény's argument has groups as the underlying sets. The transition from groups to semigroups is, however, not automatic. For example, on some semigroups, exponential functions may vanish on certain subsets; and translations of measures need not be continuous. Our analysis singles out "well-behaved" semigroups where such pathologies do not arise, and considers the ICFE on them. The necessary machinery is set up in Sections 9.1–9.3. The principal auxiliary results are developed in Section 9.4, and the solution of the ICFE is established in Section 9.5. An appendix to this chapter provides a brief description of the Choquet theorem, and some topological results needed in this chapter.

## 9.1. EXPONENTIAL FUNCTIONS ON SEMIGROUPS

Throughout, $(S, +)$ is assumed to be a locally compact semigroup of $\mathbb{R}^d$. It follows that $S$ is an $F_\sigma$ subset of $\mathbb{R}^d$, and hence a Borel subset of $\mathbb{R}^d$. A real-valued function $f$ on $S$ is called an *exponential function* if $f \not\equiv 0$ is nonnegative, continuous and satisfies the Cauchy functional equation

$$f(x+y) = f(x)f(y) \qquad \forall\, x, y \in S.$$

We first exhibit examples of exponential functions on some special semigroups.

***Example 9.1.1.*** If $S = \mathbb{R}^d$, or $\mathbb{R}^d_+$, or any subgroup of $\mathbb{R}^d$, then the exponential functions $f$ are of the form

$$f(x) = e^{\langle \alpha, x \rangle} \qquad \forall\, x \in S,$$

where $\alpha \in \mathbb{R}^d$. If $S = \mathbb{Z}^d$ or $\mathbb{Z}^d_+$, then every exponential function reduces to the form

$$f(r) = t_1^{r_1} \cdots t_d^{r_d} \qquad \forall\, r = (r_1, \ldots, r_d) \in S,$$

where $t_1, \ldots, t_d \in \mathbb{R}_+$.

***Example 9.1.2.*** Let $v = (1, 1)$ and let $S = \{tv: t \in \mathbb{R}\} \subseteq \mathbb{R}^2$. Then, the exponential functions on $S$ are obviously of the form

$$f(tv) = e^{\alpha t}, \qquad t \in \mathbb{R}, \text{ for some } \alpha \in \mathbb{R}.$$

If $f$ is a nonnegative, continuous function on $\mathbb{R}^2$ and satisfies

$$f(x + tv) = f(x)f(tv), \qquad x \in \mathbb{R}^2,\ t \in \mathbb{R}, \tag{9.1.1}$$

then

$$f(x) = p((x_1 - x_2)/2)e^{\alpha(x_1+x_2)/2} \qquad \forall\, x = (x_1, x_2) \in \mathbb{R}^2,$$

where $\alpha \in \mathbb{R}$ and $p$ is a nonnegative continuous function on $\mathbb{R}$. Indeed, using the transformation $x_1 = u_1 + u_2$, $x_2 = u_1 - u_2$, and writing

$$\bar{f}(u_1, u_2) = f(u_1 + u_2, u_1 - u_2),$$

(9.1.1) reduces to

$$\bar{f}(u_1 + t, u_2) = \bar{f}(u_1, u_2)\bar{f}(t, 0), \qquad t \in \mathbb{R},\ (u_1, u_2) \in \mathbb{R}^2,$$

so that

$$\bar{f}(u_1, u_2) = \bar{f}(u_1 - t, u_2)\bar{f}(t, 0), \qquad t \in \mathbb{R},\ (u_1, u_2) \in \mathbb{R}^2,$$

and so, in particular (for $t = u_1$),

$$\bar{f}(u_1, u_2) = \bar{f}(0, u_2)\bar{f}(u_1, 0).$$

Noting that $\bar{f}(u_1, 0) = f(u_1, u_1)$ is, by (9.1.1), an exponential on $S$ and so of the form $\exp(\alpha u_1)$ for some real $\alpha$, and defining $\bar{f}(0, u_2) = p(u_2)$, we have proven the stated representation for $f$.

***Example 9.1.3.*** Let $v = (1, 1)$, and let $f$ be a nonnegative continuous function on $\mathbb{R}_+^2$ satisfying

$$f(x + tv) = f(x)f(tv), \qquad x \in \mathbb{R}_+^2,\ t \in \mathbb{R}_+.$$

Then, by an argument similar to the preceding, we have

$$f(x) = \begin{cases} p_1\left(\dfrac{x_1 - x_2}{2}\right)e^{\alpha(x_1+x_2)/2} & \text{if } x_1 \geq x_2 \geq 0, \\ p_2\left(\dfrac{x_2 - x_1}{2}\right)e^{\alpha(x_1+x_2)/2} & \text{if } x_2 \geq x_1 \geq 0, \end{cases}$$

where $\alpha \in \mathbb{R}$, and $p_1, p_2$ are nonnegative continuous functions on $\mathbb{R}_+$. This example is important and is related to a multivariate extension of exponential distributions in reliability theory (see Marshall and Olkin, 1967).

***Example 9.1.4.*** Let $S = \{0\} \cup [1, \infty)$. Then, the exponential functions on $S$ are either $f(x) = e^{\alpha x}$ for all $x \in S$, for some $\alpha \in \mathbb{R}$, or

$$f(x) = \begin{cases} 1 & \text{if } x = 0, \\ 0 & \text{if } x \neq 0. \end{cases}$$

Indeed, suppose $f$ is not of the second form. Then, there exists $x_0 \in [1, \infty)$ such that $f(x_0) \neq 0$. Let ($Q$ denoting the set of rationals, as usual)

$$A = \{rx_0 \in [1, \infty) : r \in Q\}.$$

The Cauchy functional equation implies that, for $rx_0 \in A$, $r \in Q$, we have $f(rx_0) = f(x_0)^r$, so that $f(x) = e^{\alpha x}$ for all $x \geq 1$. That $f(1) = f(0)f(1)$ implies that $f(0) = 1$, and hence $f(x) = e^{\alpha x}$ for all $x \in S$.

Also, if $f$ is a nonnegative continuous function on $\mathbb{R}_+$ and satisfies

$$f(x + y) = f(x)f(y), \qquad \forall\, x \in \mathbb{R}_+,\ y \in S,$$

then either there exists an $\alpha \in \mathbb{R}$ such that $f(x) = e^{\alpha x}$ for all $x \in S$, or

$$f(x) = \begin{cases} p(x) & \text{if } x \in [0, 1], \\ 0 & \text{if } x \geq 1, \end{cases}$$

where $p(\cdot)$ is an arbitrary nonnegative continuous function on $[0, 1]$ with $p(0) = 1$, $p(1) = 0$.

Similarly, we have the following example.

***Example 9.1.5.*** Let $S = S_1 \cup S_2$, where $S_1 = \{0\} \times \mathbb{R}_+$, $S_2 = [1, \infty) \times \mathbb{R}_+$. Then, the exponential functions on $S$ are either (i) $f(x) = e^{\langle \alpha, x\rangle}$ $\forall\, x \in S$, or (ii):

$$f(x) = \begin{cases} e^{\alpha_2 x_2} & \text{if } x_1 = 0,\ x_2 \in \mathbb{R}_+, \\ 0 & \text{if } x_1 \in [1, \infty),\ x_2 \in \mathbb{R}_+, \end{cases}$$

where $\alpha = (\alpha_1, \alpha_2) \in \mathbb{R}^2$.

If $f$ is a nonnegative continuous function on $\mathbb{R}^2_+$ such that

$$f(x + y) = f(x)f(y), \qquad \forall\, x \in \mathbb{R}^2_+,\ y \in S,$$

then $f$ is of the form either (i), or (ii′):

$$f(x) = \begin{cases} p(x_1)e^{\alpha_2 x_2} & \text{if } x_1 \in [0, 1],\ x_2 \in \mathbb{R}_+, \\ 0 & \text{if } x_1 \in [1, \infty),\ x_2 \in \mathbb{R}_+, \end{cases}$$

where $p(\cdot)$ is an arbitrary nonnegative function on $[0, 1]$ with $p(0) = 1$, $p(1) = 0$.

We now introduce a condition on the semigroup $S$ that will ensure that an exponential function on $S$ vanishes nowhere on $S$, thus ruling out the pathological phenomena encountered in the last two examples.

**Definition 9.1.6.** *A semigroup $S$ of $\mathbb{R}^d$ is said to have the* component-generating property *if, for every subsemigroup $T$ of $S$ that is both open and closed in $S$, $S - T$ is dense in $S - S$.*

It is clear that if $S$ is a subgroup, or a connected semigroup of $\mathbb{R}^d$, then $S$ has the preceding property (in fact, for every such $T$, $S - T = S - S$). Also, the same is true if every such $T$ has nonvoid intersection with $S^\circ$, the interior of $S$ in $\mathbb{R}^d$ (in particular, when $S$ is an open semigroup): For, in this case, $S - T$ contains an open neighborhood of 0, and the semigroup property implies that $S - T$ actually equals $\mathbb{R}^d$.

On the other hand, the semigroups $S$ in Examples 9.1.4 and 9.1.5 do not have such a property. We also remark that the denseness of $S - T$ in $S - S$ required in Definition 9.1.6 cannot be replaced by equality in general; e.g., let $S$ be the semigroup generated by $\{1, \sqrt{2}\}$ in $\mathbb{R}_+$, and let $T$ be the subsemigroup generated by $\{1\}$. Then, $S - T$ is dense in $\mathbb{R}$ (apply Proposition 8.1.1 to the set $\{\sqrt{2}, 1, -1\}$), and hence dense in $S - S$, but they are not equal.

**Theorem 9.1.7.** *Suppose $S \subseteq \mathbb{R}^d$ has the component-generating property. If $f$ is an exponential function on $S$, then there exists an $\alpha \in \mathbb{R}^d$ such that*

$$f(x) = e^{\langle \alpha, x \rangle} \qquad \forall\, x \in S.$$

***Proof.*** Let $f$ be an exponential function on $S$, and let

$$T = \{x \in S : f(x) \neq 0\}.$$

We claim that $T$ is an open and closed subsemigroup of $S$. Indeed, the continuity of $f$ implies that $T$ is open, and the Cauchy functional equation guarantees that $T$ is a semigroup. To show that $T$ is closed, we first observe that the domain of $f$ can be extended to $T - T$ by defining $f(u) = f(x)/f(y)$, where $u = x - y$ with $x, y \in T$. Since $f$ satisfies the Cauchy functional equation on $T$, it is well-defined on $T - T$, and

$$f(u + v) = f(u)f(v) \qquad \forall\, u, v \in T - T.$$

That $f \geq 0$, and that $T - T$ is a subgroup of $\mathbb{R}^d$, imply that there exists $\alpha \in \mathbb{R}^d$ such that

$$f(x) = e^{\langle \alpha, x \rangle} \qquad \forall\, x \in T - T$$

(Example 9.1.1), in particular for all $x \in T$. It follows that the same is true for $x \in \bar{T} \cap S$.

By using the preceding argument again, we can extend the domain of $f$ from $S$ to $S - T$, with $f$ satisfying the equation

$$f(x + y) = f(x)f(y) \qquad \forall\, x, y \in S - T.$$

Since $0 \in T - T$, $f(0)$ is necessarily equal to 1, and hence $f(x) > 0$ on a neighborhood $V$ of 0 in $S - T$. By the component-generating property of $S$, $V$ is dense in a neighborhood of 0 in $\mathbb{R}^d$, so that $U = \bigcup_{n=1}^{\infty} nV$ is a dense semigroup of $\mathbb{R}^d$ and $f > 0$ on $U$. Repeating the preceding extension argument, we see that $f$ can actually be defined on the dense subgroup $U - U$ of $\mathbb{R}^d$, and satisfies the Cauchy functional equation. Hence, $f$ can be continuously extended to $\mathbb{R}^d$ and $f(x) = e^{\langle \alpha, x \rangle}$ $\forall\, x \in \mathbb{R}^d$. In particular, $f(x) = e^{\langle \alpha, x \rangle}$ $\forall\, x \in S$.

To conclude this section, we will consider an important special case of the ICFE on a semigroup $S$ of $\mathbb{R}^d$. The general case is more complicated, and is developed in the next three sections. At one crucial step in the proof of Lemma 9.4.3, we will need to apply this special case.

For any subset $A$ of $\mathbb{R}^d$, we use $\langle A \rangle$ to denote the closed subgroup generated by $A$. If $\sigma$ is a measure on $S$, we use $S(\sigma)$ to denote the semigroup generated by $\operatorname{supp} \sigma$.

**Theorem 9.1.8.** *Let $\sigma$ be a probability measure on $S$. Suppose $f$ is a bounded uniformly continuous function on $S$, and satisfies*

$$f(x) = \int_S f(x + y)\, d\sigma(y), \qquad \forall\, x \in S. \tag{9.1.2}$$

*Then,*

$$f(x + y) = f(x), \qquad \forall\, x \in S,\ y \in S(\sigma).$$

*In particular, if $S(\sigma)$ has the component-generating property, and if $S \subseteq \langle S(\sigma)\rangle$, then $f$ is a constant function on $S$.*

***Proof.*** Let $a \in \operatorname{supp} \sigma$ be fixed, and define

$$g(x) = f(x) - f(x + a), \qquad x \in S.$$

Then, $g$ is also bounded and uniformly continuous. Let

$$\alpha = \sup_{x \in S} g(x),$$

and let $\{x_i\}$ be a sequence in $S$ such that $\lim_{i\to\infty} g(x_i) = \alpha$. The translations $\{g_i\}$ of $g$ defined by $g_i(x) = g(x + x_i)$ are equicontinuous on $S$. A subsequence of $\{g_i\}$ can be selected that converges uniformly on compact sets to a bounded, continuous function $g_0$ on $S$, that satisfies (9.1.2), and, further, so that

$$\alpha = g_0(0) = \int_S g_0(x)\, d\sigma(x).$$

Since $\sigma$ is a probability measure, we have $g_0(x) = \alpha$ for all $x \in \operatorname{supp} \sigma$, and hence for all $x \in S(\sigma)$ as well. Suppose now that $\alpha > 0$; let $k \in \mathbb{N}$ be such that $|f(x)| \leq k\alpha/4$ for all $x \in S$. Then, the fact that $g_0(x) = \alpha$ for $x = ja$, $j = 1, \ldots, k$, implies that, for some $i$,

$$g_i(ja) > \frac{\alpha}{2} \qquad \text{for } j = 1, \ldots, k.$$

For this $i$, we have

$$f(x_i + ja) - f(x_i + (j + 1)a) > \frac{\alpha}{2} \qquad \text{for } j = 1, \ldots, k,$$

so that

$$f(x_i + a) - f(x_i + (k + 1)a) > \frac{k\alpha}{2},$$

contradicting our choice of $k$. Thus, $\alpha > 0$ is ruled out, so that $\alpha \leq 0$. Similarly, we also have $\sup_x\{f(x + a) - f(x)\} \leq 0$. Hence, $g(x) = 0$, i.e., $f(x) = f(x + a)$ for all $x \in S$; since $a \in \operatorname{supp} \sigma$ is arbitrary, we have

$$f(x + y) = f(x), \qquad \forall\, x \in S,\ y \in S(\sigma).$$

The last conclusion of the theorem follows directly from Theorem 9.1.7 with $S$ replaced by $S(\sigma)$.

If $S$ is a subgroup instead of a semigroup, then the requirement of the uniform continuity of $f$ can be replaced by that of the continuity of $f$. For, in this case, we can convolve $f$ with a continuous function with compact support and reduce the discussion to the uniformly continuous case.

## 9.2. TRANSLATIONS OF MEASURES

By a Radon measure $\mu$ on a locally compact Hausdorff space $X$, we mean a regular Borel measure $\mu$ on $X$, with $\mu(K) < \infty$ for every compact subset $K$ of $X$. Let $M(X)$ denote the class of Radon measures on $X$ and let $M^+(X)$ be the subclass of positive measures. Let $C_c(X)$ be the space of real-valued continuous functions on $X$ with compact support. The topology we use on $M(X)$ is the weak topology generated by $C_c(X)$; if, in addition, $X$ is separable and metrizable, then $M^+(X)$ is metrizable (see Theorem 9.A.1), and hence, for $\{\mu_n\}$, $\mu$ in $M^+(X)$,

$$\mu_n \to \mu \quad \text{if and only if} \quad \mu_n(\phi) \to \mu(\phi), \qquad \forall\, \phi \in C_c(X),$$

where $\mu(\phi)$ means $\int_X \phi(x)\, d\mu(x)$.

Let $S$ be a locally compact semigroup of $\mathbb{R}^d$ as before, and let $\mu \in M^+(S)$; we define $\mu_x$ the *x-translation* of $\mu$, by

$$\mu_x(E) = \mu(x + E),$$

for every Borel subset $E$ of $S$. It is known that if, in addition, $S$ is a group, then for any $\mu_n$, $\mu$ in $M^+(S)$, $x_n$, $x$ in $S$,

$$\mu_n \to \mu \;\Rightarrow\; (\mu_n)_x \to \mu_x; \qquad x_n \to x \;\Rightarrow\; \mu_{x_n} \to \mu_x.$$

If $S$ is not a group, then this assertion is not true in general.

***Example 9.2.1.*** Let $S = [0, \infty)$. Suppose $\mu_n = \delta_{1-1/n}$, $\mu = \delta_1$ are the point mass measures at $1 - 1/n$ and 1, respectively. Then, for any Borel subset $E$ of $S$, and for $x = 1$,

$$(\mu_n)_x(E) = \delta_{1-1/n}(1 + E) = 0, \qquad \mu_x(E) = \delta_1(1 + E) = \delta_0(E).$$

Hence, $\mu_n \to \mu$, but $(\mu_n)_x \nrightarrow \mu_x$.

Also, if we let $x_n = 1 + 1/n$, $x = 1$, and $\mu = \delta_1$, then $x_n \to x$, but $\mu_{x_n} = 0 \nrightarrow \mu_x = \delta_0$.

We first consider therefore a preliminary, simple restriction of the domain of definition of our measures, which will ensure that translations of measures thereon will be continuous. Let $S^\circ$ denote the interior of $S$ in $\mathbb{R}^d$. We will henceforth assume that $S^\circ \neq \phi$.

**Lemma 9.2.2.** *$S^\circ$ is an ideal of $S$ (i.e., if $x \in S$, $y \in S^\circ$, then $x + y \in S^\circ$), and in particular a subsemigroup of $S$.*

***Proof.*** $x + S^\circ$ is then an open subset of $\mathbb{R}^d$, which contains $x + y$ and is contained in $S$.

**Theorem 9.2.3.** *The map $S \times M^+(S^\circ) \to M^+(S^\circ)$ defined by $(x, \mu) \to \mu_x$ is jointly continuous. In particular, for $\mu_n$, $\mu \in M^+(S^\circ)$, and $x_n$, $x \in S$:*

(i) $\mu_n \to \mu \Rightarrow (\mu_n)_x \to \mu_x$;
(ii) $x_n \to x \Rightarrow \mu_{x_n} \to \mu_x$;
(iii) *if $\mu$ is "S-shift-bounded," i.e., if the set $\{\mu_x : x \in S\}$ of translates of $\mu$ (by the elements of $S$) is a weakly bounded subset of $M^+(S^\circ)$, then the convergence in* (ii) *is uniform; i.e., for any $\phi \in C_0(S^\circ)$ and any $\varepsilon > 0$, there exists a $\delta > 0$ (depending only on $\phi$ and $\varepsilon$) such that, if $|x_n - x| < \delta$, then $|\mu_{x_n}(\phi) - \mu_x(\phi)| < \varepsilon$.*

***Remark.*** An important example of a shift-bounded but not bounded (i.e., finite) measure is Lebesgue measure on $S = \mathbb{R}^d$. A necessary and sufficient condition for $\mu$ to be $S$-shift-bounded is that, for every fixed compact $K \subset S^\circ$, the set $\{\mu(x + K) : x \in S\}$ of real numbers be bounded. We shall use this fact in what follows.

***Proof.*** Note that if $E$ is a Borel subset of $S^\circ$, and $x \in S$, then $x + E \subseteq S^\circ$ (Lemma 9.2.2); so $\mu_x(E) = \mu(x + E)$ is defined, and $\mu_x \in M^+(S^\circ)$. We need to show that $x_n \to x$ and $\mu_n \to \mu$ imply that $(\mu_n)_{x_n} \to \mu_x$. Applying a criterion equivalent to the weak convergence of measures (Theorem 9.A.2), it suffices to show that

$$\limsup_{n\to\infty} \mu_n(x_n + K) \le \mu(x + K),$$

and

$$\liminf_{n\to\infty} \mu_n(x_n + U) \ge \mu(x + U),$$

for any compact subset $K \subseteq S^\circ$, and any open subset $U$ of $S^\circ$ with compact closure.

For $\varepsilon > 0$, by the regularity of $\mu$, there exists an open set $O \subseteq S^\circ$ such that $K \subseteq O$, $\mu(\partial(x + O)) = 0$ ($\partial A$ denotes the boundary of $A$) (*cf.* Billingsley, 1968, p. 13), and

$$\mu(x + O) \le \mu(x + K) + \varepsilon.$$

For $n$ sufficiently large, we have

$$x_n + K \subseteq x + O,$$

and hence

$$\limsup_{n\to\infty} \mu_n(x_n + K) \le \limsup_{n\to\infty} \mu_n(x + O) = \mu(x + O) \le \mu(x + K) + \varepsilon$$

(the equality holds because $\mu(\partial(x + O)) = 0$). Since $\varepsilon > 0$ is arbitrary, this completes the proof of the first inequality. To prove the second inequality, we choose a compact set $K \subseteq U$ such that

$$\mu(x + U) \le \mu(x + K) + \varepsilon$$

and $\mu(\partial(x + K)) = 0$, then proceed dually to the preceding argument.

Statements (i) and (ii) then follow immediately. As for (iii), let $\mu \in M^+(S^\circ)$ be $S$-shift-bounded. Since $\phi$ is uniformly continuous, let $V$ be a closed ball around the origin of $\mathbb{R}^d$ such that $|\phi(y) - \phi(y')| < \varepsilon$ for all $y, y'$ such that $y - y' \in V$. Then, for all large $n$ (with $x_n - x \in V$),

$$\left| \int_{S^\circ} \phi(y)\, d\mu_{x_n}(y) - \int_{S^\circ} \phi(y)\, d\mu_x(y) \right|$$

$$\le \int_{S^\circ} |\phi(y - x_n + x) - \phi(y)|\, d\mu_x(y) \le \varepsilon \mu_x(K + V \cap S^\circ) < C\varepsilon,$$

where $K \subset S^\circ$ is the support of $\phi$, and $C$ is a positive constant independent of $x \in S$, as per the previous remark. Hence, the theorem is proven.

We will call a measure $\mu \in M^+(S^\circ)$, not identically zero, an *exponential measure* if there exists a function $g\colon S^\circ \to \mathbb{R}_+$ such that $\mu_x = g(x)\mu$ for all $x \in S^\circ$.

**Proposition 9.2.4.** *Let $\mu \in M^+(S^\circ)$ be an exponential measure, and let $g$ be the associated function. Then, there exists an $\alpha \in \mathbb{R}$ such that*

$$g(x) = e^{\langle \alpha, x\rangle} \qquad \forall\, x \in S^\circ,$$

*and* $\operatorname{supp} \mu = S^\circ$.

***Proof.*** For any $x, y \in S^\circ$,

$$g(x + y)\mu = \mu_{x+y} = (\mu_y)_x = g(x)g(y)\mu.$$

Since $\mu$ is not identically zero, $g(x + y) = g(x)g(y)$ for all $x, y \in S^\circ$. To show that $g$ is continuous, we let $x_n, x \in S^\circ$ with $\lim_{n\to\infty} x_n = x$; then, by Theorem 9.2.3(ii),

$$\lim_{n\to\infty} g(x_n)\mu = \lim_{n\to\infty} \mu_{x_n} = \mu_x = g(x)\mu.$$

This implies that $\lim_{n\to\infty} g(x_n) = g(x)$, so that $g$ is continuous, and hence an exponential function on $S^\circ$. Note that $S^\circ$ (as an open set of $\mathbb{R}^d$) has the component-generating property (see the remarks after Definition 9.1.6). Theorem 9.1.7 implies that $g$ has the asserted form.

To show that $\operatorname{supp} \mu = S^\circ$, let $y \in \operatorname{supp} \mu \subseteq S^\circ$ be fixed. For any $x \in S^\circ$, and for any neighborhood $U$ of $x$, $y + U$ is a neighborhood of $x + y$. There exists a neighborhood $V$ of $y$ in $S^\circ$ such that $x + V \subseteq y + U$. Hence,

$$0 < \mu(V) = g(x)^{-1}\mu(x + V) \leq g(x)^{-1}\mu(y + U) = g(x)^{-1}g(y)\mu(U).$$

This implies that $x \in \operatorname{supp} \mu$, i.e., $S^\circ \subseteq \operatorname{supp} \mu$.

**Theorem 9.2.5.** *Let $\mu \in M^+(S^\circ)$. Then, $\mu$ is an exponential measure if and only if $d\mu = cg\, d\omega$ on $S^\circ$ for some exponential function $g$ on $S^\circ$ and some $c > 0$ ($\omega$ denotes the Lebesgue measure on $\mathbb{R}^d$).*

***Proof.*** We need only prove the necessity. By Proposition 9.2.4, $g(x) > 0$; let $d\tau = g^{-1}\, d\mu$; then, $\tau_x = \tau$ for all $x \in S^\circ$. We will extend $\tau$ to a translation-invariant measure $\bar{\tau}$ on $S^\circ - S^\circ = \mathbb{R}^d$; the uniqueness of such a measure to within a positive multiplicative constant then implies that $\bar{\tau} = c\omega$, and $\mu$ will have the asserted form.

Let $x \in \mathbb{R}^d$; define $\bar{\tau}$ on $S^\circ + x$ by

$$\bar{\tau}(E) = \tau(E - x),$$

for any Borel subset $E \subseteq S^\circ + x$. We observe that $\bar{\tau}(E) = \tau(E)$ if $E \subseteq S^\circ \cap (S^\circ + x)$: Choose $a \in S^\circ$ such that $a + x \in S^\circ$ (this is possible since $S^\circ - S^\circ = \mathbb{R}^d$); then,

$$\bar{\tau}(E) = \tau(E - x) = \tau((E - x) + (a + x)) = \tau(E + a) = \tau(E).$$

We can therefore extend the measure $\tau$ to $\bar{\tau}$ on $S^\circ \cup (S^\circ + x)$. Furthermore, if $E \subseteq S^\circ \cup (S^\circ + x)$, write $E$ as a disjoint union of $E_1 \subseteq S^\circ$ and $E_2 \subseteq S^\circ + x$. Then, for any $s \in S^\circ$,

$$\begin{aligned} \bar{\tau}_s(E) &= \bar{\tau}((E_1 \cup E_2) + s) = \bar{\tau}(E_1 + s) + \bar{\tau}(E_2 + s) \\ &= \tau(E_1) + \tau(E_2 + s - x) = \tau(E_1) + \tau(E_2 - x) \\ &= \bar{\tau}(E_1) + \bar{\tau}(E_2) = \bar{\tau}(E). \end{aligned}$$

Inductively, we can define $\bar{\tau}$ on $\bigcup_{k=0}^{n} (S^\circ + kx)$ satisfying (for $n \geq 1$):

(i) if $E \subseteq S^\circ + nx$, then $\bar{\tau}(E) = \bar{\tau}(E - x)$; and
(ii) if $E \subseteq \bigcup_{k=0}^{n} (S^\circ + kx)$, $z \in \bigcup_{k=0}^{n-1} (S^\circ + kx)$, then $\bar{\tau}_z(E) = \bar{\tau}(E)$.

Using (i), we can extend $\bar{\tau}$ to the open subsemigroup $S_x = \bigcup_{k=0}^{\infty} (S^\circ + kx)$. It can be checked, by using compact sets and (ii), that $\bar{\tau}_z = \bar{\tau}$ for $z \in S_x$.

Let $\{x_n\}$ be a countable dense subset of $\mathbb{R}^d$. We can repeat the preceding argument to extend $\bar{\tau}$ to a translation-invariant measure on $\bigcup_{n=1}^{\infty} S_{x_n} = \mathbb{R}^d$.

**Corollary 9.2.6.** *Let $T$ be a subsemigroup of $S$ such that $S^\circ \subseteq T - T$. Suppose $\mu \in M^+(S^\circ)$, $\mu_x = g(x)\mu$, and $g(x) > 0$ for all $x \in T$. Then, $d\mu = cg\, d\omega$ on $S^\circ$.*

***Proof.*** Since $g$ is positive on $T$, $g$ can be extended to $T - T$ by defining $g(x - y) = g(x)/g(y)$, $x, y \in T$, and $g$ satisfies the Cauchy functional equation on $T - T$. It follows from the preceding theorem that $d\mu = cg\, d\omega$.

## 9.3. THE SKEW CONVOLUTION

Let $\mathcal{K}$ denote the family of compact subsets of $S$. For $\mu, \nu \in M^+(S)$, we recall that the convolution $\mu * \nu$ can be obtained as follows: Define

$$\mu * \nu(K) = (\mu \times \nu)\{(x, y): x, y \in S, x + y \in K\}, \qquad K \in \mathcal{K}.$$

Then, $(\mu * \nu)$ is finitely additive on $\mathcal{K}$, and if $(\mu * \nu)(K) < \infty$ for all $K \in \mathcal{K}$, then $\mu * \nu$ can be extended to a Radon measure on $S$. It is easy to check that

$$(\mu * \nu)(\phi) = \int_S \int_S \phi(x + y)\, d\mu(x)\, d\nu(y),$$

for all $\phi$ on $S$ integrable with respect to $\mu * \nu$.

Analogously to the standard convolution just given, we also define, for $\mu, \nu \in M^+(S)$,

$$\mu \bullet \nu(K) = (\mu \times \nu)\{(x, y): x, y \in S, x \in y + K\}, \qquad K \in \mathcal{K};$$

it follows that $\mu \bullet \nu$ is finitely additive. If $(\mu \bullet \nu)(K) < \infty$ for all $K \in \mathcal{K}$, then $\mu \bullet \nu$ can be extended to a Radon measure on $S$. We will call such a convolution the *skew convolution* (*s-convolution*) of $\mu$ and $\nu$. Note that for any Borel subset $E$ of $S$, we have

$$\begin{aligned} \mu \bullet \nu(E) &= \int_{S\times S} \chi_{y+E}(x)\, d(\mu \times \nu)(x, y) \\ &= \int_S \left( \int_S \chi_E(x)\, d\mu_y(x) \right) d\nu(y) \qquad (9.3.1) \end{aligned}$$

$$= \int_S \mu_y(E)\, d\nu(y)$$

$$= \int_S \mu(y + E)\, d\nu(y). \qquad (9.3.2)$$

It follows from (9.3.1) and the monotone convergence theorem that if $\phi \in C_c(S)$ and $\phi \geq 0$, then

$$\mu \bullet \nu(\phi) = \int_S \left( \int_S \varphi(x)\, d\mu_y(x) \right) d\nu(y). \qquad (9.3.1')$$

The same holds for arbitrary $\phi \in C_c(S)$ in view of $\phi = \phi^+ - \phi^-$.

The s-convolution does not obey the same algebraic rules as the $*$-convolution, as shown by the following example and proposition.

***Example 9.3.1.*** Let $S = [0, \infty)$. Then, for any Borel subset $E$ of $[0, \infty)$,

$$\delta_0 \bullet \delta_1(E) = (\delta_0 \times \delta_1)\{(x, y): x \in y + E\} = 0,$$

but

$$\delta_1 \bullet \delta_0(E) = (\delta_1 \times \delta_0)\{(x, y): x \in y + E\} = \delta_1(E).$$

Also, a similar calculation shows that

$$(\delta_0 \bullet \delta_1) \bullet \delta_2 = 0, \qquad \text{but} \qquad \delta_0 \bullet (\delta_1 \bullet \delta_2) = \delta_1.$$

The s-convolution is hence neither commutative nor associative.

**Proposition 9.3.2.** *Let $\mu, \nu, \sigma \in M^+(S)$, and assume that all the involved convolutions exist. Then*:

(i) $(\mu \bullet \nu) \bullet \sigma = \mu \bullet (\nu * \sigma) = (\mu \bullet \sigma) \bullet \nu$;

(ii) *if $S$ is assumed to be a group, then*

$$\mu \bullet \nu = \mu * \nu^{\sim} \qquad \text{and} \qquad \mu \bullet (\nu \bullet \sigma) = (\mu * \sigma) \bullet \nu,$$

*where $\nu^{\sim}(E) = \nu(-E)$.*

***Proof.*** (i) For any Borel subset $E$ of $S$, by (9.3.2),

$$((\mu \bullet \nu) \bullet \sigma)(E) = \int_S (\mu \bullet \nu)(z + E)\, d\sigma(z)$$

$$= \int_S \int_S \mu(y + z + E)\, d\nu(y)\, d\sigma(z)$$

$$= \int_S \mu(x + E)\, d(\nu * \sigma)(x)$$

$$= (\mu \bullet (\nu * \sigma))(E).$$

Hence,

$$(\mu \bullet \nu) \bullet \sigma = \mu \bullet (\nu * \sigma) = \mu \bullet (\sigma * \nu) = (\mu \bullet \sigma) \bullet \nu.$$

(ii) For any Borel subset $E$ of $S$,

$$\begin{aligned}
\mu \bullet \nu(E) &= \int_S \int_S \chi_{y+E}(x)\, d\mu(x)\, d\nu(y) \\
&= \int_S \int_S \chi_E(x - y)\, d\mu(x)\, d\nu(y) \\
&= \int_S \int_S \chi_E(x + y)\, d\mu(x)\, d\nu^{\sim}(y) \\
&= (\mu * \nu^{\sim})(E).
\end{aligned}$$

For the second identity, we observe that

$$\begin{aligned}
\mu \bullet (\nu \bullet \sigma) &= \mu * (\nu \bullet \sigma)^{\sim} = \mu * (\nu * \sigma^{\sim})^{\sim} = \mu * (\sigma * \nu^{\sim}) \\
&= (\mu * \sigma) * \nu^{\sim} = (\mu * \sigma) \bullet \nu.
\end{aligned}$$

We remark that if $S$ is not a group, then $\nu^{\sim}$ in the first identity of (ii) is not defined, and the second identity may not hold in general; e.g., let $S = [0, \infty)$, $a, b \in S$ with $a > b$; then

$$\delta_a \bullet (\delta_b \bullet \delta_a) = 0, \qquad (\delta_a * \delta_b) \bullet \delta_a = \delta_b.$$

We have seen in Section 9.2 that, to consider a convergence property related to translations of measures, we have to restrict the domain of definition of the measures to $S^\circ$. For such problems related to convolutions and s-convolutions, we have to restrict one of the components likewise. Note that for $\mu \in M^+(S^\circ)$, $\sigma \in M^+(S)$, and for any Borel set $E$ in $S^\circ$, the property of $S^\circ$ of being an ideal in $S$ (Lemma 9.2.2) implies that if $y \in S$, then $y + E \subseteq S^\circ$, so that $\mu(y + E)$ is defined; so is $\mu \bullet \sigma(E)$ in view of (9.3.2). If $\mu \bullet \sigma(K) < \infty$ for all compact sets $K$ in $S^\circ$, we can then extend $\mu \bullet \sigma$ to a Radon measure on $S^\circ$ so that $\mu \bullet \sigma \in M^+(S^\circ)$.

**Theorem 9.3.3.** *Let $\{\mu_n\}$, $\mu$ be in $M^+(S^\circ)$ with $\mu_n \to \mu$, and let $\sigma \in M^+(S)$. Suppose either:*

(i) *$\{\mu_n\} \nearrow \mu$, and $\mu \bullet \sigma \in M^+(S^\circ)$; or*
(ii) *there exists $\nu \in M^+(S^\circ)$ such that $\mu_n \leq \nu$ for all $n$, and $\nu \bullet \sigma \in M^+(S^\circ)$. Then, $\mu_n \bullet \sigma \to \mu \bullet \sigma$ in $M^+(S^\circ)$.*

***Proof.*** For any $\phi \in C_c(S^\circ)$ with $\phi \geq 0$, we observe that

$$(\mu_n \bullet \sigma)(\phi) = \int_S \left( \int_{S^\circ} \phi(x)\, d(\mu_n)_y(x) \right) d\sigma(y),$$

and

$$(\mu \bullet \sigma)(\phi) = \int_S \left( \int_{S^\circ} \phi(x)\, d\mu_y(x) \right) d\sigma(y).$$

Assume (i) holds; we can apply Theorem 9.2.3(ii) and have

$$\lim_{n\to\infty} \int_S \phi(x)\, d(\mu_n)_y(x) = \int_S \phi(x)\, d\mu_y(x).$$

The monotone convergence theorem then implies that

$$\lim_{n\to\infty} (\mu_n \bullet \sigma)(\phi) = \mu \bullet \sigma(\phi),$$

for all $\phi \in C_c(S^\circ)$, $\phi \geq 0$. This yields $\mu_n \bullet \sigma \to \mu \bullet \sigma$.

If (ii) is assumed, then the conclusion follows from the same proof by applying the bounded convergence theorem instead.

**Theorem 9.3.4.** *Let $\mu \in M^+(S^\circ)$ and let $\{\sigma_n\}$, $\sigma$ be in $M^+(S)$ with supports contained in a compact set $K$. Suppose $\sigma_n \to \sigma$ in $M^+(S)$. Then, $\mu \bullet \sigma_n \to \mu \bullet \sigma$ in $M^+(S^\circ)$.*

***Proof.*** Let $\phi \in C_c(S^\circ)$; then, $\int_S \phi(x)\, d\mu_y(x)$ is a continuous function of $y$ (Theorem 9.2.3(i)), and

$$\begin{aligned}\lim_{n\to\infty} (\mu \bullet \sigma_n)(\phi) &= \lim_{n\to\infty} \int_K \left( \int_{S^\circ} \phi(x)\, d\mu_y(x) \right) d\sigma_n(y) \\ &= \int_K \left( \int_{S^\circ} \phi(x)\, d\mu_y(x) \right) d\sigma(y) \\ &= (\mu \bullet \sigma)(\phi).\end{aligned}$$

This implies that $\mu \bullet \sigma_n \to \mu \bullet \sigma$.

**Theorem 9.3.5.** *Let $\{\mu_n\}$, $\{\nu_n\}$, $\mu$, $\nu$ be in $M^+(S^\circ)$ with $\mu_n \to \mu$, and $\nu_n \to \nu$. Let $\sigma \in M^+(S)$, and $\mu_n \bullet \sigma \leq \nu_n$ for all $n$. Then, $\mu \bullet \sigma$ exists with $\mu \bullet \sigma \leq \nu$.*

***Proof.*** Let $\phi \in C_c(S^\circ)$ with $\phi \geq 0$; then,

$$\begin{aligned}
(\mu \bullet \sigma)(\phi) &= \int_S \left( \int_{S^\circ} \phi(x)\, d\mu_y(x) \right) d\sigma(y) \\
&\leq \liminf_{n \to \infty} \int_S \left( \int_{S^\circ} \phi(x)\, d(\mu_n)_y(x) \right) d\sigma(y) \\
&\qquad \text{(by (9.3.1') and the Fatou Lemma)} \\
&= \liminf_{n \to \infty} \int_{S^\circ} \phi(x)\, d(\mu_n \bullet \sigma)(x) \\
&\leq \liminf_{n \to \infty} \int_{S^\circ} \phi(x)\, d\nu_n(x) \\
&= \int_{S^\circ} \phi(x)\, d\nu(x) < \infty.
\end{aligned}$$

This implies that $\mu \bullet \sigma$ exists and $\mu \bullet \sigma \leq \nu$.

## 9.4. THE CONES DEFINED BY CONVOLUTIONS, AND THEIR EXTREME RAYS

We assume, as before, that $S$ is a subsemigroup of $\mathbb{R}^d$ such that $S^\circ \neq \emptyset$. Let $0 \not\equiv \sigma \in M^+(S)$ be fixed. Let

$$H_\sigma = \{\mu \in M^+(S^\circ) : \mu \bullet \sigma = \mu\},$$

and

$$C_\sigma = \{\mu \in M^+(S^\circ) : \mu \bullet \sigma \leq \mu\}.$$

A subset $A$ of a linear space $V$ (over the real field) is called a (*convex*) *cone* if $A \cap (-A) = \{0\}$, and if $u,\ v \in A$, $\alpha,\ \beta \geq 0$, implies that $\alpha u + \beta v \in A$. We use the notation $u \leq v$ to mean $v - u \in A$. Note that $\leq$ defines a partial ordering on $V$.

**Proposition 9.4.1.** *Let $H_\sigma$ and $C_\sigma$ be defined as in the preceding. Then:*

(i) *$C_\sigma$ and $H_\sigma$ are cones in $M(S^\circ)$;*
(ii) *if $\mu \in H_\sigma$ (respectively, $C_\sigma$), then for any $\nu \in M^+(S)$ such that $\mu \bullet \nu \in M^+(S^\circ)$, we have $\mu \bullet \nu \in H_\sigma$ (respectively, $C_\sigma$);*
(iii) *if $\mu \in H_\sigma$ (respectively, $C_\sigma$), then $\mu \bullet \sigma^n \in H_\sigma$ (respectively, $C_\sigma$), where $\mu \bullet \sigma^n$ is defined as $(\mu \bullet \sigma^{n-1}) \bullet \sigma$;*
(iv) *if $\mu \in H_\sigma$ (respectively, $C_\sigma$), then so does $\mu_x$ for all $x \in S$.*

***Proof.*** (i) is clear. To prove (ii), we let $\mu \in H_\sigma$; then, by Theorem 9.3.2(i),

$$(\mu \bullet \nu) \bullet \sigma = (\mu \bullet \sigma) \bullet \nu = \mu \bullet \nu,$$

and hence $\mu \bullet \nu \in H_\sigma$. (iii) follows from a simple induction. (iv) is also immediate from the definition.

**Proposition 9.4.2.** *Any $\mu \in C_\sigma$ admits a unique representation*

$$\mu = \tau + \eta$$

*with $\tau \in C_\sigma$, $\eta \in H_\sigma$, where $\eta$ is maximal among such measures: If $\mu = \tau' + \eta'$ is another decomposition with $\tau' \in C_\sigma$, $\eta' \in H_\sigma$, then $\eta' \leq \eta$. (We call this the Riesz decomposition of $\mu$.)*

***Proof.*** Let $\xi = \mu - \mu \bullet \sigma$, so $\xi$ is $\geq 0$; then,

$$\sum_{n=1}^{k} \xi \bullet \sigma^n = \mu - \mu \bullet \sigma^{k+1} \leq \mu,$$

so that $\sum_{n=1}^{\infty} \xi \bullet \sigma^n$ converges. Denote the limit by $\tau$, and let $\eta = \mu - \tau$; then, $\tau \in C_\sigma$, $\eta \in H_\sigma$, and $\eta = \lim_{n \to \infty} \mu \bullet \sigma^n$. If $\mu = \tau' + \eta'$ for some $\tau' \in C_\sigma$, $\eta' \in H_\sigma$, then

$$\eta' = \lim_{n \to \infty} \eta' \bullet \sigma^n \leq \lim_{n \to \infty} \mu \bullet \sigma^n = \eta.$$

The uniqueness of the decomposition follows directly from this.

Let $A$ be a cone and, for $u \in A$, let

$$\rho_u = \{ru : r \geq 0\}$$

be the ray generated by $u$. $\rho_u$ is called an *extreme ray* if, for any $v \in A$, $v \leq u$ implies that $v \in \rho_u$; we also say that $u$ is extremal in $A$. Let $\partial_r A$ denote the set of all such elements of $A$. Our goal in this section is to characterize the extreme rays of $H_\sigma$; this characterization is used in the next section to obtain the solution of the ICFE($\sigma$) on $S$.

**Lemma 9.4.3.** *Let $\sigma$ be a probability measure on $S$, and let $\mu \in H_\sigma$ be such that $\{\mu_x : x \in S\}$ is a weakly bounded subset of $M^+(S^\circ)$. Then, $\mu_y = \mu$ for all $y \in S(\sigma)$.*

***Proof.*** Assume without loss of generality that $0 \in S$ (otherwise, consider $S' = S \cup \{0\}$, and $\sigma' = (\sigma + \delta_0)/2$). Let $\phi \in C_c(S^\circ)$ and, for $x \in S$, let

$$\Phi(x) = \mu_x(\phi) = \int_{S^\circ} \phi(z)\, d\mu_x(z).$$

By Theorem 9.2.3(iii), $\Phi$ is a uniformly continuous function on $S$, and the definition of shift-boundedness itself implies that $\Phi$ is also bounded. Also, for $x \in S$,

$$\int_S \Phi(x + y)\, d\sigma(y) = \int_S \left[ \int_{S^\circ} \phi(z)\, d\mu_{x+y}(z) \right] d\sigma(y)$$

$$= (\mu_x \bullet \sigma)(\phi) = \mu_x(\phi) = \Phi(x).$$

Theorem 9.1.8 implies that $\Phi(x + y) = \Phi(x)$ for all $x \in S$, $y \in S(\sigma)$. Let $x = 0$; then, $\mu_y(\phi) = \mu(\phi)$ for all $y \in S(\sigma)$, $\phi \in C_c(S^\circ)$, which implies that $\mu_y = \mu$ on $S^\circ$, for all $y \in S(\sigma)$.

**Proposition 9.4.4.** *Let $\sigma \in M^+(S)$ with $\langle S(\sigma) \rangle = \mathbb{R}^d$, and let $\mu \in H_\sigma$ be such that $d\mu = g\, d\omega$, where $g$ is a positive exponential on $S$. Then, $\mu$ is extremal in $H_\sigma$, and*

$$\int_S g(x)\, d\sigma(x) = 1.$$

***Proof.*** We recall that $S(\sigma)$ is the semigroup generated by supp $\sigma$, and $\langle A \rangle$ is the group generated by $A$. Let $\nu \in H_\sigma$ and $\nu \leq \mu$. We will show that $\nu = c\mu$, which will imply that $\mu$ is extremal in $H_\sigma$. Define $h = g^{-1}$, $d\hat{\nu} = h\, d\nu$, and $d\hat{\sigma} = g\, d\sigma$. Then, for $\phi \in C_c(S^\circ)$,

$$\begin{aligned}
(\hat{\nu} \bullet \hat{\sigma})(\phi) &= \int_S \left[ \int_{S^\circ} \phi(x)\, d\hat{\nu}_y(x) \right] d\hat{\sigma}(y) \\
&= \int_S \left[ \int_{S^\circ} \phi(x) h(x + y)\, d\nu_y(x) \right] g(y)\, d\sigma(y) \\
&= \int_S \left[ \int_{S^\circ} \phi(x) h(x)\, d\nu_y(x) \right] d\sigma(y) \\
&= \int_{S^\circ} \phi(x) h(x)\, d\nu \bullet \sigma(x) \\
&= \int_{S^\circ} \phi(x) h(x)\, d\nu(x) \\
&= \hat{\nu}(\phi);
\end{aligned}$$

i.e., $\hat{\nu} \bullet \hat{\sigma} = \hat{\nu}$, so that $\hat{\nu} \in H_{\hat{\sigma}}$. Moreover, $d\mu = g\, d\omega$ and $\mu \bullet \sigma = \mu$ imply that, for any compact set $K \subset S^\circ$, $\mu(K) = \int_S g(y)\, d\sigma(y) \times \mu(K)$, whence

$$\hat{\sigma}(S) = \int_S g(y)\, d\sigma(y) = 1,$$

so that $\hat{\sigma}$ is a probability measure. Note that

$$d\hat{\nu} = h\,d\nu \leq h\,d\mu = d\omega$$

implies that $\hat{\nu}$ is $S$-shift-bounded: $\{\hat{\nu}_x : x \in S\}$ is weakly bounded, and by Lemma 9.4.3, $\hat{\nu}_x = \hat{\nu}$ for all $x \in S(\hat{\sigma}) = S(\sigma)$. Since $\langle S(\sigma) \rangle = \mathbb{R}^d \supseteq S^\circ$ by assumption, Corollary 9.2.6 implies that $\hat{\nu} = c\omega$ on $S^\circ$. We hence conclude that

$$d\nu = g\,d\hat{\nu} = cg\,d\omega = c\,d\mu$$

on $S^\circ$, and $\mu$ is extremal in $H_\sigma$.

**Proposition 9.4.5.** *Let $\mu$ be extremal in $H_\sigma$. Then, there exists a function $g \geq 0$ such that $\mu_x = g(x)\mu$ for all $x \in S(\sigma)$, and $g$ satisfies*

$$g(x + y) = g(x)g(y), \qquad x, y \in S(\sigma).$$

***Proof.*** Let $\mu \in H_\sigma$ be extremal. For any $x \in \operatorname{supp} \sigma$, and for any neighborhood $V$ of $x$ with compact closure, let $\sigma_V$ be the restriction of $\sigma$ to $V$. Then, $\mu - \mu \bullet \sigma_V = \mu \bullet \sigma_{S \setminus V}$ is in $H_\sigma$ by Proposition 9.4.1(ii). Since $\mu$ is extremal in $H_\sigma$, we obtain, from $\mu - \mu \bullet \sigma_v \leq \mu$, that $\mu \bullet \sigma_V = \alpha_V \mu$ for some $\alpha_V > 0$. Let

$$\tau_V = \sigma_V / \sigma(V), \qquad \beta_V = \alpha_V / \sigma(V).$$

Then, $\mu \bullet \tau_V = \beta_V \mu$. Let $V_n = \{y \in S : |y - x| < 1/n\}$. Then, for any $\phi \in C_c(S)$,

$$\lim_{n \to \infty} \frac{1}{\sigma(V_n)} \int_S \phi(y)\,d\sigma_{V_n}(y) = \phi(x).$$

This implies that $\lim_{n \to \infty} \tau_{V_n} = \delta_x$. By Theorem 9.3.4,

$$\mu_x = \mu \bullet \delta_x = \lim_{n \to \infty} \mu \bullet \tau_{V_n} = \lim_{n \to \infty} \beta_{V_n} \mu.$$

Let $g(x) = \lim_{n \to \infty} \beta_{V_n}$. Then, the preceding equality can be rewritten as $\mu_x = g(x)\mu$, $x \in \operatorname{supp} \sigma$. Now, for any $x$, $y \in \operatorname{supp} \sigma$,

$$\mu_{x+y} = (\mu_y)_x = g(x)g(y)\mu.$$

We can define $g$ on $S(\sigma)$ (unambiguously) by $g(x + y) = g(x)g(y)$ for all $x, y \in S(\sigma)$. Then, $\mu_x = g(x)\mu$ for all $x \in S(\sigma)$ (unambiguously).

**Theorem 9.4.6.** *Suppose that $S(\sigma)$ has the component-generating property, and that $\langle S(\sigma) \rangle = \mathbb{R}^d$. Let $\mu \in H_\sigma$; then, $\mu$ is extremal in $H_\sigma$ if and only if $d\mu = cg\,d\omega$ for some $c > 0$, where $g(x) = e^{\langle \alpha, x \rangle}$ for all $x \in S$, for some $\alpha \in \mathbb{R}^d$, and*

$$\int_S e^{\langle \alpha, x \rangle}\,d\sigma(x) = 1. \tag{9.4.1}$$

***Proof.*** The sufficiency follows from Proposition 9.4.4. To prove the necessity, we conclude from Proposition 9.4.5 that

$$\mu_x = g(x)\mu, \qquad x \in S(\sigma), \tag{9.4.2}$$

and $g$ is an exponential function on $S(\sigma)$. Since $S(\sigma)$ has the component-generating property by assumption, Theorem 9.1.7 implies that, for some $\alpha \in \mathbb{R}^d$, $g(x) = e^{\langle \alpha, x \rangle}$ for all $x \in S(\sigma)$; the assumption that $\langle S(\sigma) \rangle = \mathbb{R}^d$ implies that this holds for all $x \in \mathbb{R}^d$. Finally, (9.4.1) follows as in Proposition 9.4.4.

We conclude this section by considering some properties of the set

$$A(\sigma) = \left\{ \alpha \in \mathbb{R}^d : \int_S e^{\langle \alpha, y \rangle}\, d\sigma(y) = 1 \right\},$$

which has an obvious association with the extremal elements of $H_\sigma$ in view of the preceding result.

**Proposition 9.4.7.** *Suppose $S(\sigma)$ has the component-generating property, and $\langle S(\sigma) \rangle = \mathbb{R}^d$. Then, $\mathbb{R}_+ \times A(\sigma)$ is homeomorphic to $\partial_r H_\sigma$, the set of extremal elements of $H_\sigma$.*

***Proof.*** Let $i: \mathbb{R}_+ \times A(\sigma) \to \partial_r H_\sigma$ be defined by $i(r, \alpha) = \mu$, where $d\mu = r g_\alpha\, d\omega$, and $g_\alpha(x) = e^{\langle \alpha, x \rangle}$. Suppose $r_n \to r$, $\alpha_n \to \alpha$; then, $r_n g_{\alpha_n} \to r g_\alpha$ uniformly on compact sets. Hence, for any $\phi \in C_c(S^\circ)$, the bounded convergence theorem implies that

$$\lim_{n \to \infty} \int_S r_n g_{\alpha_n}(x) \phi(x)\, d\omega(x) = \int_S r g_\alpha(x) \phi(x)\, d\omega(x),$$

so that $i(r_n, x_n) \to i(r, x)$, and the continuity of $i$ follows.

To prove that $i^{-1}$ is continuous, we let $d\mu_n = r_n g_{\alpha_n}\, d\omega$, $d\mu = r g_\alpha\, d\omega$, and $\mu_n \to \mu$. Then, for any $x \in S$, and $\phi \in C_c(S)$ such that $\int_S \phi(y) g_\alpha(y)\, dy \neq 0$,

$$\begin{aligned}
&\lim_{n \to \infty} (r_n g_{\alpha_n}(x)) \int_S \phi(y) g_{\alpha_n}(y)\, d\omega(y) \\
&\quad = \lim_{n \to \infty} r_n \int_S \phi(y) g_{\alpha_n}(x + y)\, d\omega(x + y) \\
&\quad = \lim_{n \to \infty} (\mu_n)_x(\phi) = \mu_x(\phi) \\
&\quad = r g_\alpha(x) \int_S \phi(y) g_\alpha(y)\, d\omega(y)
\end{aligned}$$

(the third equality follows from Theorem 9.2.3(i)). This implies that

$$\lim_{n\to\infty} r_n g_{\alpha_n}(x) = r g_\alpha(x), \qquad \forall\, x \in S.$$

Observing that $g_{\alpha_n}(x) = e^{\langle \alpha_n, x\rangle}$, $g_\alpha(x) = e^{\langle \alpha, x\rangle}$, we have $r_n \to r$ and $\alpha_n \to \alpha$.

**Proposition 9.4.8.** *For any $\sigma \in M^+(S)$, let*

$$B(\sigma) = \left\{\alpha \in \mathbb{R}^d : \int_S e^{\langle \alpha, x\rangle}\, d\sigma(x) \le 1\right\} \ne \varnothing.$$

*Then, $B(\sigma)$ is convex and closed, and $A(\sigma)$ is contained in (but may not equal) its boundary.*

***Proof.*** Define

$$\psi(\alpha) = \int_S e^{\langle \alpha, x\rangle}\, d\sigma(x), \qquad x \in \mathbb{R}^d$$

($\psi(\alpha)$ may be $\infty$). By the convexity of exponential functions, we have, for $0 < \lambda < 1$, $\alpha_1, \alpha_2 \in \mathbb{R}^d$,

$$\begin{aligned}\psi(\lambda\alpha_1 + (1-\lambda)\alpha_2) &= \int_S e^{\langle \lambda\alpha_1 + (1-\lambda)\alpha_2, x\rangle}\, d\sigma(x)\\ &\le \lambda \int_S e^{\langle \alpha_1, x\rangle}\, d\sigma(x) + (1-\lambda)\int_S e^{\langle \alpha_2, x\rangle}\, d\sigma(x)\\ &= \lambda\psi(\alpha_1) + (1-\lambda)\psi(\alpha_2).\end{aligned}$$

This implies that $B(\sigma)$ is convex, and the continuity of $\psi$ on

$$\{\alpha \in \mathbb{R}^d : \psi(\alpha) < \infty\}$$

implies that $B(\sigma)$ is closed. That $A(\sigma)$ is contained in the boundary of $B(\sigma)$ is obvious. To show that $A(\sigma)$ may not be equal to $B(\sigma)$, we let $d\sigma(x) = \chi_{[2,\infty)}\, x^{-2}\, dx$ on $S = [0, \infty)$; then, a direct calculation shows that

$$\int_0^\infty e^{\alpha x} x^{-2}\, dx \begin{cases} \le \frac{1}{2} & \text{if } \alpha \le 0,\\ = \infty & \text{if } \alpha > 0.\end{cases}$$

This implies that $B(\sigma) = (-\infty, 0]$, but $A(\sigma) = \varnothing$.

## 9.5. THE ICFE ON SEMIGROUPS OF $\mathbb{R}^d$

We first establish some topological and measure-theoretic properties of the cones $C_\sigma$ and $H_\sigma$ in order to apply the Choquet theorem (see Theorem 9.A.5 and Corollary 9.A.6).

**Lemma 9.5.1.** *Let $\mu \in H_\sigma$. Then, $\mu$ is extremal in $H_\sigma$ if and only if it is extremal in $C_\sigma$.*

***Proof.*** The sufficiency is clear. To prove the necessity, let $\mu$ be extremal in $H_\sigma$, and let $\nu \in C_\sigma$ with $\nu \le \mu$ (i.e., $\mu - \nu \in C_\sigma$). Then,

$$\mu = \nu \bullet \sigma + (\mu - \nu) \bullet \sigma \le \nu + (\mu - \nu) = \mu$$

implies that $\nu \bullet \sigma = \nu$, i.e., $\nu \in H_\sigma$. That $\mu$ is extremal in $H_\sigma$ implies that $\nu = c\mu$, and hence $\mu$ is extremal in $C_\sigma$.

**Lemma 9.5.2.** *$C_\sigma$ is closed in $M^+(S^\circ)$ and is metrizable, and $H_\sigma$, $\partial_r C_\sigma$, and $\partial_r H_\sigma$ are Borel subsets of $C_\sigma$.*

***Proof.*** Since $M^+(S^\circ)$ is metrizable, so is the subset $C_\sigma$. To show that $C_\sigma$ is closed, we let $\{\mu_n\}$ be a sequence in $C_\sigma$ with $\mu_n \to \mu$. Then, $\mu_n \bullet \sigma \le \mu_n$ and, by Theorem 9.3.5, $\mu \bullet \sigma \le \mu$, whence $\mu \in C_\sigma$.

Since $C_c(S^\circ)$ is separable, let $\{\phi_n\}$ be a dense sequence in $C_c(S^\circ)$. Then,

$$\begin{aligned} H_\sigma &= \bigcap_{n=1}^{\infty} \{\mu \in C_\sigma : \mu \bullet \sigma(\phi_n) = \mu(\phi_n)\} \\ &= \bigcap_{n=1}^{\infty} \left\{\mu \in C_\sigma : \lim_{k\to\infty} \mu \bullet \sigma_k(\phi_n) = \mu(\phi_n)\right\} \\ &= \bigcap_{n=1}^{\infty} \bigcap_{i=1}^{\infty} \bigcup_{j=1}^{\infty} \bigcap_{k=j}^{\infty} \{\mu \in C_\sigma : |\mu \bullet \sigma_k(\phi_n) - \mu(\phi_n)| < 1/i\}, \qquad (9.5.1) \end{aligned}$$

where $\sigma_k$ is the restriction of $\sigma$ on $F_k$, and $\{F_k\}$ is an increasing sequence of compact subsets of $S$ such that $\bigcup_k F_k = S$. Since the set

$$\{\mu \in C_\sigma : |\mu \bullet \sigma_k(\phi_n) - \mu(\phi_n)| < 1/i\}$$

is open in $C_\sigma$, $H_\sigma$ is a Borel subset. That $\partial_r C_\sigma$ is a Borel subset follows from the first part of Theorem 9.A.5 and the fact that $C_\sigma$ is metrizable. By Lemma 9.5.1, $\partial_r H_\sigma = \partial_r C_\sigma \cap H_\sigma$, and hence $\partial_r H_\sigma$ is also a Borel subset of $C_\sigma$.

We remark that $H_\sigma$ may not be closed, and this fact is the main reason to introduce the auxiliary cone $C_\sigma$ (which is closed), to which we can apply the Choquet theorem.

**Theorem 9.5.3.** *Suppose $S(\sigma)$ has the component-generating property and $\langle S(\sigma)\rangle = \mathbb{R}^d$. Let $\mu \in M^+(S^\circ)$. Then, $\mu$ satisfies*

$$\mu \bullet \sigma = \mu$$

*if and only if $d\mu = h\, d\omega$ on $S^\circ$, where*

$$h(x) = \int_{A(\sigma)} e^{\langle \alpha, x\rangle}\, dP(\alpha), \qquad x \in S^\circ \quad a.e.,$$

*and where $P$ is a positive Radon measure on*

$$A(\sigma) = \left\{\alpha \in \mathbb{R}^d : \int_S e^{\langle \alpha, x\rangle}\, d\sigma(x) = 1\right\}.$$

***Proof.*** Since $C_\sigma$ is closed and metrizable, we can apply Corollary 9.A.6 to conclude the existence of a probability measure $Q$ on $C_\sigma$ with $Q(\partial_r C_\sigma) = 1$, and

$$\mu = \int_{\partial_r C_\sigma} v\, dQ(v). \tag{9.5.2}$$

Recalling that $\partial_r H_\sigma = \partial_r C_\sigma \cap H_\sigma$, we claim that $Q(\partial_r H_\sigma) = 1$. For, otherwise, $Q(\partial_r C_\sigma \backslash \partial_r H_\sigma) > 0$, and by (9.5.1) there exists $\phi \in C_c(S^\circ)$ such that

$$Q\{v \in \partial_r C_\sigma : v \bullet \sigma(\phi) < v(\phi)\} > 0,$$

and hence

$$\mu \bullet \sigma(\phi) = \int_{\partial_r C_\sigma} v \bullet \sigma(\phi)\, dQ(v) < \int_{\partial_r C_\sigma} v(\phi)\, dQ(v) = \mu(\phi).$$

This contradicts $\mu \bullet \sigma = \mu$, and the claim is proven.

We can now write (9.5.2) as

$$\mu = \int_{\partial_r H_\sigma} v\, dQ(v).$$

In view of Proposition 9.4.7, we have, for each $\phi \in C_c(S^\circ)$,

$$\mu(\phi) = \int_{\mathbb{R}_+ \times A(\sigma)} \left(\int_S \phi(x)(r g_\alpha(x))\, dx\right) d\tilde{Q}(r, \alpha),$$

where $g_\alpha(x) = e^{\langle \alpha, x\rangle}$, and $\tilde{Q}$ is the induced measure on $\mathbb{R}_+ \times A(\sigma)$. If we define $P$ by

$$dP(\alpha) = \int_{\mathbb{R}_+} r \, d\tilde{Q}(r, \alpha),$$

then

$$\mu(\phi) = \int_{A(\sigma)} \left( \int_S \phi(x) g_\alpha(x) \, dx \right) dP(\alpha)$$

for all $\phi \in C_c(S^\circ)$, which implies that $d\mu = h \, d\omega$, where

$$h(x) = \int_{A(\sigma)} e^{\langle \alpha, x \rangle} \, dP(\alpha) \qquad \forall \, x \in S^\circ.$$

**Corollary 9.5.4.** *Suppose $\sigma \in M^+(S)$ satisfies the conditions in Theorem 9.5.3, and in addition $A(\sigma)$ is a bounded subset of $\mathbb{R}^d$. Let $\mu \in M^+(S^\circ)$. Then, $\mu$ satisfies $\mu \bullet \sigma = \mu$ if and only if $d\mu = h \, d\omega$ on $S^\circ$, where*

$$h(x) = c \int_{A(\sigma)} e^{\langle \alpha, x \rangle} \, d\bar{P}(\alpha), \qquad x \in S^\circ \quad a.e.,$$

*$c \geq 0$ is a constant, and $\bar{P}$ is a probability measure on $A(\sigma)$.*

***Proof.*** Let $h$ be defined as in Theorem 9.5.3, let

$$\gamma = \sup\{|\alpha| : \alpha \in A(\sigma)\},$$

and let $x_0$ be such that $h(x_0) < \infty$. Then, using the notation in Theorem 9.5.3, we have

$$0 < e^{-\gamma |x_0|} \int_{\mathbb{R}_+ \times A(\sigma)} r \, d\tilde{Q}(r, \alpha) \leq \int_{A(\sigma)} e^{\langle \alpha, x_0 \rangle} \, dP(\alpha) < \infty.$$

Let $c$ be the value of the first integral here, and define $\bar{P}$ by

$$d\bar{P} = c^{-1} \int_{\mathbb{R}_+} r \, d\tilde{Q}(r, \alpha).$$

Then, $\bar{P}$ is a probability measure on $A(\sigma)$.

The solution of the ICFE($\sigma$) follows directly:

**Theorem 9.5.5.** *Suppose that $S(\sigma)$ has the component-generating property, and that $\langle S(\sigma) \rangle = \mathbb{R}^d$. Then, $f$ is a nonnegative locally integrable solution of*

$$f(x) = \int_S f(x + y) \, d\sigma(y), \qquad x \in S^\circ \quad a.e., \tag{9.5.3}$$

*if and only if*

$$f(x) = \int_{A(\sigma)} e^{\langle \alpha, x \rangle}\, dP(\alpha), \qquad x \in S^\circ \quad a.e.,$$

where $A(\sigma) = \{\alpha \in \mathbb{R}^d : \int_S e^{\langle \alpha, x \rangle}\, d\sigma(x) = 1\}$, *and* $P$ *is a positive Radon measure on* $A(\sigma)$.

*If, in addition,* $A(\sigma)$ *is a bounded set, then* $f$ *can be reduced to the form*

$$f(x) = c \int_{A(\sigma)} e^{\langle \alpha, x \rangle}\, d\bar{P}(\alpha),$$

*where* $\bar{P}$ *is a probability measure on* $A(\sigma)$.

***Proof.*** Define $d\mu = f\, d\omega$ on $S^\circ$. Then, (9.5.3) reduces to $\mu \bullet \sigma = \mu$ on $S^\circ$. Theorem 9.5.3 and Corollary 9.5.4 hence apply.

**Corollary 9.5.6.** *Suppose that* $0 \in S \subseteq \overline{S^\circ}$ *and* $\sigma \in M^+(S)$ *are such that* $S(\sigma)$ *has the component-generating property, and that* $\langle S(\sigma) \rangle = \mathbb{R}^d$. *Then,* $f$ *is a nonnegative continuous solution of*

$$f(x) = \int_S f(x + y)\, d\sigma(y), \qquad \forall\, x \in S,$$

*if and only if*

$$f(x) = c \int_S e^{\langle \alpha, x \rangle}\, d\bar{P}(\alpha), \qquad \forall\, x \in S,$$

*where* $\bar{P}$ *is a probability measure on* $A(\sigma)$.

***Proof.*** It follows from Theorem 9.5.5, the continuity of $f$, and $S \subseteq \overline{S^\circ}$ that

$$f(x) = \int_{A(\sigma)} e^{\langle \alpha, x \rangle}\, dP(\alpha), \qquad x \in S.$$

In particular, $f(0) = \int_{A(\sigma)} dP(\alpha)$, and the conclusion follows by taking $\bar{P} = (f(0))^{-1} P$.

**Theorem 9.5.7.** *Suppose* $S = \mathbb{R}^d$ *and* $\langle S(\sigma) \rangle = \mathbb{R}^d$. *Then,* $f$ *is a non-negative, locally integrable function on* $\mathbb{R}^d$ *such that*

$$f(x) = \int_{\mathbb{R}^d} f(x + y)\, d\sigma(y), \qquad x \in \mathbb{R}^d \quad a.e., \tag{9.5.4}$$

*if and only if*

$$f(x) = c \int_{A(\sigma)} e^{\langle \alpha, x \rangle}\, d\bar{P}(\alpha), \qquad x \in \mathbb{R}^d \quad a.e.,$$

*where* $c \geq 0$ and $\bar{P}$ *is a probability measure on* $A(\sigma)$.

***Proof.*** We first observe that the use of the component-generating property in Theorem 9.4.6 can be replaced by the following argument if $S$ is a group: Let $x \in S$ be such that $-x \in S(\sigma)$; then, (9.4.2) implies that

$$\mu = (\mu_x)_{-x} = g(-x)\mu_x = g(-x)g(x)\mu,$$

so that $g(x) = g(-x)^{-1}$ if $x \in -S(\sigma)$. From this, it follows that $g(x + y) = g(x)g(y)$ for all $x, y \in S(\sigma) - S(\sigma)$.

Now let $f \not\equiv 0$ on $\mathbb{R}^d$ be a solution of (9.5.4); by convolving $f$ with a continuous function supported by a compact neighborhood of 0, we may assume that $f$ is continuous and

$$f(0) = \int_{\mathbb{R}^d} f(y)\, d\sigma(y) > 0.$$

By letting $d\mu = f\, d\omega$ on $\mathbb{R}^d$, and by applying Theorem 9.4.6 (with the modification mentioned previously) and Corollary 9.5.6, we can express $f$ as stated.

If $S$ is a semigroup of $\mathbb{Z}^d$, all the previous arguments go through (actually, all the continuity properties in Section 9.3 are trivial), and we have the following.

**Theorem 9.5.8.** *Suppose $S$ is a subsemigroup of $\mathbb{Z}^d$, and $\sigma: S \to \mathbb{R}^+$ is any nonnegative function such that $\langle S(\sigma)\rangle = \mathbb{Z}^d$. Suppose further* (i) *$S(\sigma)$ has the component-generating property, or* (ii) *$S = \mathbb{Z}^d$. Then, $f \geq 0$ is a nonnegative solution of*

$$f(r) = \sum_{s \in S} f(r + s)\sigma(s), \qquad r \in S,$$

*if and only if*

$$f(r) = c\int_{A(\sigma)} t^r\, dP(t), \qquad r \in S,$$

*where $c > 0$, $t^r = t_1^{r_1}, \ldots, t_1^{r_d}$, and $P$ is a probability measure on*

$$A(\sigma) = \left\{t \in \mathbb{R}^d_+ : \sum_{r \in S} t^r = 1\right\}.$$

For the case where $\langle S(\sigma)\rangle \neq \langle S\rangle$ ($=\mathbb{R}^d$, or $\mathbb{Z}^d$), the general principle is to reduce the ICFE($\sigma$) on $S$ to $S(\sigma)$ as indicated in the following.

**Proposition 9.5.9.** *Let $\sigma \in M^+(S)$ and let $T = \langle S(\sigma)\rangle \cap S$. Suppose there exists a set $D \subseteq S$ such that $\{x + T : x \in D\}$ is a disjoint family whose union* is $S$. Then, *the nonnegative continuous solutions of*

$$f(x) = \int_S f(x+y)\, d\sigma(y), \qquad x \in S,$$

*are of the form* $f(x+t) = g_x(t)$, $x \in D$, $t \in T$, *where* $g_x(t)$ *satisfies*

$$g_x(t) = \int_T g_x(t+s)\, d\sigma(s), \qquad t \in T.$$

As an illustration, we consider the following.

**Proposition 9.5.10.** *Let* $S = \mathbb{R}_+^2$, *and let* $\sigma \in M^+(S)$ *be such that* $S(\sigma)$ *is a nonlattice subset of* $\{(x,x): x \in \mathbb{R}_+\}$. *Then, nonnegative continuous solutions of*

$$f(x) = \int_S f(x+y)\, d\sigma(y), \qquad x \in S,$$

*are of the form*

$$f(x) = \begin{cases} p_1\left(\dfrac{x_1 - x_2}{2}\right) e^{\alpha(x_1+x_2)/2} & \text{if } x_1 \geq x_2 \geq 0, \\ p_2\left(\dfrac{x_2 - x_1}{2}\right) e^{\alpha(x_1+x_2)/2} & \text{if } x_2 > x_1 \geq 0, \end{cases}$$

*where* $p_1$, $p_2$ *are continuous functions and*

$$\int_{\mathbb{R}_+} e^{\alpha x}\, d\sigma(x,x) = 1.$$

***Proof.*** Let $T = \langle S(\sigma) \rangle \cap S$ and $D = (\mathbb{R}_+ \times \{0\}) \cup (\{0\} \times \mathbb{R}_+)$. Then, $T = \{(t,t): t \in \mathbb{R}_+\}$, and $S$ is the disjoint union of $x + T$, $x \in D$. We can consider the equation

$$g_x(t) = \int_T g_x(t+s)\, d\sigma(s), \qquad t \in T.$$

By applying the solution of the ICFE($\sigma$) on $\mathbb{R}_+$, we obtain

$$g_x(t_1 + t_2) = g_x(t_1) g_x(t_2), \qquad t_1, t_2 \in T,$$

which is the same as

$$f(x + tv) = f(x) f(tv), \qquad x \in \mathbb{R}_+^2,\ t \in \mathbb{R}_+,$$

where $v = (1,1)$ as in Example 9.1.3. The desired expression for $f$ hence follows.

## APPENDIX: WEAK CONVERGENCE OF MEASURES; CHOQUET'S THEOREM

Let $X$ be a locally compact Hausdorff space, and let $\mathcal{B}$ be the family of Borel subsets of $X$. A Borel measure $\mu$ on $X$ is called a *Radon measure* if it satisfies:

(i) $\mu(K) < \infty$ for all compact subsets $K$ of $X$;
(ii) $\mu$ is regular; i.e.,

$$\mu(E) = \inf\{\mu(O): E \subseteq O,\ O \text{ open in } X\},$$

for every $E \in \mathcal{B}$ and

$$\mu(E) = \sup\{\mu(K): K \subseteq E,\ K \text{ compact}\},$$

for every $E$ open in $X$ with $\mu(E) < \infty$.

Let $M(X)$ denote the class of Radon measures on $X$, and let $M^+(X)$ be the subclass of positive measures. We will give $M^+(X)$ the topology generated by the following subbasic open neighborhoods of 0 (the zero measure):

$$U(\phi, \varepsilon) = \{\mu \in M(X) : |\mu(\phi)| < \varepsilon\},$$

where $\phi \in C_c(X)$, the space of continuous functions with compact supports, and $\mu(\phi)$ denotes $\int_X \phi \, d\mu$. We call this topology the *weak topology* generated by $C_c(X)$. It follows from the definition that, for any net $\{\mu_\alpha\}$, and $\mu \in M(X)$,

$$\mu_\alpha \to \mu \quad \text{if and only if} \quad \mu_\alpha(\phi) \to \mu(\phi), \qquad \forall\, \phi \in C_c(X).$$

**Theorem 9.A.1.** *Let $X$ be a locally compact Hausdorff space. Then:*

(i) *$M^+(X)$ is complete;*
(ii) *$M^+(X)$ is metrizable and separable if and only if $X$ has a countable base.*

A proof is given in Choquet (1969), Vol. I, p. 209 and p. 220. Note that if $X$ is metrizable and separable, then $X$ has a countable base, so that $M^+(X)$ will be complete and metrizable. The following theorem formulates conditions equivalent to weak convergence, and forms part of what is sometimes called the *portmanteau* theorem. The proof restricted to probability measures is given in Billingsley (1968).

**Theorem 9.A.2.** *Let $X$ be a metrizable, separable, locally compact Hausdorff space. Then, the following are equivalent*:

(i) $\mu_n \to \mu$ *weakly*;
(ii) $\overline{\lim}_{n\to\infty} \mu_n(K) \leq \mu(K)$, $\underline{\lim}_{n\to\infty} \mu_n(U) \geq \mu(U)$ *for all compact subsets $K$, and for all open subsets $U$ with compact closure*;
(iii) $\lim_{n\to\infty} \mu_n(E) = \mu(E)$ *for all Borel subsets $E$ with compact closure, and with $\mu(\partial E) = 0$ ($\partial E$ is the boundary of $E$).*

***Proof.*** (i) implies (ii). Let $\mu_n \to \mu$ weakly, and let $K$ be a compact subset of $X$. The regularity of $\mu$ implies that, for each $\varepsilon > 0$, there exists an open set $U$ with compact closure such that $K \subseteq U$ and $\mu(U) < \mu(K) + \varepsilon$. Since $X$ is a metric space, there exists a continuous function $\phi \in C_c(X)$ with $0 \leq \phi \leq 1$ and

$$\phi(x) = \begin{cases} 1, & x \in K, \\ 0, & x \in U^c; \end{cases} \tag{9.A.1}$$

hence,

$$\overline{\lim_{n\to\infty}} \mu_n(K) \leq \overline{\lim_{n\to\infty}} \int_X \phi\, d\mu_n = \int_X \phi\, d\mu \leq \mu(U) < \mu(K) + \varepsilon,$$

and the first inequality in (ii) follows. To prove the second inequality, we can find, for $\varepsilon > 0$ and for such $U$, a compact $K \subseteq U$ such that $\mu(U) < \mu(K) + \varepsilon$. Let $\phi \in C_c(X)$ be as in (9.A.1); then,

$$\mu(U) - \varepsilon < \mu(K) \leq \int_X \phi\, d\mu = \lim_{n\to\infty} \int_X \phi\, d\mu_n \leq \underline{\lim}_{n\to\infty} \mu_n(U).$$

(ii) implies (iii). Let $E$ be a Borel subset with compact closure, and with $\mu(\partial E) = 0$. Then,

$$\mu(E) = \mu(\bar{E}) \geq \overline{\lim_{n\to\infty}} \mu_n(\bar{E}) \geq \overline{\lim_{n\to\infty}} \mu_n(E)$$

$$\geq \underline{\lim}_{n\to\infty} \mu_n(E) \geq \underline{\lim}_{n\to\infty} \mu_n(E^\circ) \geq \mu(E^\circ) = \mu(E),$$

and hence $\lim_{n\to\infty} \mu_n(E) = \mu(E)$.

(iii) implies (i). Let $\phi \in C_c(X)$; without loss of generality, we may assume that $0 \leq \phi \leq 1$. Let $K$ be a compact set containing the support of $\phi$, and with $\mu(\partial K) = 0$. Then, $\lim_{n\to\infty} \mu_n(K) = \mu(K)$, so that $\{\mu_n(K)\}$ and $\mu(K)$ are bounded, by $M$, say. For $\varepsilon > 0$, let $0 = t_0 < t_1 < \cdots < t_m = 1$ be such that $0 < t_i - t_{i-1} < \varepsilon$, $i = 1, \ldots, m$, and

$$\mu\{x \in X : f(x) = \alpha_i\} = 0, \qquad i = 0, 1, \ldots, m,$$

where $\alpha_i = f(t_i)$. Let $U_i = \{x \in X : \alpha_{i-1} < f(x) < \alpha_i\}$, $i = 1, \ldots, m$. Then, $\mu(\partial U_i) = 0$, so that $\lim_{n\to\infty} \mu_n(U_i) = \mu(U_i)$, $i = 1, \ldots, m$. Let $\phi_\varepsilon = \sum_{i=1}^m \alpha_i \chi_{U_i}$. The inequality

$$\left|\int_X \phi \, d(\mu_n - \mu)\right| \le \int_K |\phi - \phi_\varepsilon| \, d|\mu_n - \mu| + \left|\int_X \phi_\varepsilon \, d(\mu_n - \mu)\right|$$

$$\le 2M\varepsilon + \sum_{i=1}^m \alpha_i |(\mu_n - \mu)(U_i)|$$

implies that $\lim_{n\to\infty} (\mu_n - \mu)(\phi) = 0$ for any $\phi \in C_c(X)$, so that $\mu_n \to \mu$.

In the following, we will outline the basic ideas of Choquet's extremal representation theory. For details, the reader may refer to Choquet (1969), Vol. II, Chapter 6, or Phelps (1966). Let $K$ be a convex subset of a vector space $V$; $x \in K$ is called an *extreme point* of $K$ if $x = \frac{1}{2}(x_1 + x_2)$, $x_1, x_2 \in K$, implies that $x = x_1 = x_2$. Let $\partial_e K$ denote the set of extreme points of $K$.

**Theorem 9.A.3** (Krein–Milman). *Let $K$ be a compact convex subset in a locally convex Hausdorff topological vector space $V$. Then, $K$ equals the closed convex hull of $\partial_e K$, the set of its extreme points.*

Choquet's extension of the Krein–Milman theorem replaces the closed convex hull of $\partial_e K$ by probability measures supported on $\partial_e K$.

**Theorem 9.A.4** (Choquet). *Let $V$ be as before, and let $K$ be a metrizable compact convex subset of $V$. Then, $\partial_e K$ is a Borel subset of $K$, and, for any $x \in K$, there exists a probability measure $P$ on $K$, supported on $\partial_e K$, such that $x = \int_{\partial_e K} y \, dP(y)$, equivalently, such that*

$$\langle \phi, x \rangle = \int_{\partial_e K} \langle \phi, y \rangle \, dP(y), \tag{9.A.2}$$

*for any continuous linear functional $\phi$ on $V$.*

We say then that $x$ is *represented* by the probability measure $P$. The extension of the theorem from compact convex sets to cones is as follows: A *ray* $\rho$ in a cone $C$ is a half-line of the form $\rho = \{\lambda u : \lambda \ge 0\}$ for some $u \in C$. A ray $\rho$ is called an *extreme ray* of $C$ if, for any $x \in \rho$, $y, z \in C$, $x = y + z$ implies that $y, z \in \rho$. We will use $\partial_r C$ to denote the set of extreme rays of $C$.

**Theorem 9.A.5** (Choquet). *Let $V$ be as before, and let $C \subseteq V$ be a weakly complete, metrizable cone. Then, $\partial_r C$ is a Borel subset of $C$; also, for any $x \in C$, there exists a probability measure $P$ on $C$, supported on $\partial_r C$, which represents $x$ in the sense of* (9.A.2).

The special version we use in Theorem 9.5.3 is the following consequence of Theorems 9.A.1 and 9.A.5.

**Corollary 9.A.6.** *Let $X$ be a locally compact Hausdorff space with a countable base, and let $C$ be a closed cone in $M^+(X)$. Then, $C$ is complete and metrizable, and $\partial_r C$ is a Borel subset of $C$; also, for any $\mu \in C$, there exists a probability measure $P$ supported on $\partial_r C$ such that*

$$\mu(\phi) = \int_{\partial_r C} \nu(\phi)\, dP(\nu)$$

*for all $\phi \in C_c(X)$.*

## NOTES AND REMARKS

Choquet and Dény (1960) was the first paper to study the convolution equation $\mu = \mu * \sigma$; this was done in the setting of a locally compact abelian group $G$. They assumed that $\sigma$ is a probability measure and $\mu$ is shift-bounded (i.e., $\mu * \phi$ is a bounded function for any continuous $\phi$ on $G$ with compact support). Their proof is simplified in Dény (1960); the latter's is essentially that given for Theorem 9.1.8; Doob *et al.* (1960) proved this by a martingale argument, reproduced in Meyer (1966). The general case where $\sigma$ and $\mu$ are only assumed nonnegative is proved in Dény (1960), and the solutions $\mu$ are called $\sigma$-harmonics.

An extension of Dény's theorem to semigroups was given by Davies and Shanbhag (1987), with the basic equation taken in the form $f(x) = \int_S f(x + y)\, d\sigma(y)$, for $x \in S$, with $\sigma \geq 0$, a given measure, and $f$ a nonnegative continuous function to be solved for, using a martingale approach; this extended considerably the Doob–Snell–Williamson argument. The proof presented here is from Lau and Zeng (1990), using Dény's extreme-point method, and restricting attention to subsemigroups of $\mathbb{R}^d$ for simplicity. An application to a characterization of multivariate stable distributions is considered in Zeng (1990).

The convolution equation $\mu = \mu * \sigma$ was also studied on semisimple Lie groups by Furstenberg (1963) and Azencott (1970) in the restricted case where $\sigma$ is a probability measure and $\mu$ is bounded.

For the case where $\sigma$ is not assumed positive, the equation $\mu * \sigma = \mu$ can be recast into the form $\mu * \sigma = 0$. Schwartz (1947) (see also Kahane, 1959) considered $\sigma$ defined on $\mathbb{R}$ with compact support, and characterized all continuous functions $f$ on $\mathbb{R}$ that satisfy $f * \sigma = 0$; he called these $f$ *mean-periodic functions*. His theorem is used by Brown *et al.* (1973) to study the Pompeiu problem: Let $D$ be a bounded region in $\mathbb{R}^2$, and let $T$ denote the group of all rigid motions on the plane. Determine $D$ so that

$$\iint_{\tau(D)} f(x, y)\, dx\, dy = 0, \qquad \forall\, \tau \in T,$$

will imply $f(x, y) = 0$ for all $(x, y) \in \mathbb{R}^2$ for $f$ continuous on $\mathbb{R}^2$. In particular, is the unit disk of $\mathbb{R}^2$ essentially the only domain $D$ *without* this property? For more recent developments on this problem and mean-periodic functions, the reader may refer to Bagchi and Sitaram (1979, 1989), Berenstein (1987), and Zalcman (1980).

# Bibliography

(Diacritical marks in spelling proper names have been mostly omitted.)


Ahsanullah, M. (1978). Record values and the exponential distributions, *Ann. Inst. Statist. Math.*, **30**, 429–433.

Alzaid, A., Lau, K. S.., Rao, C. R., and Shanbhag, D. N. (1988). Solution of Deny convolution equations restricted to a half-line via a random walk approach, *J. Mult. Anal.*, **24**, 304–329.

Alzaid, A., Rao, C. R., and Shanbhag, D. N. (1986). An application of Perron–Frobenius theorem to a damage model problem, *Sankhya A*, **48**, 43–53.

Alzaid, A., Rao, C. R., and Shanbhag, D. N. (1987). Solution of the integrated Cauchy functional equation using exchangability, *Sankhya A*, **49**, 189–194.

Azencott, R. (1970). Espaces de Poisson des groups localement compacts, Springer-Verlag Lecture Notes, No. 148, New York.

Azlarov, T., and Volodin, N. A. (1986). Characterization problems associated with the exponential distribution (translation from Russian), Springer-Verlag, New York.

Bagchi, S. C., and Sitaram, A. (1979). Spherical mean-periodic functions on semi-simple Lie groups, *Pacific J. Math.*, **84**, 241–250.

Bagchi, S. C., and Sitaram, A. (1989). The Pompeiu problem revisited, preprint.

Berenstein, C. A. (1987). Spectral synthesis on symmetric spaces, *Contemporary Math.*, *Integral Geometry*, Vol. **63**, 1–25.

Berg, C., Christensen, J., and Ressel, P. (1984). Harmonic analysis on semigroups, theory of positive-definite and related functions, Springer-Verlag, New York.

Bergström, H. (1963). Limit theorems for convolutions, Almqvist and Wiksell, Stockholm.

Billingsley, P. (1968). Convergence of probability measures, John Wiley, New York.

Boas, R. P. (1954). Entire functions, Academic Press.

Bondesson, L. (1977). The sample variance, properly normalized, is $\chi^2$-distributed for normal laws only, *Sankhya A*, **39**, 303–304.

Brandhofe, T., and Davies, L. (1980). On a functional equation in the theory of linear statistics, *Ann. Inst. Statist. Math.*, **32**, 17–23.

Brown, L., Schreiber, B. M., and Taylor, B. A. (1973). Spectral synthesis and the Pompeiu problem, *Ann. l'Inst. Fourier, Grenoble*, **23**, 125–154.

Choquet, G. (1969). Lectures on analysis, Vols. I, II, Benjamin, New York.

Choquet, G., and Deny, J. (1960). Sur l'equation de convolution $\mu = \mu * \sigma$, *C.R. Acad. Sci. Paris*, **250**, 799–801.

Chung, K. L. (1974). A course in probability theory, Second edition, Academic Press, New York.

Cramér, H. (1946). Mathematical methods of statistics, Princeton University Press, Princeton, New Jersey.

Davies, L., and Shanbhag, D. N. (1987). A generalization of a theorem of Deny with application in characterization problems, *Quart. J. Math., Oxford*, **38**, 13–34.

Davies, L., and Shimizu, R. (1976). On identically distributed linear statistics, *Ann. Inst. Statist. Math.*, **28**, 469–489.

Davies, L., and Shimizu, R. (1980). On a functional equation in the theory of linear statistics, *Ann. Inst. Statist. Math.*, **32**, 17–23.

Deny, J. (1960). Sur l'equation de convolution $\mu * \sigma = \mu$, *Semin. Theor. Potent. M. Brelot.*, Fac. Sci. Paris, 4 ann.

Doetsch, G. (1974). Introduction to the theory and application of the Laplace transformation, Springer-Verlag, New York.

Doob, J. L. (1953). Stochastic Processes, John Wiley, New York.

Doob, J. L., Snell, J. L., and Williamson, R. E. (1960). Application of boundary theory to sums of independent random variables. *Contributions to Probability and Statistics*, Stanford Univ. Press, 182–197.

Dunford, N., and Schwartz, J. (1953). Linear operators, Vol. I, John Wiley, New York.

Feller, W. (1968). An introduction to probability theory and its applications, Vol. 1, Third edition, John Wiley, New York.

Feller, W. (1971). An introduction to probability theory and its applications, Vol. 2, Second edition, John Wiley, New York.

Furstenberg, H. (1963). Poisson formula for semi-simple Lie groups, *Annals of Math.*, **77**, 335–386.

Galambos, J., and Kotz, S. (1978). Characterizations of probability distributions, Springer-Verlag Lecture Notes, No. 75, New York.

Geisser, S. (1973). Normal characterizations via the squares of random variables, *Sankhya A*, **35**, 492–494.

Gnedenko, B. V., and Kolmogorov, A. N. (1964). Limit distributions for sums of independent random variables, Second edition, translation from Russian by K. L. Chung, Addison-Wesley, Reading, Massachusetts.

Gu, H. M., and Lau, K. S. (1984). Integrated Cauchy functional equation with an error term and the exponential law, *Sankhya A*, **46**, 339–354.

Hewitt, E., and Savage, L. J. (1955). Symmetric measures on cartesian products, *Trans. Amer. Math. Soc.*, **80**, 470–501.

Huang, J. S. (1981). On a "lack of memory" property, *Ann. Inst. Statist. Math.*, **33**, 131–134.

Huang, J. S., and Lin, G. D. (1988). Characterization of distributions via two expected values of order statistics, *Metrika*, **35**, 329–338.

Kagan, A. M., Linnik, Yu. V., and Rao, C. R. (1973). Characterization problems in mathematical statistics, translation from Russian by B. Ramachandran, John Wiley, New York.

Kahane, J.-P. (1959). Lectures on mean periodic functions, Tata Inst., Bombay.

Krishnaji, N. (1970). Characterization of the Pareto distribution through a model of unreported incomes, *Econometrica*, **38**, 251–256.

Laha, R. G., and Lukacs, E. (1965). On a linear form whose distribution is identical with that of a monomial, *Pacific J. Math.*, **15**, 207–214.

Lau, K. S., and Rao, C. R. (1982). Integrated Cauchy functional equation and characterizations of the exponential law, *Sankhya A*, **44**, 72–90.

Lau, K. S., and Rao, C. R. (1984). Solution to the integrated Cauchy functional equation on the whole line, *Sankhya A*, **46**, 311–318.

Lau, K. S., and Zeng, W. B. (1990). The convolution equation of Choquet and Deny on semigroups, *Studia Math.*, to appear.

Levin, B. Ja. (1964). The distribution of zeros of entire functions, *Amer. Math. Soc. Translations*, Vol. 5.

Lévy, P. (1937). Théorie de l'addition des variables aléatoires, Gauthier-Villars, Paris.

Lin, G. D. (1987a). On characterizations of distributions via moments of record values, *Prob. Th. Rel. Fields*, **74**, 479–483.

Lin, G. D. (1987b). Characterizations of distributions via moments of order statistics: A survey and comparison of methods, statistical data analysis and inference, Elsevier Science Publishers, ed. Y. Dodge, 297–307.

Linnik, Yu, V. (1953a, b). Linear forms and statistical criteria (in Russian), *Ukrain, Mat. Zhurnal*, **5**, 207–243 and 247–290; *Amer. Math. Soc., Selected Transl. in Math. Statist. Prob.*, Vol. **3** (1962), 1–90.

Linnik, Yu. V., and Ostrovskii, I. V. (1977). Decomposition of random variables and vectors, *Amer. Math. Soc., Translations of Math. Monographs*, Vol. 48.

Loève, M. (1977). Probability theory, Vols. I and II, Fourth edition, Springer-Verlag.

Lukacs, E. (1968). Stochastic convergence, D. C. Heath. Second edition (1975), Academic Press.

Lukacs, E. (1969). A characterization of stable processes, *J. Appl. Prob.*, **6**, 409–418.

Lukacs, E. (1970). Characteristic functions, Second edition, Griffin, London.

Marsaglia, G., and Tubilla, A. (1975). A note on the lack of memory of the exponential distribution, *Ann. Prob.*, **3**, 352–354.

Marshall, A. W., and Olkin, I. (1967). A multivariate exponential distribution, *J. Amer. Statist. Assoc.*, **62**, 30–44.

Meyer, P. A. (1966). Probability and potentials, Blaisdell, Waltham, Massachusetts.

Norton, R. M. (1975). On properties of the arc-sine law, *Sankhya A*, **37**, 306–308.

Norton, R. M. (1978). Moment properties and the arc-sine law, *Sankhya A*, **40**, 192–198.

Phelps, R. R. (1966). Lectures on Choquet's theorem, Van Nostrand, Princeton.

Puri, P. S., and Rubin, H. (1970). A characterization based on the absolute difference of two i.i.d. random variables, *Ann. Math. Statist.*, **41**, 251–255.

Ramachandran, B. (1967). Advanced theory of characteristic functions, Statist. Publ. Soc., Calcutta.

Ramachandran, B. (1969). On characteristic functions and moments, *Sankhya A*, **31**, 1–12.

Ramachandran, B. (1975). On a conjecture of Gleisser's, *Sankhya A*, **37**, 423–427.

Ramachandran, B. (1977a). On some fundamental lemmas of Linnik's, *Colloq. Math. Soc. Janos Bolyai*, Vol. 21, *Analytic function methods in probability theory*, North-Holland, 293–306.

Ramachandran, B. (1977b). On the strong Markov property of the exponential laws, *ibid.*, 277–292.

Ramachandran, B. (1979). On the "strong memorylessness property" of the exponential and geometric probability laws, *Sankhya A*, **49**, 244–251.

Ramachandran, B. (1982a). On the equation $f(x) = \int_{[0,\infty)} f(x+y)\,d\mu(y)$, $x \geq 0$, *Sankhya A*, **44**, 364–371.

Ramachandran, B. (1982b). An integral equation in probability theory and its implications, *Statistics and Probability: Essays in honor of C. R. Rao*, eds. G. Kallianpur, P. R. Krishnaiah, and J. K. Ghosh, North-Holland, 609–616.

Ramachandran, B. (1984). Renewal-type equations on $\mathbb{Z}$, *Sankhya A*, **46**, 319–325.

Ramachandran, B. (1987). On the equation $f(x) = \int_{-\infty}^{\infty} f(x+y)\,d\mu(y)$, $x \in \mathbb{R}$, *Sankhya A*, **49**, 195–198.

Ramachandran, B. (1991a). Identically distributed stochastic integrals, stable processes, and semi-stable processes, *Sankhya A*, **53**.

Ramachandran, B. (1991b). Characteristic functions and shift-symmetry, *Gujarat Statist. Rev.*, Prof. C. G. Khatti Mem. Vol.

Ramachandran, B., and Lau, K. S. (1990). Integrated Cauchy functional equation with an error term on $[0, \infty)$, *Proceedings of the R. C. Bose Memorial Symposium on Analysis and Probability*, New Delhi.

Ramachandran, B., Lau, K. S., and Gu, H. M. (1988). On characteristic functions satisfying a functional equation and related classes of simultaneous integral equations, *Sankhya A*, **50**, 190–198.

Ramachandran, B., and Prakasa Rao, B. L. S. (1984). On the equation $f(x) = \int_{-\infty}^{\infty} f(x + y)\, d\mu(y)$, *Sankhya A*, **46**, 326–338.

Ramachandran, B., Ramaswamy, V., and Balasubramanian, K. (1976). The stable laws revisited, *Sankhya A*, **38**, 300–303.

Ramachandran, B., Ramaswamy, V., and Balasubramanian, K. (1980). The stable laws revisited, addendum and corrigendum, *Sankhya A*, **42**, 298–299.

Ramachandran, B., and Rao, C. R. (1968). Some results on characteristic functions and characterizations of the normal and generalized stable laws, *Sankhya A*, **30**, 125–140.

Ramachandran, B., and Rao, C. R. (1970). Solutions of functional equations arising in some regression problems, and a characterization of the Cauchy law, *Sankhya A*, **32**, 1–30.

Rao, C. R. (1973). Linear statistical inference and its applications, John Wiley, New York.

Rao, C. R., and Rubin, H. (1964). On a characterization of the Poisson distribution, *Sankhya A*, **26**, 294–298.

Rao, C. R., and Shanbhag, D. N. (1986). Recent results on characterizations of probability distributions: A unified approach through an extension of Deny's theorem, *Adv. Appl. Prob.*, **18**, 660–678.

Riedel, M. (1980a). Representation of the characteristic function of a stochastic integral, *J. Appl. Prob.*, **17**, 448–455.

Riedel, M. (1980b). Characterization of stable processes by identically distributed stochastic integrals, *Adv. Appl.. Prob.*, **12**, 689–709.

Riedel, M. (1985). On a characterization of the normal distribution by means of identically distributed linear forms, *J. Mult. Anal.*, **16**, 241–252.

Rossberg, H.-J. (1972). Characterization of the exponential and Pareto distributions by means of some properties of the distributions which the differences and quotients of order statistics are subject to, *Math. Operationsforsch. u. Statist.*, **3**, 207–216.

Rossberg, H.-J., Jesiak, B., and Siegel, G. (1985). Analytic methods of probability theory, Akademie-Verlag, Berlin.

Ruben, H. (1978). On quadratic forms and normality, *Sankhya A*, **40**, 156–173.

Rudin, W. (1974). Real and complex analysis, Second edition, McGraw-Hill.

Schwartz, L. (1947). Théorie génerale des fonctions moyennes-périodiques, *Annals of Math.*, **48**, 857–929.

Shanbhag, D. N. (1977). An extension of the Rao–Rubin characterization of the Poisson distribution, *J. Appl. Prob.*, **14**, 640–646.

Shantaram, R. (1978). A characterization of the arc-sine law, *Sankhya A*, **40**, 199–207.

Shantaram, R. (1980). On a conjecture of Norton's, *SIAM J. Appl. Math.*, **42**, 923–925.

Shimizu, R. (1968). Characteristic functions satisfying a functional equation—I, *Ann. Inst. Statist. Math.*, **20**, 187–209.

Shimizu, R. (1978). Solution to a functional equation and its application to some characterization problems, *Sankhya A*, **40**, 319–332.

Shimizu, R. (1980). Functional equation with an error term and the stability of some characterizations of the exponential distribution, *Ann. Inst. Statist. Math.*, **32**, 1-16.

Shimizu, R., and Davies, L. (1981). General characterization theorems for the Weibull and the stable distributions, *Sankhya A*, **43**, 282-310.

Shohat, J. A., and Tamarkin, J. D. (1950). The problem of moments, *Amer. Math. Soc. Colloquium Publ. Surveys*, No. 1.

Sohobov, O. M., and Geshev, A. A. (1974). Characteristic property of the exponential distribution, *Natur. Univ. Plovdiv.*, **7**, 25-28.

Titchmarsh, E. C. (1939). The theory of functions, Second edition, Oxford University Press.

Widder, D. V. (1946). The Laplace transform, Princeton Univ. Press.

Witte, H.-J. (1988). Some characterizations of distributions based on the ICFE, *Sankhya A*, **50**, 59-63.

Zalcman, L. (1980). Offbeat integral geometry, *Amer. Math. Monthly*, **87**, 165-175.

Zeng, W. B. (1990). On a characterization of multivariate stable distributions via random linear statistics, preprint.

Zinger, A. A. (1975). On samples with identically distributed linear statistics, *Theor. Prob. Applns.*, **20**, 668-672.

Zinger, A. A. (1977). On a characterization of the normal law by identically distributed linear statistics, *Sankhya A*, **39**, 232-242.

# Index

(References to authors, made in the Bibliography, are omitted here.)

# PROBABILITY AND MATHEMATICAL STATISTICS

Thomas Ferguson, *Mathematical Statistics: A Decision Theoretic Approach*
Howard Tucker, *A Graduate Course in Probability*
K. R. Parthasarathy, *Probability Measures on Metric Spaces*
P. Révész, *The Laws of Large Numbers*
H. P. McKean, Jr., *Stochastic Integrals*
B. V. Gnedenko, Yu. K. Belyayev, and A. D. Solovyev, *Mathematical Methods of Reliability Theory*
Demetrios A. Kappos, *Probability Algebras and Stochastic Spaces*
Ivan N. Pesin, *Classical and Modern Integration Theories*
S. Vajda, *Probabilistic Programming*
Robert B. Ash, *Real Analysis and Probability*
V. V. Fedorov, *Theory of Optimal Experiments*
K. V. Mardia, *Statistics of Directional Data*
H. Dym and H. P. McKean, *Fourier Series and Integrals*
Tatsuo Kawata, *Fourier Analysis in Probability Theory*
Fritz Oberhettinger, *Fourier Transforms of Distributions and Their Inverses: A Collection of Tables*
Paul Erdös and Joel Spencer, *Probabilistic Methods of Statistical Quality Control*
K. Sarkadi and I. Vincze, *Mathematical Methods of Statistical Quality Control*
Michael R. Anderberg, *Cluster Analysis for Applications*
W. Hengartner and R. Theodorescu, *Concentration Functions*
Kai Lai Chung, *A Course in Probability Theory, Second Edition*
L. H. Koopmans, *The Spectral Analysis of Time Series*
L. E. Maistrov, *Probability Theory: A Historical Sketch*
William F. Stout, *Almost Sure Convergence*
E. J. McShane, *Stochastic Calculus and Stochastic Models*
Robert B. Ash and Melvin F. Gardner, *Topics in Stochastic Processes*
Avner Friedman, *Stochastic Differential Equations and Applications, Volume 1, Volume 2*
Roger Cuppens, *Decomposition of Multivariate Probabilities*
Eugene Lukacs, *Stochastic Convergence, Second Edition*
H. Dym and H. P. McKean, *Gaussian Processes, Function Theory, and the Inverse Spectral Problem*
N. C. Giri, *Multivariate Statistical Inference*
Lloyd Fisher and John McDonald, *Fixed Effects Analysis of Variance*
Sidney C. Port, and Charles J. Stone, *Brownian Motion and Classical Potential Theory*

Konrad Jacobs, *Measure and Integral*
K. V. Mardia, J. T. Kent, and J. M. Biddy, *Multivariate Analysis*
Sri Gopal Mohanty, *Lattice Path Counting and Applications*
Y. L. Tong, *Probability Inequalities in Multivariate Distributions*
Michel Metivier and J. Pellaumail, *Stochastic Integration*
M. B. Priestly, *Spectral Analysis and Time Series*
Ishwar V. Basawa and B. L. S. Prakasa Rao, *Statistical Inference for Stochastic Processes*
M. Csörgö and P. Révész, *Strong Approximations in Probability and Statistics*
Sheldon Ross, *Introduction to Probability Models, Second Edition*
P. Hall and C. C. Heyde, *Martingale Limit Theory and Its Application*
Imre Csiszár and János Körner, *Information Theory: Coding Theorems for Discrete Memoryless Systems*
A. Hald, *Statistical Theory of Sampling Inspection by Attributes*
H. Bauer, *Probability Theory and Elements of Measure Theory*
M. M. Rao, *Foundations of Stochastic Analysis*
Jean-Rene Barra, *Mathematical Basis of Statistics*
Harald Bergström, *Weak Convergence of Measures*
Sheldon Ross, *Introduction to Stochastic Dynamic Programming*
B. L. S. Prakasa Rao, *Nonparametric Functional Estimation*
M. M. Rao, *Probability Theory with Applications*
A. T. Bharucha-Reid and M. Sambandham, *Random Polynomials*
Sudhakar Dharmadhikari and Kumar Joag-dev, *Unimodality, Convexity, and Applications*
Stanley P. Gudder, *Quantum Probability*
B. Ramachandran and Ka-Sing Lau, *Functional Equations in Probability Theory*